Studies in Logic
Logic and Argumentation
Volume 104

Argument, Sex and Logic

Studies in Logic Series Editor
Dov Gabbay   dov.gabbay@kcl.ac.uk

Argument, Sex and Logic

Dov Gabbay
King's College London, University of
Luxembourg, and Bar-Ilan University

Gadi Rozenberg
Ashkelon Academic College, and

Merhavim – Medical Center for Treatment of
Brain and Mind (Maba'n)

Lydia Rivlin
Author and Broadcaster

ISBN 978-1-84890-454-5

College Publications
Scientific Director: Dov Gabbay
Managing Director: Jane Spurr

http://www.collegepublications.co.uk

CONTENTS

Sex Offender Practice

INTRODUCTION TO THIS BOOK

This book is intended first for the formal argumentation community and second to the sex offender therapist community.

The argumentation community develops logics and systems modelling argumentation and dialogues. The community is in search of major applications areas for their models. One such application area, for example, is Law.

The message of this book is that there is another major application area for formal argumentation. There is an international community of sex offender therapists which is well established, well funded, and their therapy methods use (methods which can be modelled by) formal argumentation and logic. This community presents a natural application area for formal argumentation. We thus describe in this book how the sex offender therapists work, and give the formal argumentation researcher a view of this application area. At the same time we hope the sex offender therapist will get interested in understanding of how they are using logic in their practices. What is especially important about this application area is that in order to model it and learn from it, the formal argumentation community have to evolve their formal methods and adapt to this new application. More than that ; they have to open their minds and stop thinking locally about more and more little formal modifications to their models and think more on the global picture. Part of this enhancement is to modify and import certain methods from other areas of Logic, for example, from Non-Monotonic logic.

The members of the formal argumentation community are not familiar, on average, with other areas of logic, and so we also describe in this book, what we need from neighbouring logics.

This makes this book of interest also to sex offender therapist as well. They may already be familiar with some intuitive logic in their own practices, but the additional logics described in this book will be of interest to them.

Our message to the argumentation community, expressed in their own terms, is that there is a bigger picture of logic use to be formalised and studied.

We view an argumentation network as a system $(S, R$, Tools, Algorithm,Intended Application Area). The (S, R) part is a Dung like formal argumentation graph, (see Chapter 1), the Tools are Enhancement to (S, R), the algorithm is a specific algorithm for computing extensions for $(S, R$, Tools), and the application area is where we want to apply $(S, R$, Tools, Algorithm).

In practice we start with the application area and see what aspects we want to model in the area. Then we design the system $(S, R$, Tools, Algorithm) which can be used as a model for what we want to achieve. The system $(S, R$, Tools, Algorithm) can also be studied in the abstract without knowing from where it originated.

So following the above expanded point of view, we are saying to the argumentation community take a look at this wide and important application area of the sex offender therapy.

There are four independent international communities dealing with reasoning and arguments, namely ((1), (2), (3) and (4) below), (1) and (2) never interacted and have not been interacting until recently and in fact have not even been aware of their common interest, while (2) and (3) were aware of one another. These are:

1. The logic and argumentation community, studying and modelling reasoning and argumentation [1, 2].

2. The sex offender forensic therapy community, using reasoning and argumentation methods in their therapy.

3. The Logic Based Therapy (LBT), see [6, 7, 8]. The integration of logic with clinical practice is not new and already exists in the professional literature, for example in Cohen, E.D. 2022 and others [5, 10, 12, 13], which proposes the integration of logic in cognitive behavioral intervention and Therapy, (LBT). LBT recognises that the need for therapy arises from cognitive distortions in the mind of their patient which in turn causes various maladies. see the LBT list of Fallacies quoted in [3].

4. The Rational Emotive Behavior Therapy (REBT). Quoting from [16], REBT is a form of psychotherapy that helps you identify self-defeating thoughts and feelings, challenge the nature of irrational and unproductive feelings, and replace them with healthier, more productive beliefs. See [11, 9, 14, 13].

From the point of view of Formal logic, the fallacies (and there are maybe 400 of them) can be treated/corrected by pointing out the formal mistakes. This is pure theory . The Therapist cannot just tell the patient he/she are logically/mathematically wrong. The therapist needs to use different more human sensitive approach to help the patient. These Cognitive Distortions/Fallacious reasoning may be different in the LBT patient case as compared with the sex offender patient case and the respective therapy approach may be different. The present article recognises the achievement and wealth of experience of LBT.We can see that LBT deals with Cognitive Distortions and we refer the readers from the formal logic communities and the sex offender communities to look at for example book [5] and paper [4] but seeks to focus on the treatment of sex offenders.

The challenges facing this book in its attempt to interest the formal argumentation community in the sex offender therapy area are fourfold:

1. Some phenomena of reasoning occurring and treated in the sex offender community need to be modelled in logic. For example there is no formal logical model for universal reasoning distortions.[1] Having become aware of this phenomena we need to study it in general logic, the way we study other reasoning

---

[1]This footnote is intended to clarify the our use of the concept Universal Distortion in Reasoning. By a universal reasoning distortion (URD) we mean here a complete distortion of a reasoning system across all of its component. We do not regard a local error of reasoning as a system distortion. For example a reasoner using one or two fallacies such as generalisation or exaggeration is not distorted in his thinking. He is just making an error. (This error may lead the patient to depression but still it is not a Universal distortion). By comparison, a sex offender who would dismiss any argument and commit any fallacy in order to make himself look good and reject any criticism, is universally distorted in his thinking. More examples of a Universal distortion is a person who is drunk. Such a person is

mechanisms like fallacies, fake news, bias, attack, support, etc., etc.

2. We need to identify the kind of distortions sex offenders practice and compare them with other kind of "distorted" reasoning.

3. We need to collect data from sex offender therapists to figure out how they try to correct/address sex offender distortion. This is important to General Logic. (If a good reasoner puts forward an argument which is a fallacy, our response is to point out that it is a fallacy, but doing this to a sex offender may not be the correct response.

4. We need to give a simplified exposition/survey of distortions and therapy in such a way that we get the formal argumentation and the sex offender therapy communities to work together.

There is a serious challenge here as many of the active argumentation researchers are relative beginners and many of the sex offender therapists can also be considered beginners in logic we need to have a lot of repetition in the book so as to make it easier to follow any particular thread locally and not having to read the entire book. For example the definition of the Caminada labelling is used in 3 different chapters, including the background Chapter 1, and reproducing it three times in each context is better for the reader than having the reader run back and forth to Chapter 1.

---

distorted in his functionality across the board. His ability to think straight, his reaction time, his digestion, his balance, etc.

Sex offenders have typical thinking distortions for their group. There are egosyntonic thinking distortions (acceptable to the ego) and the individual is sure that these perceptions are correct and sees no need to change them. Thus, for example, a person with pedophilic sexual deviation often believes that his thoughts are correct and true. For example, the child enjoys sexual contact with an adult. He knows what he wants, and he agrees. This is an example of a universal distortion of a sex offender who will reject any argument, facts, assumptions, etc which threatens his egosyntonic belief. By comparison in the egodystonic case the sex offender has no reasoning distortion. He knows he is wrong, and his mind is open to listening to therapists.

Acknowledgements

We are grateful to Steve Spurr, retired social worker, for contributing Section 3 of Chapter 7.

References

[1] Baroni, P, Gabbay, D., and Giacomin, M., eds. *Handbook of Formal Argumentation*, College Publications, 2018.

[2] Baroni, P, Gabbay, D., and Giacomin, M., eds. *Handbook of Formal Argumentation, Volume 2*, College Publications, 2018.

[3] https://psychology.fandom.com/wiki/Cognitive_Behaviour_Therapy

[4] Carrasco, N., & Garza-Louis, D. (1997). Hispanic sex offenders: Cultural characteristics and implications for treatment. In B. K. Schwartz (Ed.), Handbook of sex offender treatment (pp. 45-1 – 45-10). Civic Research Institute, Inc.

[5] Elliot D. Cohen. *Cognitive behavior interventions for self defeating thoughts. Helping clients to overcome the tyranny of "I can't"*. Routledge, 2022.

[6] Elliot D. Cohen. *Logic-Based Therapy and Everyday Emotions: A Case-Based Approach* (Lanham, MD: Lexington Books, 2016).

[7] Elliot D. Cohen. *Making Peace with Imperfection: Discover Your Perfectionism Type, End the Cycle of Criticism, and Embrace Self-Acceptance* (Oakland, CA: Impact, 2019).

[8] Elliot D. Cohen. *Critical Thinking Unleashed (Elements of Philosophy)*, Rowman & Littlefield Publishers, 2009.

[9] Windy Dryden. Rational Emotive Behavior Therapy. *Encyclopedia of Cognitive Behavior Therapy*. Springer US. 321–324.

[10] Albert Ellis. *Reason and emotion in Psychotherapy*. Lyle Stuart, 1962.

[11] Albert Ellis. Rational Emotive Behavior Therapy. *Encyclopedia of Psychotherapy*. Vol. 2. p. 483–487. 2002 Elsevier Science USA.

[12] Stefan G. Hofmann. *An Introduction to Modern CBT Psychological Solutions to Mental Health Problems*. John Wiley & Sons, 2012.

[13] Jerome Neu. *Emotion, thought and Therapy. A Study of Hume and Spinoza and the Relationship of Philosophical Theories of Emption to Psychological Theories of Therapy*. Routledge and Kegan Paul, 1977.

[14] Rational Emotive Behavior Therapy (REBT), Irrational and Rational Beliefs, and the Mental Health of Athletes, Frontiers in Psychology. 50 years of rational-emotive and cognitive-behavioral therapy: A systematic review and meta-analysis. *Journal of Clinical Psychology*, 2018. `https://www.psychologytoday.com/intl/therapy-types/ rational-emotive-behavior-therapy#:~:text=Rational% 20Emotive%20Behavior%20Therapy%20is,with%20healthier%2C% 20more%20productive%20beliefs.AddressedFebruary09,2024`

PART ONE

CHAPTER 1
A PRIMER TO ARGUMENTATION BASED ON CLASSICAL LOGIC

This chapter gives background from argumentation for this book as the logic and distortion in thinking of a sex offender is modelled in formal argumentation.

We give a short survey of different types and different views of argumentation networks. Our advice is that it will benefit not only the young researcher in argumentation but also the sex offender therapist who can get a feel for argumentation without necessarily following the technical details.

1 Beginning discussion and orientation

Classical logic is the basic workhorse of the area of logic pure and applied. It serves as:

1. Basic logic to teach to the beginning student;

2. Basic logic for modelling application areas;

3. Basic logic into which other logics can be translated and studied and unified and interact;

4. It has many diverse proof systems and machines and algorithms for theorem proving which can be used as initial programs to work for applications;

5. It is strong for formalising mathematics.

For this reason we present the elements of argumentation directly formalised in classical logic.

The presentation is geared towards the following:

1. Model Dung semantics, Caminada labelling, weak admissibility and a multitude of new semantics;

2. Model some algorithms for finding semantics for applications;

3. Model universal distortions of argumentation networks as used by sex offenders;

4. Model certain dialogue and monologue systems as used by sex offenders' group therapy discussions;

5. We specially draw motivating examples from sex offender reasoning and therapy and from the problem of breaking reasoning loops.

2 Formal language

Definition 2.1. *Our language relates to models containing a domain $S \neq \varnothing$ and several binary relations as follows: $(x, y, \ldots)$ variables.*

> *$xR_a y$ attack relation $\subseteq S \times S$*
> *$xR_s y$ support relation $\subseteq S \times S$*
> *$I(x)$, in predicate*
> *$O(x)$ out predicate*
> *$U(x)$ undecided predicate*
> *$L(x)$ a label predicate meaning x is abnormal*
> *$E(x)$ E is an extension.*

Other labels $L_i(x), i = 2, 3, \ldots.$

Definition 2.2. *The following are basic axioms relating to the predicates.*

Axiom Group 1. *Axioms relating R_a and R_s. These depend on the application area and the interpretation of "attack" and "support" in the application.*

For example, in deductive support we have

$$xR_s y \wedge zR_a y \rightarrow zR_a x$$

For other types of support, this axiom may not be taken. If xR_sy means x recommends y, then z attacking y does not imply z attacking x.

Another possible axiom is to reduce support to attack: x supports y iff x attacks all z which attack y.

Axioms Group 2. Relating *In, Out, Undecided* and Extensions E. *All agree that*

1. *If x is* in *and x attacks y and x is* not *abnormal, then y is* out.

2. *If all attackers of y are either abnormal or* out *then y is* in.

3. *Other options involving undecided may depend on the application. Yet other options relating to the other L label:*

   (a) *The Dung axiom for undecided is: If all attackers of y are either* out *or undecided and at least one attacker is undecided then y is undecided.*

4. *Another axiom for abnormal nodes*

   (a) $L(x) = xR_ax$ *or another axiom* $\forall x \neg L(x)$ *Nothing is abnormal.*

5. *Axioms about E (Extensions).*

   For example an axiom for E to be a Dung preferred extension:

   $$E(x) \land E(y) \rightarrow \neg xRy$$
   $$E(x) \rightarrow in(x)$$
   $$\forall z[(\neg \exists x(E(x) \land zRx) \land \forall u(uRz \land \neg zRu$$
   $$\Rightarrow \exists v(E(v) \land vRu))$$
   $$\Rightarrow E(z)]$$

   If E is not maximal extension the axiom will be violated.

Remark 2.3 (Background theory). *We may have a background theory Δ using other predicates $\Pi_1, \Pi_2, \ldots$ which is used to define the $L(x)$, xR_ay and xR_sy predicates.*

Changing Δ to Δ^* this changes the values/meaning of $L(x), R_a, R_s$, to also change, which causes a universal distortion. See remark below.

Remark 2.4.

1. *What is an argumentation system?*
   It has the form $\mathbf{M} = (S, R_a, R_s, \text{in}, \text{out}, \text{und}, E, \mathcal{B} = \text{basic axioms}, \Delta).$

2. *What is semantics?*
   Semantics is a set of additional axioms $\mathcal{SEM}$ involving in, out, und, L *and* E. R_a, R_s.

3. *What is an extension of the semantics $\mathcal{SEM}$? How to find extensions?*
   Take a model of $\mathbf{M} \cup \Delta \cup \mathcal{B} \cup \mathcal{SEM}$ *with domain S and look at the set*

   $$E_{\mathbf{M}} = \{x \in S \,|\, \mathbf{M} \models E(x)\}.$$

   Then $E_{\mathbf{M}}$ is one extension for $\mathcal{SEM}$. All the extensions are obtained by looking at all the models $\mathbf{M}$ of $\mathcal{B} \cup \mathcal{SEM}$.

Remark 2.5. *It is also possible to take many labels*

$$L_1(x), L_2(x) \ldots$$

This allows us to extend the semantics with strength, etc.

3 Argumentation through dialogue

Note that this approach allows us to model dialogues with a changing database and universal distortion, as practiced in Sex Offender Therapy sessions.

Such modelling runs contrary to what we find in the literature on argumentation dialogues, where the database is fixed and the arguments are minimal pieces of the database and an atom they prove for example Definition 2.1 of the paper "Properties and Complexity of Some Formal Inter-agent Dialogues".

This generalizes ASPIC and structured argumentation, which we will not elaborate on in our current book. See chapter 2 of our book 691 in progress.

In this book we model sex offender therapy dialogues. But to motivate this we need first to look at an example.

Example 3.1 (Example from a sex offender therapy dialogue). *We need to look at Figure 1. Here we use → for attack and —□ for support.*

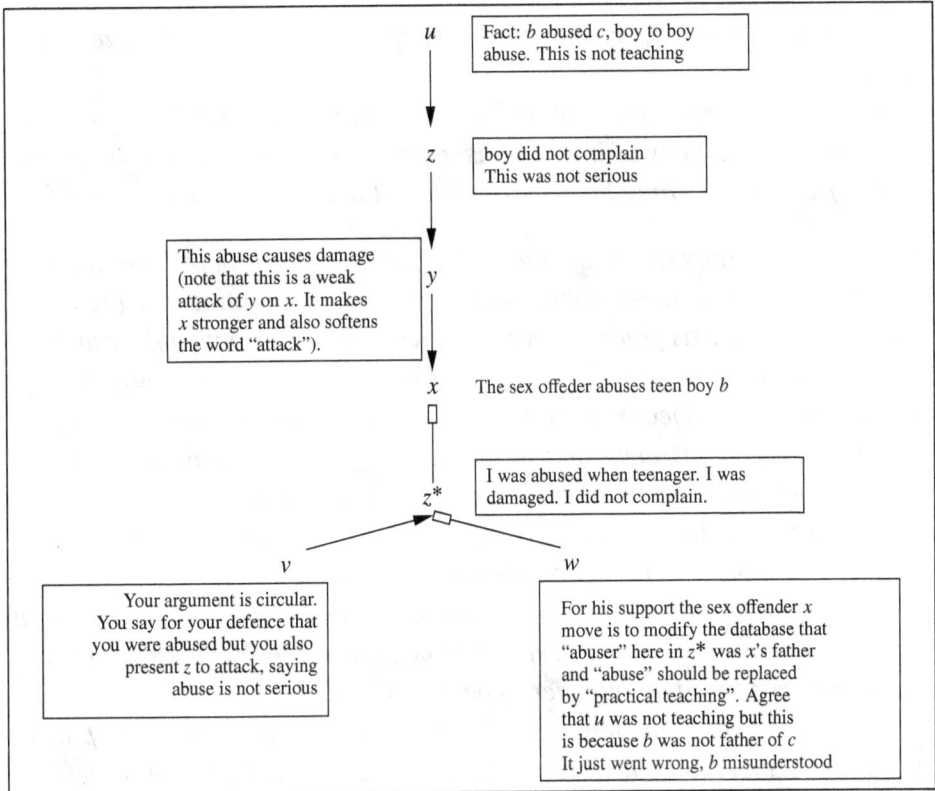

Figure 1

The situation here is typical with the sex offender narrative. The dialogue is as follows (Figure 1):
Node x = x abused a teenager b
Node y: The therapist attacked by saying it is damaging.
The sex offender said two things for his defence:
Node z, (attacking node y): It could not have been serious because the boy b did not complain. In fact it was good he taught b practical sex!
The schools are not doing a good job.
Node z: Also don't blame me if I was abused myself when I was a teenager.*
Node v: The therapist says you x are contradicting yourself. You say b did

not complain, this is not serious and at the same time you bring to your defence the fact that you were abused.

Now the sex offender can change the theory Δ and introduce a distortion:
Node w: The person who "abused" me was my father. He did not abuse me he taught me the facts of life. This is good. I taught the facts of life to the boy b. But it went wrong because b could not teach b is not the father of c.

Remark 3.2. *Many subjects try to find different ways to control the situation of the clinical risk assessment. They try to regain control and break the hierarchy that exists from the very difference between patient/therapist by trying to dictate the setting of risk assessment or the therapeutic setting. For example: a subject who is a recidivist and a high sexual risk offender came to a risk assessment and when the conversation began he refused to answer, took out a page and wrote on it that he was mute following a recent violent arrest by the police and is only willing to be tested on the condition that allows him to write his answers on a page. This is an example of a person who is trying to control the situation, trying to dictate the rules and trying to emphasize his own feelings of being a victim by himself and shift the focus from being an offender to being a victim.*

This practice occurs many times in both treatment and risk assessment, when the subjects/patients are mainly preoccupied with feeling of being a victim and they are blaming external factors, and this pattern allows them to avoid introspection and looking at their offensive behaviour, but to remain preoccupied with their filings of being a victim. A person who feels pitiful, hapless and a victim often allows himself a behaviour that hurts another and explains it to others and to himself from the place of the victim, such as, for example, "It doesn't matter that I hurt 10 victims, in one of the cases the police were not accurate and wrote something wrong about me... They are all criminals and therefore I am also allowed to be a criminal". This is a defense mechanism that helps a person continue to hurt and it is also a severe thought distortion that helps a person live at peace with himself.

Remark 3.3. *We offer further discussion about dialogues. Dialogues have two components. One is the subject matter itself and the other is a matter* *of procedure. When a therapist conducts a dialogue with a patient it is a subject matter questions and answers dialogues with a view of offering some solutions to problems with the patient. However, when the therapist gives testimony in court and conducts a dialogue with the prosecution and the defence attorneys in front of a judge, this dialogue has also a procedural component.*

In Chapter 8, we see how such dialogues are conducted in Israel. The expert therapist submits a report and the prosecution and defence can question his qualifications and ask the expert questions about his report. The expert gives his answers. The procedure does not allow the expert to modify or change his report. Moreover, the back and forth dialogue has only two steps. The expert is questioned and he answers. In fact, even the dialogue questions in this case are controlled (see Chapter 8).

When an expert writes an assessment of how dangerous a sex offender is, to submit to a court, the expert must write down all the parameters and all the arguments that led him to the decision he formulated in his report and present his argument of how according to the data he determined the level of dangerousness of the sex offender.

In cases where the court or a lawyer representing the sex offender invites the expert to testify in court, the risk assessor must defend and explain what is written in the document he submitted. He cannot add additional/new parameters that did not appear in his document, and if there are important parameters that do not appear in his document, it harms the professionalism and prestige of the writer.

In fact, the field in which the dialogue involving arguments for and against and the subtleties which can be conducted is very stylized and focuses on what was written in the initial document

The expert author even declares at the end of each assessment point that this opinion constitutes a testimony as if it were given in court before a judge.

There are many examples of dialogues in this book: Figure 1, Chapter 1. In Chapter 3, Examples 4.3 and 5.6; Section 5.1 and Appendix A.

4 Argumentation systems with attack only

This Section presents, for the convenience of the reader, some basic concepts of what we called traditional argumentation theory. Such systems contain attacks only. We refer to such system as Argumentation with Attack only. One can also add support to the system and in this case we get systems of Argumentation with Attack and Support. We shall then explain in what way the systems required for this chapter depart from the traditional ones.

There are two ways to present the semantics for argumentation with attack, the traditional set theoretical approach and the Caminada labelling approach. For the mapping connections between the two approaches, see [1]. Let us briefly quote the traditional set theoretic approach:

Definition 4.1.

1. *We begin with a pair (S, R), where S is a nonempty set of points (arguments) and R is a binary relation on S (the "attack" relation).*

2. *Given (S, R), a subset E of S is said to be conflict free if for no x, y in E do we have xRy.*

3. *E protects an element $a \in S$, if for every x such that xRa, there exists a $y \in E$ such that yRx holds.*

4. *E is admissible if E is conflict free and protects all of its elements.*

5. *E is a complete extension if E is admissible and contains every element which it protects.*

6. *A subset E is a stable extension if E is a complete extension and for each $y \notin E$ there exists $x \in E$ such that xRy.*

7. *E is the grounded extension if it is the unique minimal extension (it exists, see Lemma 4.2).*

8. *E is a preferred extension, if E is a maximal (with respect to set inclusion) complete extension.*

9. *A Semantics is a (metalevel) property $\mathbb{S}$ of extensions, such as being stable, or being grounded or being preferred. Thus we can talk about $\mathbb{S}$-Semantics, (stable semantics, grounded semantics and preferred semantics) where we consider only $\mathbb{S}$- extensions.*

Lemma 4.2. *For any network (S, R) there exists a grounded extension (which may be empty).*

Proof. This can be proved, using set theoretical methods, see [1, 2]. A proof can be obtained from the proof of Lemma 2.2 of Chapter 5 for the case of V giving all arguments life 1. See also the general construction of Section 8.3 of Chapter 5. ∎

We can also present the complete extensions of $A = (S, R)$, using the Caminada labelling approach, see [1].

Definition 4.3. *A Caminada labelling of S is a function $\lambda : S \mapsto \{in, out, und\}$ such that the following holds.*

(C1) $\lambda(x) = in$, *if for all y attacking x, $\lambda(y) = out$.*

(C2) $\lambda(x) = out$, *if for some y attacking x, $\lambda(y) = in$.*

(C3) $\lambda(x) = und$, *if for all y attacking x, $\lambda(y) \neq in$, and for some z attacking x, $\lambda(z) = und$.*

Lemma 4.4.

1. *A consequence of (C1) is that if x is not attacked at all, then $\lambda(x) = in$.*

2. *Given an extension E let λ_E be defined by $\lambda_E(x) = \{$ in if $x \in E$, out if for some $y \in E$ we have yRx, and undecided otherwise$\}$. Conversely given a λ, define E_λ to be $\{x | \lambda(x) = in \}$.*

3. *Any Caminada labelling yields a complete extension and vice versa.*

4. *Any $\{$in, out$\}$ Caminada labelling (i.e. with no "und" value) yields a stable extension and vice versa.*

5. *Set theoretic minimality or maximality conditions on extensions E correspond to the respective conditions on the "in" parts of the corresponding Caminada labellings.*

Proof. See [1]. ∎

Example 4.5. *It is useful to introduce a familiar story as an example, the story of the party.*

Story. The Party: *We are planning a party and we have a set S which is the maximal set of all relatives friends, colleagues, etc. who can be invited to the party. The problem is that some of them do not get along/hate some others. So we have a relation R, where xRy (which we might denote by x ⇸ y) means that if x is invited, y must not be invited. We get here a traditional argumentation network with attack relation R. The complete extensions are possible groups of people we can invite, provided we invite all those to whom no other invitee objects.*

Remark 4.6 (Translation into classical logic). *The network (S,R) can be viewed as a classical predicate model for a binary relation symbol **R**. The domain of the model is S and the extension of **R** is R.*

*With this point of view, we can add additional predicates symbols to the language to be able to talk in classical logic syntax about extensions. Let us add the predicates **E**, unary for subsets of S, **In** for points in S that are in, **Out** for points which are out and **Und** for points which are undecided.*

*We can write axiomatically the conditions for **E** being a complete extension and for the conditions of the vector (**In**, **Out**, **Und**) to be a Caminada labelling.*

1. *Conditions on **E** (where "..." indicates a formula of predicate logic):*

   (a) *"**E** is conflict free": $\forall x,y[\mathbf{E}(x) \wedge \mathbf{E}(y) \rightarrow \neg xRy]$.*

   (b) *"**E** protects the element a": $\forall x[xRa, \rightarrow \exists y(\mathbf{E}(y) \wedge yRx)]$.*

   (c) *"**E** is admissible (i.e. **E** is conflict free and **E** protects all of its elements)": "**E** is conflict free" $\wedge \forall a[\mathbf{E}(a) \rightarrow \forall x[xRa, \rightarrow \exists y(\mathbf{E}(y) \wedge yRx)]]$*

18

(d) "**E** *is a complete extension (i.e* **E** *is admissible and contains every element which it protects)":* "**E** *is admissible"* $\wedge \forall a$ *[*"**E** *protects the element a"* $\rightarrow$ **E**(a)*]*

2. *Conditions for Caminada labelling:*

(a) $\forall x[\mathbf{In}(x) \vee \mathbf{Out}(x) \vee \mathbf{Und}(x)])$

(b) $\forall x[\neg(\mathbf{In}(x) \wedge \mathbf{Out}(x) \wedge \neg(\mathbf{Und}(x) \wedge \mathbf{In}(x)) \wedge \neg(\mathbf{Out}(x \wedge \mathbf{Und}(x))]$

(c) $\forall x[\exists y(y\mathbf{R}x \rightarrow \mathbf{Out}(x)]$

(d) $\forall x[[\forall y(y\mathbf{R}x \rightarrow \mathbf{Out}(y)] \rightarrow \mathbf{In}(x)]$

4.1 Adding valuations or preferences

Given an argumentation network (S, R), consider a node $s \in S$ and its set of attackers $A(s) = \{y \in S \,|\, y R s\}$. Following considerations in Subsection 4, the question of whether s is "in" or not is basically algorithmic based on the geometry of (S, R). The arguments $\{y \,|\, y \in A(s)\}$ themselves are atomic, and no considerations are available about their nature, such as "who put them forward", "why we think they are true", "how strong they are compared with other arguments", "are they independent of each other", etc., etc.

If we wish to protect node s from its attackers $A(s)$, we might wish to identify various properties $V_1(y), \ldots, V_k(y)$, of nodes $y \in A(s)$ and then argue, that given these properties, we want to reject some or all the attackers of $A(s)$. The predicates $V_i(y)$ are called valuations. They could be qualitative (true or false of y) or numerical ($V_i(y) \in [0, 1]$). These predicates are meta-level to the arguments in S, and they compensate for the abstract atomic nature of the elements of S. Their purpose is to mitigate the attacks of $A(s)$ on s. We can thus consider a formula $\mathbb{B}^s$ involving the predicates $\{V_i(y), V_i(s) \,|\, i = 1, \ldots, k, y \in A(s)\}$ which says something about $\{y \,|\, y \in A(s)\}$ and use it to modify clauses (C1)–(C3) involved in the Caminada labelling of Subsection 4. For example, we can say that if $\mathbb{B}^s$ does not hold then certainly s must be "in" because we should ignore all the attacks on s.

19

The idea of valuation is helpful in modelling reasoning distortions. We shall see our sex offender of our case study in Section 2.2, Chapter 7 when defending the attack on the claim

$s = $ " I have not offended"

against the testimony of child y about the offence, the offender added

$V(y) = $ "y has not complained about the offence for a long time".

This is supposed to mitigate the seriousness of the offence. The predicate "V" is not part of the language of (S, R), it is a valuation added to it, and seems useful in modelling distortions introduced by sex offenders.

So if we have a system $(S, R, V_1, \ldots, V_k)$, a distortion can be affected either by modifying $\{V_i\}$ or by tinkering with $\{\mathbb{B}^s\}$.

Definition 4.7 (Networks with Bench-Capon type valuations).

1. *We say that networks of the form $(S, R, V_1, ..., V_n)$ are of Bench - Capon type if all V_i are subsets of S.*

2. *We define Caminada labelling for such networks in terms of the translation into classical logic of item 2 of Remark 4.6.*

3. *Let $\mathbb{B}(y, s)$ be defined as the formula $\mathbb{B}(y, s) = \bigwedge_{i=1 \to n}[V_i(s) \to V_i(y)]$*

4. *Conditions for Caminada labelling:*

   (a) *$\forall x[\mathbf{In}(x) \vee \mathbf{Out}(x) \vee \mathbf{Und}(x)])$*

   (b) *$\forall x[\neg(\mathbf{In}(x) \wedge \mathbf{Out}(x) \wedge \neg(\mathbf{Und}(x) \wedge \mathbf{In}(x)) \wedge \neg(\mathbf{Out}(x \wedge \mathbf{Und}(x))]$*

   (c) *$\forall x[\exists y(y\mathbf{R}x \wedge \mathbb{B}(y, s) \to \mathbf{Out}(x)]$*

   (d) *$\forall x[[\forall y(y\mathbf{R}x \to \mathbf{Out}(y) \vee \neg\mathbb{B}(y, s)] \to \mathbf{In}(x)$*

20

4.2 Abstract dialectical frameworks (ADF)

ADF was introduced to the argumentation community in 2010 in [6]. It was originally introduced as (in our humble opinion) a mathematical extension of argumentation, giving it the strength of the classical propositional calculus in an explicit form. Such moves are common and useful in mathematics, and indeed ADF has evolved, overcame various difficulties and gained respectable grounds in the argumentation community since its introduction and is now a powerful tool, see [7]–[11], [12], [13]. To be able to discuss ADF, let us just give a definition of some simplified version of ADF.

Definition 4.8 (Boolean ADF networks). *We formally define the notion of a Boolean ADF network. This is a notion just for this chaptre so the reader can have a formal definition of the type of ADF we are talking about.*

Let (S, R) be an argumentation network and for each s in S let $\mathbb{C}(s)$ be a Boolean combination of variables in the set of elements of $\{y|yRs\}$. Consider the set of equations of the form

$$s \leftrightarrow \mathbb{C}(s), s \in S.$$

Any solution of these equations in Kleene 3 valued logic is considered an Extension for the system $(S, R, \mathbb{C}(s), s \in S)$.

For Kleene 3 valued logic, see Figure 22 of Chapter 6 (1=in, 0=out, 1/2 = undecided).

Remark 4.9 (Some ADF variations). *We can add some predicates to (S, R), say a family of n_i-place predicate P_i, and allow them to be used in the formulas $\mathbb{C}(s)$ above. We need to regard each $P(y_j)$, as propositional atomic, (a trick well known from classical logic model theory)*

Models of the sets of equations (equivalences) above in this language in Kleene 3 valued logic will give us the complete extensions, if they exist.

For the practical sex offender therapist, however, the mathematics is less important than the intuition behind it, and in this case, the ADF is intuitively capable, mathematically powerful, especially when generalised to a matrix form or a more general form as will be indicated in Subsections 8.2 and 8.3 of Chapter 6. We do require, however, some generalisations and some restrictions on ADF, if we want to use it for the sex offender case. First we need to use predicate logic formulas, not propositional logic formulas. Second, the argumentation/logical attacks on sex offenders are monotonic. The more attackers you have the stronger is the case. So if the set of arguments E can kill x, and E is a subset of E', then also E' can kill x. This property does not hold for a general ADF. We now explain.

Note that the traditional Dung notion of attack is lost in the framework of ADF. We start with (S, R), but regard, for each $s \in S$ the set $A(s) = \{y|yRs\}$ not as a set of attackers, but as a linked set of related nodes. We say that s is 'in" iff some Boolean combination $\mathbb{C}_s(\{y|y \in A(s)\})$ holds. So s is out when $\mathbb{C}_s$ does not hold, and then the set $A(s)$ can mount a successful attack on s.Thus we need to have a $\mathbb{C}_s$ associated with each s. This is not a monotonic attack condition. $\neg\mathbb{C}_s$ might say, for example, that the set $A(s)$ contains an even number of "in" elements. So if another attacker becomes "in", the attack fails. There is no monotonicity here. We may consider only monotonic ADF predicates. But even for this case ADF have their own way of calculating extensions, which is still incompatible with the sex offender case (we omit details here). Suppose still that we adopt this approach (of taking monotonic $\mathbb{C}_s$) but calculate the extensions using Boolean equations, namely, we take as our (AVF) complete extensions of the network as all Kleene three-valued models of the theory $\{s \leftrightarrow \mathbb{C}_s|s \in S\}$.

We still cannot use this mathematical version for two reasons.

1. Even if we restrict the formulas $\mathbb{C}_s$ to be monotonic, (and this is not easy because the formula does not have a fixed number of variables), we still need the explicit intuition of "attack" which is implicit in ADF. The sex offender creates distortion as a response to direct explicit threats, not to implicit ones. Furthermore the sex offender does not protect himself by talking about combinations of in out of his attackers; he gives other types of valuations, like "racist attack", "unimportant", etc. This requires the formulas $\mathbb{C}_s$ to be predicate logic formulas containing the predicates $V(y)$, for yRs, which can be genuinely set theoretical or also be numerical values.

2. The valuation approach can technically represent the Boolean fragment of ADF. If we add naming valuations of the form $V_s, s \in S$ such

that $V_s^y = \top$ iff $y = s$, then we can express Boolean ADF.[1] More importantly, we can be very specific and project a distortion on each and every individual argument $x \in S$ via the use of the naming predicate $V_x(y)$.

This observation is in fact important for Section 9 of Chapter 6, the comparison with the literature. There will also be a discussion at the end of Option 3 in Section 7.3 of Chapter 6.

4.3 The equational approach

This approach views (S, R) as a carrier for equations in $[0, 1]$ for the variables $\{s | s \in S\}$. We are given for each s a continuous function $\mathbf{f}_s(A(s))$ in the variable $A(s) = \{y | yRs\}$. We look at the equations $\{s = \mathbf{f}_s(A(s))\}$. These have solutions in $[0, 1]$. Any such solution generates a complete extension for (S, R) in the sense of Subsection 4, provided $\mathbf{f}_s$ satisfied the suitable properties, and provided that we let value 1 for s to mean "in", value 0 for s to mean "out" and otherwise undecided.

Two such functions are notable

$$\mathbf{f}_{\max}(s) = 1 - \max\{y | yRs\}$$

and

$$\mathbf{f}_{\mathrm{inv}}(s) = \prod_{yRs}(1 - y).$$

The extensions we get from $\mathbf{f}_{\max}$ can be proved to yield exactly the extensions in Subsection 4. See [3].

The equational approach $\mathbf{f}_{\max}$ is suitable for modelling argumentation networks with numerical valuations. If the meta-predicates $V_i(s)$ give numerical values (see Subsection 4.1), these need to be integrated with the traditional "in", "out", "undecided" values for $s \in S$ (see Subsection 4). The best way of doing this is to adopt the equational approach to (S, R) and integrate the numerical functions V_i into the equations.

---

[1]This is not a criticism of ADF. In logic there are many systems expressing one another, it is the presentation that makes the difference. Note that our machinery is not the same as ADF machinery, so although we can express Boolean formulas we do not do to it what ADF does.

5   More background and concepts from abstract argumentation

5.1   Adding support

We now add support to the attack only networks, to get networks with attack and support. Let us denote such networks by (S, R, R_S), where $R_S \subseteq S \times S$ is the support relation (notation: xRy is denoted by $x \twoheadrightarrow y$ and xR_Sy is denoted by $x \to y$).

While there is agreement on the attack networks $(S.R)$, there is no community agreement on how to handle support. See [4] for a research/survey. Systems with both attack and support are called bipolar networks. There are many proposals to give semantics to such networks. Our chapter, on attack and support for the expert witness model, requires a new approach, to enable us to model how support is handled in court in this particular case.

To see our options for support, consider Figure 2

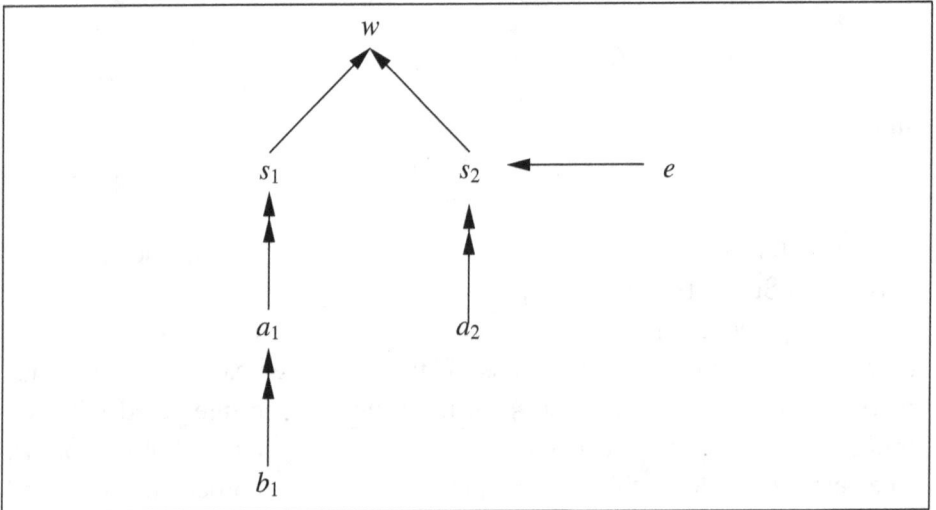

Figure 2

In this figure s_1 and s_2 support w. Consider the attack of a_1 on s_1. Assume that this attack is successful. So certainly s_1 is out/dead.

Question 1. Does this successful attack also knock out the other supporters of w, namely s_2 and e?

Question 2. Does e also support w, through its support of s_2?

Question 3.

   i. Does a_2 also attack e (because e supports s_2 and a_2 attacks s_2).

   ii. Does a_2 also attack w, (through its attack of its supporter s_2)?

Our first definition, Definition 5.1 adopts the following view for support.

1. The support relation is reflexive and transitive.

2. We "lump" together any z and all of its supporters. To attack z you can attack either z or any of its supporters. In comparison, Definition 5.1 adopts a slightly different view of support.

3. We lump together all supporters y of any z. Any attack on any y is considered an attack on all of the ys and if successful, takes out/dead all of them.

Definition 5.1. *Let $S \neq \varnothing$ be a finite set and let R_A and R_S be two binary relations on S. The system (S, R_A, R_S) is called argumentation system with attack and support, using options 1 and 2 if the following holds:*

1. *R_S is reflexive and transitive. We understand xR_Ay as x attacks y (notation also $x \twoheadrightarrow y$) and we understand xR_Sy as x supports y (notation $x \rightarrow y$).*

2. *Given a system (S, R_A, R_S) and a set $E \subseteq S$, we say that E is conflict free iff the following holds*

(*) *There are no u, x, y, z such that $x, z \in E$ and $u \rightarrow x \twoheadrightarrow y \rightarrow z$.*

3. *Let $a \in S$ and $E \subseteq S$. We say that E protects a if condition ($\sharp$) holds:*

($\sharp$) *For any $u \twoheadrightarrow y \rightarrow a$ there exist $z \in E$ and a such that $z \twoheadrightarrow v \rightarrow u$.*

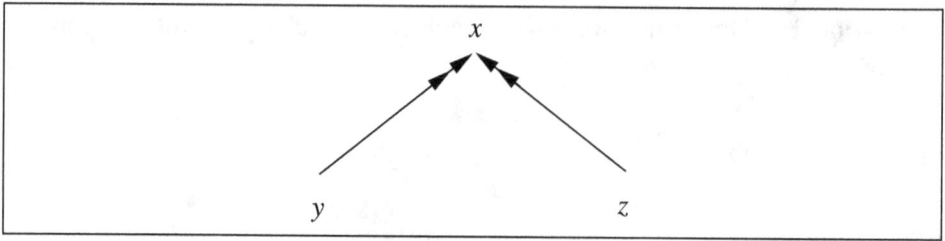

Figure 3: Informational attack

4. *E is admissible if E protects all of its elements.*

5. *E is a complete extension if E is admissible and contains every element it protects.*

6. (S, R_A, R_S) *is attack only system if R_S is identity, $(xR_S y$ iff $x = y)$ in which case we write (S, R).*

Definition 5.2. *This definition is similar to Definition 5.1 except that:*

1. *We do not require that R_S is reflexive nor transitive*

2. *We replace $(*)$ by $(**)$ and $(\natural)$ by $(\natural\natural)$ as follows:*

$(**)$ *There are no $x, z \in E$ such that $x\rho z$ holds, where $x\rho z$ iff there are u, y such that $[x \twoheadrightarrow u$ and $u \to y$ and $z \to y]$ or $x \twoheadrightarrow z$.*

$(\natural\natural)$ *For any x such that $x\rho a$, there exists $e \in E$ such that $e\rho x$.*

Definition 5.3. *We say an argumentation network (S, R_A, R_S) is a-cyclic if the transitive closure of $R_A \cup R_B$ contains no cycles with more than one element (equivalently it is anti-symmetric). See for example Figure 8 of Chapter 8.*

5.2 Examples illustrating informational attacks

In this Subsection we would like to give some material illustrating informational attacks. For a full coverage see the long version of [5].

Example 5.4. *Let us tentatively consider the situation in Figure 3.*

In this figure we assume that x, y and z are pieces of information and that y and z attack x. We understand the attack of any a on b, written as a ↠ b, to mean that we update b with information $\tau(a)$ sent from a to b. $\tau(a)$ is part of the information a, which is sent to attack b. We could have $\tau(a) = a$. We obtain a new piece of information $b \oplus \tau(a)$ (in many cases we have $\tau(x) = x$ and $\oplus = \cup$ and so the result is $b \cup a$).

Remark 5.5. *the idea of information update and the notation $\oplus$ is not new. It appears in the context of the logic and semantics for substructural logics, in connection with the semantical condition and the substructural implication ⊸.*

We envisage a language with atoms and ⊸ and with semantical models of the form $(S, \oplus, h)$, where S is a set of pieces of information and $\oplus$ is an associative and commutative binary operation (being a fancy way of writing conjunction "and" in the context of multi-sets, i.e., when we want to stress that "x and x" is strictly more than just "x") and where h is the assignment to the atoms (e.g. $h(q) \subseteq S$) and we let, for $x \in S$, satisfaction $\models$ to be defined by

- *$x \models q$, if $x \in h(q)$, for q atomic*

- *$x \models A \multimap B$ iff for every $y \models A$ we have $x \oplus y \models B$.*

Here $\tau(a) = a$, i.e. the node a sends the complete/entire information it has to its target. In practice (witness cases) only some part or a combination $\tau(a)$ derived from a is sent.

Example 5.6. *To give examples from substructural logics, consider the following possibilities.*

1. *If S is a family of multi sets and $\oplus$ is multi set union then ⊸ becomes linear implication.*

2. *If S is a family of sets and $\oplus$ is set union then ⊸ becomes relevance implication.*

3. *If we interpret $\oplus$ as equality = then ⊸ becomes S5 strict implication.*

$$x = \{a \wedge b \rightarrow \bot, \neg e \rightarrow a\}$$

$$y = \{e\} \qquad\qquad z = \{b\}$$

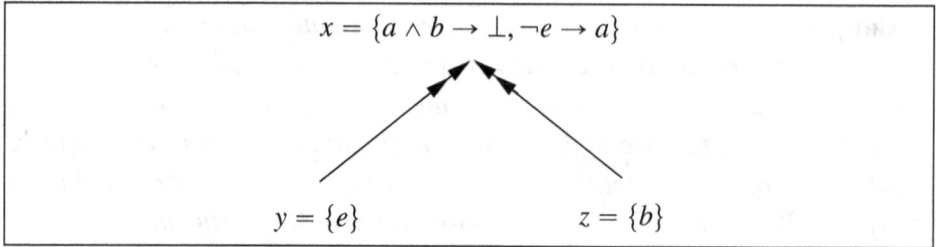

Figure 4: A logic programming example

Thus we can use $\oplus$ to explain the attacks in Figure 3 to mean that $x = \tau(y) \oplus \tau(z)$ or $x = y \oplus z$ if we let $\tau(u) = u$.

We need to give some examples of the use of $\oplus$ in attack.

Example 5.7.

1. *Consider Figure 4*

   The nodes in Figure 4 are logic programming databases where "$\neg$" is negation by failure, "$\wedge$" is conjunction, and "$\rightarrow$" is the logic programming implication. a, b, e are atoms and "$\bot$" is falsity. The database x can derive a, we write $x \vdash a$. If x gets attacked by y alone, then it gets the input e and so $\neg e$ no longer succeeds from $x \oplus \{e\}$, and so $x \oplus y$ cannot derive a. If x is attacked by z alone, then we get that x gets the input b alone so we have $x \oplus z$, which becomes inconsistent.

   If x is attacked by both $y = \{e\}$ and $z = \{b\}$, then it becomes $x \oplus y \oplus z$ which remains consistent and can derive just b.

2. *To further our understanding, note that the databases $x = \{\neg e \rightarrow a, a \wedge b \rightarrow \bot\}$ and $x1 = \{a, a \wedge b \rightarrow \bot\}$ are not the same, even though both derive $\{a, a \wedge b \rightarrow \bot\}$, because $x1 \oplus \{e\}$ derives a while $x \oplus e$ does not derive a.*

3. *Consider now Figure 5 In this figure the node u attacks the node v. The question we ask is what information does u send to v?*

   On the one hand u can derive a. So if it sends $\{a\}$ (i.e. $\tau(u) = \{a\}$), it will render $u \oplus \tau(u)$ inconsistent, because v derives $a \wedge b \rightarrow \bot$.

   But if it sends itself, namely $\tau(u) = u$, then $v \oplus u$ cannot derive a and so remains consistent.

$$v = \{e, b, a \wedge b \rightarrow \bot\}$$

$$u = \{\neg e \rightarrow a\}$$

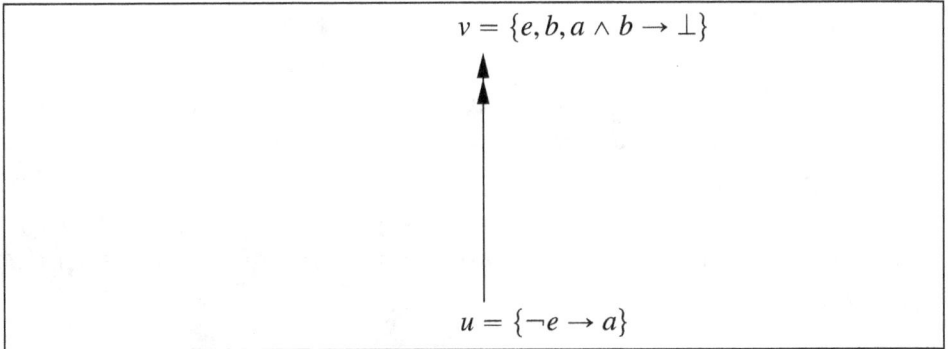

Figure 5: Second logic programming example

Definition 5.8.

1. *An information system is a set I with a binary associative and commutative operation $\oplus$ on I. Let $\mathbf{O} \in I$ be empty information with $x \oplus \mathbf{O} = x$ for all x.*

2. *Let (S, R) be a finite acyclic graph as in Definition 7.8. of Chapter 8 and let $\mathbf{f}$ be a function giving for each point x in S a value $\mathbf{f}(x) \in I$.*

   We define an information label $\mu(x)$, for each $x \in S$, in steps, using $\mathbf{f}$.

   Step 1. *Let x be any source point. Define $\mu(x)$ to be $\mathbf{f}(x)$.*

   Step $n + 1$. *Let x be any point such that $\mu(x)$ is not yet defined but for all z such that $zRx, \mu(z)$ is defined. Then let $\mu(x)$ be defined as*

   $$\mu(x) = \mathbf{f}(x) + \mu(z_1) \oplus \ldots \oplus \mu(z_m)$$

   where $z_1, \ldots, z_m$ are all the points in S such that zRx holds.

Let S_μ be the set of all $x \in S$ for which $\mu(x)$ is defined.

Proposition 5.9. *We have for S_μ of Definition 5.8 that $S_\mu = S$.*

Proof. Similar to the proof of Proposition 4.4 of Chapter 8. ∎

Example 5.10. *We now give a useful example of an information system labelling. Let (S, R) be acyclic graph. Regard the elements of S as atoms for constructing a logic programming information system. For each $x \in S$ and for $y_1, \ldots, y_m$ being* all *the nodes y such that yRx, construct the clause $C_x =_{\text{def}} \neg y_1 \wedge \ldots \wedge \neg y_m \to x$. If x is is a source point then let $C_x =_{\text{def}} x$. Consider as pieces of information as finite sets of clauses of the form C_x. Let $\oplus$ be defined on such pieces of information as set union $\cup$. Consider the function $\mathbf{f}$ on S, defined for $x \in S$ by*

$$\mathbf{f}(x) = C_x.$$

We can now consider an information system I whose elements are sets of clauses $\{C_x\}$, with $\oplus$ taken as union $\cup$. Consider μ defined from $\mathbf{f}$ as in Definition 5.8. We now got a special information system labelling for any acyclic (S, R). To fully appreciate the role of this example we need to know more about logic programming. The next definitions and theorem will tell us a bit more.

Definition 5.11. *Let $(I, \oplus)$ be an information system. Let Q be a set of atomic sentences. A function $\pi : Q \times I \mapsto \{0, 1\}$ is called a consequence function. For $q \in Q$ and $x \in I$ we also write $x \vdash_\pi q$ when $\pi(q, x) = 1$.*

There can be various restrictions on $\vdash$, but we need not go into detail here. The system $\vdash$ we use in this Chapter will be known and recognised as reasonable.

Example 5.12. *Let us go back to the logic programming information system labelling of Example 5.10, defined for any acyclic finite graph (S, R). Consider now $\mu(x)$, for $x \in S$. $\mu(x)$ is a set of logic programming clauses. Let $\vdash$ be the logic programming consequence relation. We are not defining this here, we assume it is know to the reader), see also [5]). Let λ_μ be defined on S by*

- *$\lambda_\mu(x) = 1$ if $\mu(x) \vdash x$, and*

- *$\lambda_\mu(x) = 0$ if $\mu(x) \vdash \neg x$.*

- *If neither $\mu(x) \vdash x$ nor $\mu(x) \vdash \neg x$ holds we say that the value of λ_μ is undecided.*

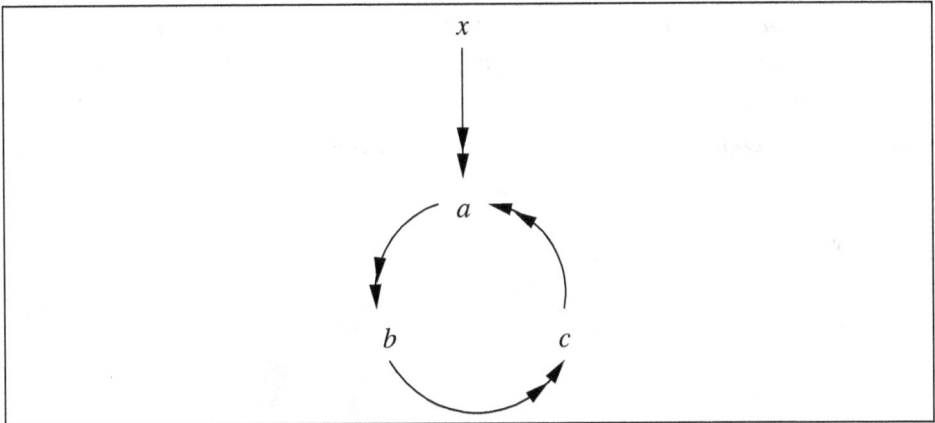

Figure 6: A cyclic graph

The above considerations introduced a very specific $\{0, 1\}$ labelling λ_μ on (S, R) obtained from a very specific logic programming information system labelling for (S, R). We now want to compare it with the traditional Caminada labelling λ for (S, R), introduced in Definition 4,3 of Chapter 8. In fact we can prove the following Proposition (see [5]):

Proposition 5.13. $\lambda = \lambda_\mu$

This shows that the informational approach can simulate the traditional approach at least for the case of acyclic networks.

A detailed investigation of the informational approach can be found in the next Subsection 5.3 and in [5].

5.3 Further discussion

This section provides more explanation of the formal model.

Example 5.14. *Consider Figure 6 which will help us make a few explanatory remarks. We considered on acyclic graphs (S, R) and compared the Caminada extension labelling with the information system labelling. The graph of Figure 6 is cyclic and so is not acceptable to us. It does, however, have a unique grounded extension with Caminada labelling $\lambda(a) = 0$, $\lambda(x) = 1$, $\lambda(b) = 1$, $\lambda(c) = 0$. We just do not need such a graph for analysing and modelling expert testimony and therefore we insisted on* *acyclic graphs for our modelling. The cycles however, could present a problem for the informational labelling because of the way these are defined. So borrowing from our paper [5], let us analyse the informational labelling approach of Definition 5.8 when applied to Figure 6. We use the same **f** as given in Example 5.10. Thus we have:*

- $\mathbf{f}(x) = x$

- $\mathbf{f}(a) = \neg x \wedge \neg c \rightarrow a$

- $\mathbf{f}(b) = \neg a \rightarrow b$

- $\mathbf{f}(c) = \neg b \rightarrow c$

We now define μ.

- $\mu(x) = x$

- $\begin{aligned} \mu(a) \ &= \mathbf{f}(a) \oplus \mu(x) \oplus \mu(c) \\ &= \{\neg x \wedge \neg c \rightarrow a\} \cup \{x\} \cup \mu(c) \end{aligned}$

- $\begin{aligned} \mu(b) \ &= \mathbf{f}(b) \oplus \mu(a) \\ &= \{\neg a \rightarrow b\} \cup \mu(a) \end{aligned}$

- $\begin{aligned} \mu(c) \ &= \mathbf{f}(c) \oplus \mu(b) \\ &= \{\neg b \rightarrow c\} \cup \{\neg a \rightarrow b\} \cup \mu(a) \end{aligned}$

Therefore

$(*)$ $\mu(c) = \{\neg b \rightarrow c\} \cup \{\neg a \rightarrow b\} \cup \{\neg x \wedge \neg c \rightarrow a, x\} \cup \mu(c)$

This can happen only if

$(**)$ $\{\neg b \rightarrow c, \neg a \rightarrow b, \neg x \wedge \neg c \rightarrow a, x\} \subseteq \mu(c)$

This actually happens only if we have equality in $(**)$ because all the players' wff are participating. We therefore get that:

$(***)$ $\mu(a) = \mu(b) = \mu(c)$, and $\mu(x) = x$

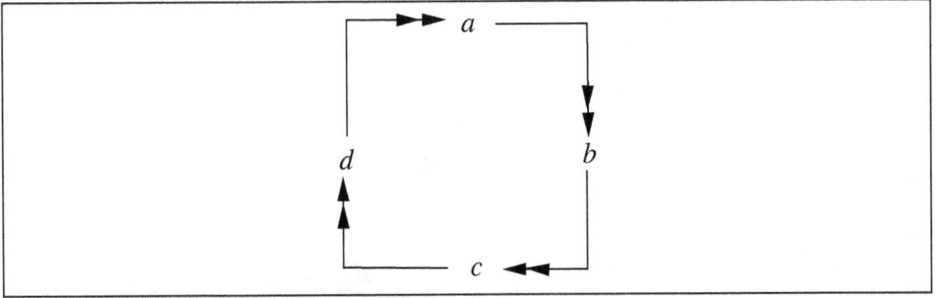

Figure 7: A four loop

Logic programming calculations will show that

$$\mu(x) \vdash x$$
$$\mu(a) \vdash \neg a$$
$$\mu(b) \vdash b$$
$$\mu(c) \vdash \neg c$$

Therefore $\lambda_\mu = \lambda$, in this case also.

Example 5.15. *Consider the four cycle graph of Figure 7. let us see what extensions we get using the logic programming informational model. As before we let:*

- $\mathbf{f}(a) = \neg d \to a$

- $\mathbf{f}(b) = \neg a \to b$

- $\mathbf{f}(c) = \neg b \to c$

- $\mathbf{f}(d) = \neg c \to d$

The function μ can be calculated as before. We get:

- $\mu(a) = \{\neg d \to a\} \cup \mu(d)$

- $\mu(b) = \{\neg a \to b\} \cup \mu(a)$

- $\mu(c) = \{\neg b \to c\} \cup \mu(b)$

33

- $\mu(d) = \{\neg c \to d\} \cup \mu(c)$

Solving these set equations we get:

$$\mu(d) = \{\neg c \to d, \neg b \to c, \neg a \to b, \neg d \to a\} \cup \mu(d)$$

Therefore

$$\mu(a) = \mu(b) = \mu(c) = \mu(d) = \Delta$$

where $\Delta = \{\neg d \to a, \neg a \to b, \neg b \to c, \neg c \to d\}$. *Since*

$$\Delta \nvdash a \qquad \Delta \nvdash \neg a$$
$$\Delta \nvdash b \qquad \Delta \nvdash \neg b$$
$$\Delta \nvdash c \qquad \Delta \nvdash \neg c$$
$$\Delta \nvdash d \qquad \Delta \nvdash \neg d$$

we get the whole undecided extension.

However there are two answer set models for Δ. *These are* $\{a, c\}$ *and* $\{b, d\}$ *and these give the other two extensions. The reader can see in [5] that in general the answer set models give all the possible extensions.*

References

[1] M. Caminada and D. Gabbay. A logical account of formal argumentation. *Studia Logica*, 93(2-3): 109-145, 2009.

[2] D. Gabbay and Co-authors, Abstract Generic Argumentation (AGA), February 2017.

[3] D. M. Gabbay An equational approach to argumentation networks. *Argument and Computation*, 3: 2–3, pp. 87–142, 2012.

[4] D. Gabbay. Logical foundations for bipolar argumentation networks, *J Logic and Computation* 26(1): 247–292, 2016. doi: 10.1093/logcom/ext027 First published online: July 22, 2013.

[5] D. Gabbay and M. Gabbay. The attack as information input. December 2015. Short version appeared in pages 311-318 of Pietro Baroni, Thomas F. Gordon, Tatjana Scheffler and Manfred Stede, *Proceedings of COMMA 2016*, IOS press, long version draft available.

[6] Gerhard Brewka and Stefan Woltran. Abstract dialectical frameworks. In *Proc. KR'10*, pages 102–111. AAAI Press, 2010.

[7] G. Brewka, S. Ellmauthaler, H. Strass, J. P. Wallner, and S. Woltran. Abstract dialectical frameworks revisited. In *Proceedings of the Twenty-Third international joint conference on Artificial Intelligence* (pp. 803-809). AAAI Press. (2013, August).

[8] S. Polberg.Extension-based semantics of abstract dialectical frameworks. In *STAIRS 2014: Proceedings of the 7th European Starting AI Researcher Symposium* (Vol. 264, p. 240). IOS Press. (2014, August).

[9] H. Strass. The relative expressiveness of abstract argumentation and logic programming. In *Proceedings of the Twenty-Ninth AAAI Conference on Artificial Intelligence (AAAI)* (S. Koenig and B. Bonet, eds.), (Austin, TX, USA), pp. 1625–1631, Jan. 2015.

[10] H. Strass. Expressiveness of two-valued semantics for abstract dialectical frameworks, *Journal of Artificial Intelligence Research*, vol. 54, pp. 193–231, 2015.

[11] H. Strass and J. P. Wallner. Analyzing the computational complexity of abstract dialectical frameworks via approximation fixpoint theory, *Artificial Intelligence*, vol. 226, pp. 34–74, 2015.

[12] Gerhard Brewka and Stefan Woltran. GRAPPA: A semantical framework for graph-based argument processing. In *ECAI 2014 - 21st European Conference on Artificial Intelligence*, 18-22 August 2014, Prague, Czech Republic - Including Prestigious Applications of Intelligent Systems (PAIS 2014), pages 153-158, 2014.

[13] Sylwia Polberg. Understanding the abstract dialectical framework. In *Proceedings of the 15th European Conference on Logics in Artificial Intelligence (JELIA'16)*, 2016.

CHAPTER 2
PROBABILISTIC ARGUMENTATION.
AN EQUATIONAL APPROACH

1 Introduction

The objective of this Chapter is to provide some orientation to underpin probabilistic semantics for abstract argumentation. We feel that a properly developed probabilistic argumentation framework cannot be obtained by simply imposing an arbitrary probability distribution on the components of an argumentation system that does not agree with the dynamic aspects of these networks. We need to find a probability distribution that is compatible with their underlying motivation.

We shall use the methodology of "Logic by Translation", which works as follows: Given a new area for which we want to study certain aspect properties AP, we translate this area to classical logic, study AP in classical logic and then translate back and evaluate what we have obtained.

Let us start by looking at interpretations of an abstract argumentation network $\langle S, R \rangle$, $S \neq \varnothing$, $R \subseteq S \times S$, into logics which already have probabilistic versions. This way we can import the probability aspect from there and it will have a meaning. We begin with translating abstract argumentation frames into classical propositional logic. In the abstract form, the elements of S are just atoms waiting to be instantiated as arguments coming from another application system. R may be defined using the source application system or may represent additional constraints. At any rate, in this abstract form, S is just a set of atoms and all we have about it is R. In translating $\langle S, R \rangle$ into classical propositional logic, we view S as a set of atomic

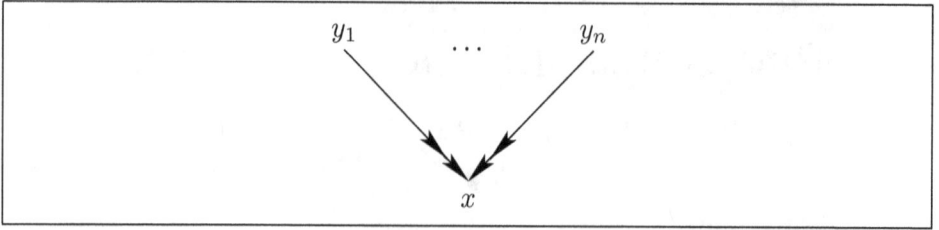

Figure 1: Basic attack formation in an argumentation network.

propositions and we use R to generate a classical theory $\Delta_{\langle S, R \rangle}$. Consider Figure 1, which describes the basic attack formation of all the attackers $Att(x) = \{y \in S \mid (y, x) \in R\} = \{y_1, \ldots, y_n\}$ of the node x in a network $\langle S, R \rangle$.

The essential logic translation of the attack on each node x is given by (E1) below, where x, y_i are propositional symbols representing the elements $x, y_i \in S$:

$$x \leftrightarrow \bigwedge_i \neg y_i \qquad \text{(E1)}$$

So $\langle S, R \rangle$ corresponds to a classical *propositional* theory $\Delta_{\langle S, R \rangle} = \{x \leftrightarrow \bigwedge_i \neg y_i \mid x \in S\}$.[1] Note that in classical logic, this theory may be inconsistent and have no models. For example, if S contains a single node x and R is $\{(x, x)\}$, i.e., the network has a single self-attacking node, then the associated theory is $\{x \leftrightarrow \neg x\}$, which has no model. For this reason it is convenient to regard these theories as theories of Kleene three-valued logic, with values in $\{0, \frac{1}{2}, 1\}$. In this 3-valued semantics, a valuation would satisfy $x \leftrightarrow \neg x$ if and only if it gives the value $\frac{1}{2}$ to x.[2]

If we consider the *equational approach* [5], then we can write

$$x = \bigwedge_i \neg y_i \qquad \text{(E2)}$$

---

[1]If there is a logical relationship between the arguments of S that can be captured by formulae, then we can alternatively instantiate $x \longmapsto \varphi_x$, giving $\Delta_{\langle S, R \rangle} = \{\varphi_x \leftrightarrow \bigwedge_i \neg \varphi_y \mid x, y \in S\}$.

[2]In Kleene's logic, one can interpret $\neg$ as complement to 1; $\wedge$ as min; and $\vee$ as max. Thus, if the values of A, B are $v(A), v(B)$, then $v(\neg A) = 1 - v(A)$, $v(A \wedge B) = \min(v(A), v(B))$ and $v(A \vee B) = \max(v(A), v(B))$.

where (E2) is a numerical equation over the real interval $[0, 1]$, with conjunction and negation interpreted as numerical functions expressing the correspondence of the values of the two sides.

A complete extension of $\langle S, R \rangle$ is a solution to the equations of the form of (E2) when they are viewed as a set of Boolean equations in Kleene's 3-valued logic with values $\left\{0, \frac{1}{2}, 1\right\}$, where

$x = 0$ means that $x = $ **out** (at least one attacker $y_i = $ **in**)   (1)

$x = 1$ means that $x = $ **in** (all attackers $y_i = $ **out**)   (2)

$x = \dfrac{1}{2}$ means that $x = $ **und** (no attacker $y_i = $ **in** and at least (3) one attacker $y_j = $ **und**)

The acceptability semantics above can be re-written in terms of the semantics of Kleene's logic as

$$v(x) = \min\{1 - v(y_i)\}$$

which in equational form can be simplified to

$$x = 1 - \max\{y_i\} \qquad \text{(E2*)}$$

The reader should note that we actually solve the equations over the unit interval $[0, 1]$ and project onto Kleene's 3-valued logic by letting

$x = 0$ mean $x = $ **out**   (at least one attacker $y_i = $ **in**)

$0 < x < 1$ mean $x = $ **und**   (no attacker $y_i = $ **in** and at least one attacker $y_j = $ **und**)

$x = 1$ mean $x = $ **in**   (all attackers $y_i = $ **out**)

Now there are probabilistic approaches to two-valued classical logic. The simplest two methods are described in Gabbay's book *Logic for Artificial Intelligence and Information Technology* [4]. Our idea is to bring the probabilistic approach through the above translation into argumentation theory.

Let us start with a description of the probabilistic approaches to classical propositional logic.

Method 1: Syntactic. Impose probability $P(q)$ on the atoms q of the language and propagate this probability to arbitrary well-formed formulas (wffs). So if $\varphi(q_1, \ldots, q_m)$ is built up from the atoms $q_1, \ldots, q_m$, we can calculate $P(\varphi)$ if we know $P(q_i)$, $i = 1, \ldots, m$.

Method 2: Semantic. Impose probability on the models of the language of $\{q_1, \ldots, q_m\}$. The totality of models is the space W of all $\{0, 1\}$-vectors in 2^m. We give values $P(\varepsilon)$, for any $\varepsilon \in 2^m$, with the restriction that $\Sigma_{\varepsilon \in 2^m} P(\varepsilon) = 1$. The probability of any wff φ is then

$$P(\varphi) = \Sigma_{\varepsilon \Vdash \varphi} P(\varepsilon) \tag{P1}$$

The motivation for the syntactical Method 1 is that the atoms $\{q_1, \ldots, q_m\}$ are all independent. So for example, the date of birth of a person (p) is independent of whether it is going to rain heavily on that person's 21st birthday (q). However, if we want to hold a birthday party r in the garden on the 21st birthday, then we have that q attacks r.

If, on the other hand, we have:

$a =$ John comes to the party

$b =$ Mary comes to the party

then a and b may be dependent, especially if some relationship exists between John and Mary. We may decide that the probability of $a \wedge b$ is 0, but the probabilities of $\neg a \wedge b$ and of $a \wedge \neg b$ are $\frac{1}{4}$ each and the probability of $\neg a \wedge \neg b$ is $\frac{1}{2}$. Assigning probability in this way depends on the likelihood we attach to a particular situation (model). This is the semantic approach.

Example 1.1 shows that these two methods are orthogonal.

Example 1.1. *What can $\Delta_{\langle S, R \rangle}$ mean in classical logic? It is a generalisation of the "Liar's paradox". x attacking itself is like x saying "I am lying": $x = \top$ if and only if $x = \bot$. Figure 1 represents y_i saying x is a lie. $\Delta_{\langle S, R \rangle}$ represents a system of lying accusations: a community liar paradox.*

Similarly, S can represent people possibly invited to a birthday party. $y \to x$ means y saying "if I come, x cannot come". So Figure 1 is saying "invite x if and only if you do not invite any of the y_i".

Suppose we instantiate $x \longmapsto \varphi_x$. Then we must have

$$P(\varphi_x) = P(\bigwedge_i \neg\varphi_{y_i}).$$

However, there may be also a connection between φ_x and some φ_{y_k}, e.g., $\varphi_x \vdash \varphi_{y_k}$. This will impose further restrictions on $P(\varphi_x)$ and $P(\varphi_{y_k})$, and it may be the case that no such probability function exists.

Remark 1.2. *The two approaches are of course, connected. If we are given a probability on each q_i, then we get probability on each $\varepsilon \in 2^m$ by letting*

$$P(\varepsilon) = \Pi_{\varepsilon \Vdash q}P(q) \times \Pi_{\varepsilon \Vdash \neg q}(1 - P(q)) \qquad (P2)$$

The q_i's are considered independent, so the probability of $\bigwedge_i \pm q_i$ is the product of the probabilities

$$P(\textstyle\bigwedge_i \pm q_i) = \Pi_i P(\pm q_i)$$

where $P(\neg q_i) = 1 - P(q_i)$ and the probability of $A \vee B$ is

$$P(A \vee B) = P(A) + P(B)$$

when $\Vdash \neg(A \wedge B)$, as is the case with disjuncts in a disjunctive normal form.

So, for example

$$
\begin{aligned}
P((a \wedge b) \vee (a \wedge \neg b)) &= P(a \wedge b) + P(a \wedge \neg b) \\
&= P(a)P(b) + P(a)(1 - P(b)) \\
&= P(a)(P(b) + 1 - P(b)) \\
&= P(a).
\end{aligned}
$$

2   The syntactical approach (Method 1)

Let us investigate the use of the syntactical approach.

Let $\langle S, R \rangle$ be an argumentation network. In the equational approach, according to the syntactical Method 1, we assign probabilities to all the atoms and are required to solve the equation (E3) below for each x, where $Att(x) = \{y_i\}$ and x and all y_i are numbers in $[0,1]$:

$$P(x) = P(\bigwedge_i \neg y_i), \tag{E3}$$

Since in Method 1, all atoms are independent, (E3) is equivalent to (E3*):

$$P(x) = \Pi_i(1 - P(y_i)). \tag{E3*}$$

Such equations always have a solution.

Let us check whether this makes sense. Let us try to identify the argument x equationally with its probability, namely we let $P(x) = x$.

If $x = $ **in**, let $P(x) = 1$

If $x = $ **out**, let $P(x) = 0$.

If $x = $ **und**, let $0 < P(x) < 1$

to be determined by the solution to the equations.

Equation (E3*) becomes, under $P(x) = x$, the following:

$$x = \Pi(1 - y_i) \text{ for } x \in S. \tag{E4}$$

This is the Eq$_{inv}$ equation in the equational approach (see [5]).

The following definition will be useful in the interpretation of values from $[0,1]$ and their counterparts in Caminada's labelling functions.

Definition 2.1. *A valuation function $\mathbf{f}$ can be mapped into a labelling function $\lambda(\mathbf{f})$ as follows.*

$$\begin{aligned} \mathbf{f}(x) = 1 &\rightarrow \lambda(\mathbf{f})(x) = \mathbf{in} \\ \mathbf{f}(x) = 0 &\rightarrow \lambda(\mathbf{f})(x) = \mathbf{out} \\ \mathbf{f}(x) \in (0,1) &\rightarrow \lambda(\mathbf{f})(x) = \mathbf{und} \end{aligned}$$

What do we know about Eq$_{inv}$? We quote the following from [5].

Theorem 2.2. *Let* **f** *be a solution to equations* (E4). *Then* $\lambda(\mathbf{f})$ *defined according to Definition 2.1 is a legal Caminada labelling (see [1]) and leads to a complete extension.*

Theorem 2.3. *Let* λ_0 *be a legal Caminada labelling leading to a preferred extension. Then there exists a solution* f_0, *such that* $\lambda_0 = \lambda(f_0)$.

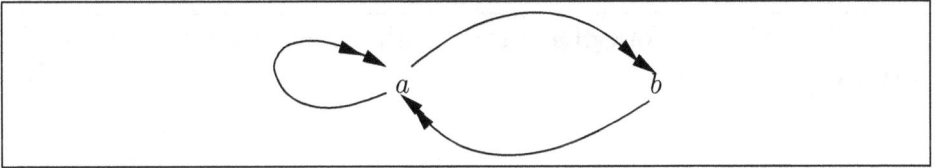

Figure 2: A sample argumentation network having a complete extension that cannot be found via Equations (E4).

Remark 2.4. *There are (complete) extensions* λ' *such that there does not exist an* f' *with* $\lambda' = \lambda(f')$.

For example, in Figure 2, the extension $a = b = $ **und** *cannot be obtained by any* f. *Only* $b = $ **in**, $a = $ **out** *can be obtained as a solution to equations* (E4).[3]

Example 2.5. *Let* $\langle S, R \rangle$ *be given and let* λ *be a complete extension which is not preferred! The reason that* λ *is not preferred, is that we have by definition, a* λ_1 *extending* λ, *which gives more* $\{$**in**, **out**$\}$ *values to points* z, *for which* λ *gives the value* **und**. *Therefore, we can prevent the existence of such an extension* λ_1, *if we force such points* z *to be undecided. This we do by attacking such points* z *by a new*

---

[3]The equations are

1. $a = (1 - a) \times (1 - b)$

2. $b = 1 - a$.

From the above two equations we get

3. $a = (1 - a) \times a$

The only possibility is $a = 0$.

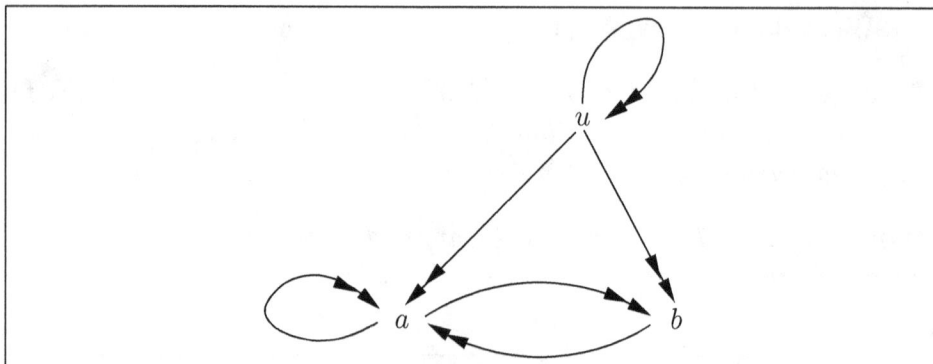

Figure 3: The network of Figure 2 with an extra undecided node u attacking all nodes.

self-attacking point u. The construction is therefore as follows. We are given $\langle S, R \rangle$ and a complete extension λ, which is not preferred. We now construct a new $\langle S', R' \rangle$ which is dependent on λ. Consider $\langle S', R' \rangle$ where $S' = S \cup \{u\}$, where $u \notin S$, is a new point. Let R' be

$$R' = R \cup \{(u, u)\} \cup \{(u, v) \mid v \in S \text{ and } \lambda(v) = \mathbf{und}\}.$$

Then $\lambda' = \lambda \cup \{(u, \mathbf{und})\}$ is a preferred extension of $\langle S', R' \rangle$ and can therefore be obtained from a function f' using the equations (E4).

Let us see what the construction above does to our example in Figure 2, and let us look at the extension $\lambda(a) = \lambda(b) = \mathbf{und}$.

Consider the network in Figure 3. Its equations (E4) are:

1. $u = 1 - u$

2. $a = (1 - u)(1 - a)(1 - b)$

3. $b = (1 - u)(1 - a)$

From (1) we get $u = \frac{1}{2}$. So we have:

2. $a = \frac{1}{2}(1 - a)(1 - b)$

3. $b = \frac{1}{2}(1 - a)$

44

$$1 - b = 1 - \tfrac{1}{2}(1 - a)$$
$$= \tfrac{2-1+a}{2}$$
$$= \tfrac{1+a}{2}$$

therefore substituting in (1) we get

$$a = \tfrac{1}{2}(1 - a)(\tfrac{1+a}{2})$$
$$= \tfrac{1}{4}(1 - a^2)$$
$$4a + a^2 - 1 = 0$$
$$(a+2)^2 - 4 - 1 = 0$$
$$(a+2)^2 = 5$$
$$a = \sqrt{5} - 2 \approx 0.236$$
$$b = \tfrac{1}{2}(1 - a)$$
$$= \tfrac{1}{2}(1 - \sqrt{5} + 2)$$
$$= \tfrac{3-\sqrt{5}}{2} \approx 0.382.$$

*The extension of the network is $a = b = $ **und**.*

Summary of the results so far for the syntactical probabilistic method. Given an argumentation network $\langle S, R \rangle$, we can find all Method 1 complete probabilistic extensions for it by solving all Eq_{inv} equations. Such complete probabilistic extensions will also be complete extensions in the traditional sense (i.e., Dung's), which will also include all preferred extensions (Theorems 2.2 and 2.3).[4]

---

[4]Note that in traditional Dung semantics a preferred extension E is maximal in the sense that there is no extension E' such that

1. If x is considered **in** (resp. **out**) by E then x is also considered **in** (resp. **out**) by E'.

2. There exists at least one node considered **in** (resp. **out**) by E' and considered **und** by E.

The above definition holds for numerical or probabilistic semantics, where the value 1 (resp. 0) is understood as **in** (resp. **out**) and values in $(0, 1)$ are understood as **und**.

However, not all complete extensions can be obtained in this manner (i.e., by Method 1, see remark 2.4 and compare with Example 3.6).

We can, nevertheless, for any complete extension E which cannot be obtained by Method 1, obtain it from the solutions of the equations generated for a larger network $\langle S', R' \rangle$ as shown in Example 2.5.

We shall say more about this in a later section.

Remark 2.6. *Evaluation of the results so far for the syntactical probabilistic method.*

1. *We discovered a formal mathematical connection between the syntactical probabilistic approach (Method 1) and the Equational Eq_{inv} approach. Is this just a formal similarity or is there also a conceptual connection?*

   The traditional view of an abstract argumentation frame $\langle S, R \rangle$, is that the arguments are abstract, some of them abstractly attack each other. We do not know the reason, but we seek complete extensions of arguments that can co-exist (i.e., being attack-free), and that protect themselves. The equational approach is an equational way of finding such extensions. Each solution $\mathbf{f}$ to the equations give rise to a complete extension. The numbers we get from such solutions $\mathbf{f}$ of the equational approach can be interpreted as giving the degree of being in the complete extension (associated with $\mathbf{f}$) or being out of it.

   Due to the mathematical similarity with the probability approach, these numbers are now interpreted as probabilities.

   To what extent is this justified? Can we do this at all?

   Let us recall the syntactical probabilistic method. We start with an abstract argumentation framework $\langle S, R \rangle$ and add the probability $P(x)$ for each $x \in S$. We can interpret $P(x)$ as the probability that x "is a player" to be considered (this is a vague statement which could mean anything but is sufficient for our purpose). The problem is how do we take into account the attack relation? Our choice was to require equation (E3). It is *this choice that allowed the connection between the syntactical probabilistic approach and the Equational approach with Eq_{inv}.*

So our syntactical probabilistic approach should work as follows.

Let P be the independent probability on each $x \in S$. This is an arbitrary number in $[0, 1]$. Such a P cannot be used for calculating extensions because it does not take into consideration the attack relation R. So modify P to a P' which does respect R via Equation (E3).

How do we modify P to find P'?

*Well, we can use a numerical iteration method. The details are not important here, the importance is in the idea, which can be applied to the traditional notion of extensions as well. Given $\langle S, R \rangle$ and an arbitrary desired assignment E of elements that are **in** (and consequently also determining elements that are **out**) for S, this E may not be legitimate in taking into account R, so we need to modify it to get the best proper extension E' nearest to E (cf. [2, 6]).*

So our syntactical probabilistic approach yielding a P satisfying Equation (E3) can be interpreted as Eq_{inv} extensions obtained from initial values which are probabilities (as opposed to, say, initial values being a result of voting) corrected via iteration procedures using R.

Alternatively, we can look at the Eq_{inv} equations as a mathematical means of finding all those syntactical probabilities P which respect the attack relation R (via Equation (E3)).

*Or we can see the solutions of the Eq_{inv} as giving probabilities for being included or excluded in the complete extension defined by these solutions (as opposed to the interpretation of the degree of being **in** or **out**).*

2. *The discussion in item 1. above hinged upon the choice we made to take account of R by respecting Equation (E3). There are other alternatives for taking R into account. We can give direct, well-motivated definitions of how to propagate probabilities along*

attack arrows. This is similar to the well-known problem of how to propagate probabilities along proofs (provability support arrows, or modus ponens, etc). Such an analysis is required anyway for instantiated networks, for example in ASPIC+ style [10]). We shall deal with this in a subsequent paper.

3 The semantical approach (Method 2)

Let us now check what can be obtained if we use Method 2, i.e., giving probability to the models of the language. In this case the equation (for $\{y_i\} = Att(x)$) (E3) $P(x) = P(\bigwedge_i \neg y_i)$ still holds, but the $\neg y_i$ are not independent. So we cannot write equation (E3*) for them and get Eq$_{\text{inv}}$. Instead we need to use the schema $P(A \vee B) = P(A) + P(B) - P(A \wedge B)$. We begin with a key lemma, which will enable us to compare later with the work of M. Thimm, see [13].

Lemma 3.1. *Let $\langle S, R \rangle$ be a network and let P be a probability measure on the space W of all models of the language whose set of atoms is S. For $x \in S$, let the following hold*

$$P(x) = P(\bigwedge_{i=1}^{n} \neg y_i)$$

where $Att(x) = \{y_1, \ldots, y_n\}$.
   Then we have

1. $P(x) \leq P(\neg y_i), 1 \leq i \leq n$

2. $P(x) \geq 1 - \Sigma_{i=1}^{n} P(y_i)$

Proof. By induction on n.

1. If $x = \neg y$ then $P(x) = 1 - P(y)$ and the above holds.

2. Assume the above holds for m, show for $m + 1$. Let $z = \bigvee_{i=1}^{m} y_i, y = y_{m+1}$. Then $x = \neg z \wedge \neg y$.

   We have by the induction hypothesis

- $P(\neg z) \leq P(\neg y_i), i = 1, \ldots, m$
- $P(\neg z) \geq 1 - \Sigma_{i=1}^{m} P(y_i)$

Consider now:

$$
\begin{aligned}
P(\neg z \wedge \neg y) &= 1 - P(y \vee z) \\
&= 1 - (P(y) + P(z) - P(y \wedge z)) \\
&= 1 - P(y) - P(z) + P(y \wedge z) \\
&= 1 - P(y) - (P(z) - P(y \wedge z))
\end{aligned}
$$

But $P(A \wedge B) \leq P(B)$ is always true.

So

$$
P(\neg z \wedge \neg y) \leq 1 - P(y) = P(\neg y)
$$

On the other hand, by our assumption

$$
1 - P(z) = P(\neg z) \geq 1 - \Sigma_{i=1}^{m} P(y_i)
$$

So

$$
\begin{aligned}
P(\neg z \wedge \neg y) &= 1 - P(y) - P(z) + P(y \wedge z) \\
&\quad (1 - P(z)) - P(y) + P(y \wedge z) \\
&\geq 1 - \Sigma P(y_i) - P(y) + P(y \wedge z) \\
&\geq 1 - \Sigma_{i=1}^{m+1} P(y_i)
\end{aligned}
$$

$\square$

Remark 3.2. *The converse of Lemma 3.1 does not hold, as we shall see in Example 3.5 below.*

Let us look at some examples illustrating the use of Method 2.

Example 3.3. *Consider the network in Figure 4. This figure is taken from Thimm's "A probabilistic semantics for abstract argumentation" [13, Figure 1]. We include it here for two reasons:*

1. *To illustrate or probabilistic semantic approach.*

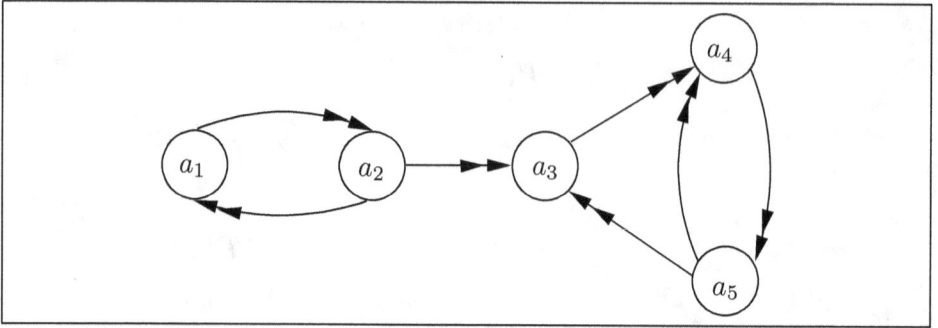

Figure 4: Figure 1 of "A probabilistic semantics for abstract argumentation" [13].

2. *To use it later to compare our work with Thimm's approach.*

Let us apply Method 2 to it and assign probabilities to the models of the propositional language with the atoms $\{a_1, a_2, a_3, a_4, a_5\}$. We assign P as follows.

$$P(a_1 \wedge \neg a_2 \wedge a_3 \wedge \neg a_4 \wedge a_5) = 0.3$$
$$P(a_1 \wedge \neg a_2 \wedge \neg a_3 \wedge a_4 \wedge \neg a_5) = 0.45$$
$$P(\neg a_1 \wedge a_2 \wedge \neg a_3 \wedge \neg a_4 \wedge a_5) = 0.1$$
$$P(\neg a_1 \wedge a_2 \wedge \neg a_3 \wedge a_4 \wedge \neg a_5) = 0.15$$
$$P(any\ other\ conjunctive\ model) = 0.$$

Let us compute $P(a_i)$, for $i = 1, \ldots, 5$.
   We have
$$P(X) = \sum_{\varepsilon \Vdash X} P(\varepsilon).$$

We get
$$P(a_1) = 0.3 + 0.45 = 0.75$$
$$P(a_2) = 0.1 + 0.15 = 0.25$$
$$P(a_3) = 0.3$$
$$P(a_4) = 0.45 + 0.15 = 0.6$$
$$P(a_5) = 0.3 + 0.1 = 0.4.$$

To be a legitimate probabilistic model P must satisfy equation (E3) relating to the attack relation of Figure 4. Namely we must have

$$P(X) = P(\bigwedge_{Y \in Att(X)} \neg Y) \tag{E3}$$

Therefore

$$P(a_1) = P(\neg a_2)$$
$$P(a_2) = P(\neg a_1)$$
$$P(a_3) = P(\neg a_2 \wedge \neg a_5)$$
$$P(a_4) = P(\neg a_3 \wedge \neg a_5)$$
$$P(a_5) = P(\neg a_4)$$

Let us calculate the P in the right hand side of the above equations.

$$P(\neg a_2) = 1 - 0.25 = 0.75$$
$$P(\neg a_1) = 1 - 0.75 = 0.25$$
$$P(\neg a_2 \wedge \neg a_5) = 0.45$$
$$P(\neg a_3 \wedge \neg a_5) = 0.45 + 0.15 = 0.6$$
$$P(\neg a_4) = 0.4$$

We see that

$$P(a_3) = 0.3 \neq P(\neg a_2 \wedge \neg a_5) = 0.45.$$

Therefore this distribution P is not legitimate according to our Method 2. It does not satisfy equations (E3) because

$$P(a_3) \neq P(\neg a_2 \wedge \neg a_5)$$

Therefore Lemma 3.1 does not apply and indeed, condition (2) of Lemma 3.1 does not hold for a_3. We have $P(a_3) = 0.3$ but $1 - P(a_2) - P(a_5) = 0.35$.

Example 3.4. *Let us look at Figure 5. This is also taken from Thimm's paper [13, Figure 2]. It shall be used later to compare our methods with Thimm's.*

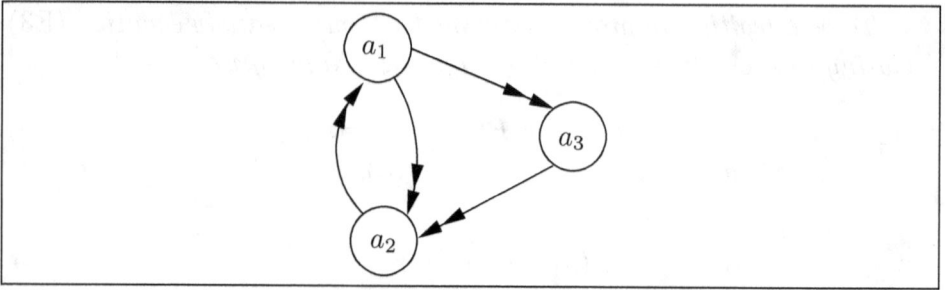

Figure 5: Figure 2 of "A probabilistic semantics for abstract argumentation" [13].

1. We use Method 2. *Consider the following probability distribution on models*

$$P(a_1 \land \neg a_2 \land \neg a_3) = 0.5$$
$$P(a_1 \land \neg a_2 \land a_3) = 0$$
$$P(a_1 \land a_2 \land \neg a_3) = 0$$
$$P(a_1 \land a_2 \land a_3) = 0$$
$$P(\neg a_1 \land a_2 \land a_3) = 0$$
$$P(\neg a_1 \land a_2 \land \neg a_3) = 0.5$$
$$P(\neg a_1 \land \neg a_2 \land a_3) = 0$$
$$P(\neg a_1 \land \neg a_2 \land \neg a_3) = 0.$$

In this model we get

$$P(a_1) = 0.5$$
$$P(a_2) = 0.5$$
$$P(a_3) = 0$$

Let us check whether this probability distribution satisfies equation (E3), namely

$$P(X) = P(\bigwedge_{Y \in Att(X)} \neg Y) \tag{E3}$$

We need to have

$$
\begin{aligned}
P(a_1) &= P(\neg a_2) \\
P(a_2) &= P(\neg a_1) \\
P(a_3) &= P(\neg a_2 \wedge \neg a_2)
\end{aligned}
$$

Indeed

$$
\begin{aligned}
P(\neg a_1) &= 1 - P(a_1) = 0.5 \\
P(\neg a_2) &= 1 - P(a_2) = 0.5 \\
P(\neg a_1 \wedge \neg a_2) &= 0.
\end{aligned}
$$

Thus we have a legitimate model.

2. We use Method 1. *Let us use Eq_{inv} on this figure, namely we try and solve the equations*

$$
\begin{aligned}
a_1 &= 1 - a_2 \\
a_2 &= 1 - a_1 \\
a_3 &= (1 - a_1)(1 - a_2)
\end{aligned}
$$

Let us use a parameter $0 \leq x \leq 1$ and let

$$
\begin{aligned}
a_1 &= x, \\
a_2 &= 1 - x, \\
a_3 &= x(1 - x)
\end{aligned}
$$

The probabilities we get with parameter x as well as for $x = 0.5$ are given below.

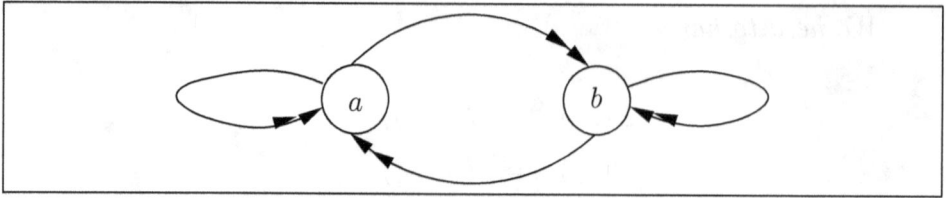

Figure 6: Network for Example 3.5.

$$P(a_1 \wedge a_2 \wedge a_3) \quad = \quad x^2(1-x)^2 \quad = \quad \tfrac{1}{16}$$

$$P(a_1 \wedge a_2 \wedge \neg a_3) \quad = \quad x(1-x)(1-x(1-x)) \quad = \quad \tfrac{3}{16}$$

$$P(a_1 \wedge \neg a_2 \wedge a_3) \quad = \quad x^3(1-x) \quad = \quad \tfrac{1}{16}$$

$$P(a_1 \wedge \neg a_2 \wedge \neg a_3) \quad = \quad x^2(1-x(1-x)) \quad = \quad \tfrac{3}{16}$$

$$P(\neg a_1 \wedge a_2 \wedge a_3) \quad = \quad x(1-x)^3 \quad = \quad \tfrac{1}{16}$$

$$P(\neg a_1 \wedge a_2 \wedge \neg a_3) \quad = \quad (1-x)^2(1-x(1-x)) \quad = \quad \tfrac{3}{16}$$

$$P(\neg a_1 \wedge \neg a_2 \wedge a_3) \quad = \quad x^2(1-x)^2 \quad = \quad \tfrac{1}{16}$$

$$P(\neg a_1 \wedge \neg a_2 \wedge \neg a_3) \quad = \quad x(1-x)(1-x(1-x)) \quad = \quad \tfrac{3}{16}$$

If we choose $x = 0.5$ we get $P(a_1) = P(a_2) = 0.5$ and $P(a_3) = \tfrac{1}{4}$.

Example 3.5. *This example shows that the converse of Lemma 3.1 does not hold. Consider the network in Figure 6.*

Any legitimate probability assigned to models would be required to satisfy the following

$$P(a) = P(\neg a \wedge \neg b)$$
$$P(b) = P(\neg a \wedge \neg b)$$

Case 1. *Try the following probability P_1.*

$$P_1(a \wedge b) = P_1(a \wedge \neg b) = P_1(\neg a \wedge b) = P_1(\neg a \wedge \neg b) = 0.25.$$

Therefore

$$P_1(a) = 0.5$$
$$P_1(b) = 0.5$$

Note that we also have

$$P_1(a) = \tfrac{1}{2} \leq 1 - P_1(b) = \tfrac{1}{2}$$
$$P_1(a) = \tfrac{1}{2} \leq 1 - P_1(a) = \tfrac{1}{2}.$$

Similarly for $P_1(b)$ by symmetry.

Also

$$P_1(a) = \frac{1}{2} \geq 1 - P_1(a) - P_1(b) = 1 - \frac{1}{2} - \frac{1}{2} = 0.$$

Thus the conditions of the conclusions of Lemma 3.1 hold. However the assumptions of Lemma 3.1 do not hold, because

$$P_1(a) = \frac{1}{2} \neq P_1(\neg a \wedge \neg b) = \frac{1}{4}.$$

Case 2. *Let us check whether we can find a probability P_2 which is indeed acceptable to Method 2. Let us try with variables y, z and create equations and solve them:*

$$P_2(a \wedge b) = y$$
$$P_2(\neg a \wedge b) = z.$$

Therefore $P_2(b) = y + z$.

$$P_2(\neg a \wedge \neg b) = y + z$$

and what is left is

$$P_2(a \wedge \neg b) = 1 - 2y - 2z$$

but we must also have

$$P_2(a) = P_2(\neg a \wedge \neg b)$$

55

and hence we must have

$$P_2(a) = 1 - 2y - 2z + y = P_2(\neg a \wedge \neg b) = y + z.$$

So we get the equation

$$1 - 2y - 3z = 0$$

$$2y + 3z = 1$$

$$y = \frac{(1-3z)}{2}$$

Since $0 \leq y, z \leq 1$ so z must be less than $\frac{1}{3}$.
    Let us choose $z = 0.2$ and so $y = 0.2$.
    We get, for example

$$P_2(a \wedge b) = 0.2$$

$$P_2(\neg a \wedge b) = 0.2$$

$$P_2(\neg a \wedge \neg b) = 0.4$$

$$P_2(a \wedge \neg b) = 0.2$$

We could also have chosen $z = \frac{1}{3}$ and $y = 0$. This would give P_3, where

$$P_3(a \wedge b) = 0$$

$$P_3(\neg a \wedge b) = \tfrac{1}{3}$$

$$P_3(\neg a \wedge \neg b) = \tfrac{1}{3}$$

$$P_3(a \wedge \neg b) = \tfrac{1}{3}$$

So we get

$$P_3(b) = P(a) = \tfrac{1}{3}$$

$$P_3(\neg a \wedge \neg b) = \tfrac{1}{3}.$$

Example 3.6. *Consider the network of Figure 2. Let us try to find a probabilistic semantics for it according to Method 2. Assume we have*

$$P(a \wedge b) = x_1$$

$$P(a \wedge \neg b) = x_2$$

$$P(\neg a \wedge b) = x_3$$

$$P(\neg a \wedge \neg b) = 1 - x_1 - x_2 - x_3.$$

We need to satisfy

$$P(a) = P(\neg a \wedge \neg b)$$
$$P(b) = P(\neg a)$$

This means we need to solve the following equations.

1. $x_1 + x_2 = 1 - x_1 - x_2 - x_3$

2. $x_1 + x_3 = 1 - x_1 - x_2$.

By adding $x_1 + x_2$ to both sides (1) can be written as

$$2(x_1 + x_2) = 1 - x_3,$$

and by swapping x_3 to the right and $-x_1 - x_2$ to the left (2) can be written as

$$2x_1 + x_2 = 1 - x_3.$$

Thus we get

3. $2x_1 + x_2 = 2x_1 + 2x_2$.

Therefore $x_2 = 0$.
There remains, therefore

4. $2x_1 = 1 - x_3$.

We can choose values for x_3.

Sample choice 1. $x_3 = 1$, so $x_1 = 0$.
We get $P_1(a \wedge b) = P(a \wedge \neg b) = P_1(\neg a \wedge \neg b) = 0$ and $P_1(\neg a \wedge b) = 1$.
This yields $P(a) = 0, P(b) = 1$. This is also the Eq$_{\text{inv}}$ solution to

$$b = 1 - a$$
$$a = (1 - a)(1 - b)$$

Sample choice 2. $x_3 = \frac{1}{2}$. So $x_1 = \frac{1}{4}$ and the probabilities are

$$P_2(a \wedge b) = \frac{1}{4}$$
$$P_2(a \wedge \neg b) = 0$$
$$P_2(\neg a \wedge b) = \frac{1}{2}$$
$$P_2(\neg a \wedge \neg b) = \frac{1}{4}.$$

P_2 is a Method 2 probability, which cannot be given by Method 1.

Sample choice 3. $x_3 = 0$. Then $x_1 = \frac{1}{2}$. We get

$$P_3(a \wedge b) = \frac{1}{2}$$
$$P_3(a \wedge \neg b) = 0$$
$$P_3(\neg a \wedge b) = 0$$
$$P_3(\neg a \wedge \neg b) = \frac{1}{2}.$$

Therefore $P_3(a) = P_3(b) = \frac{1}{2}$.

Lemma 3.7. *Let $\langle S, R \rangle$ be a network and let P be a semantic probability (Method 2) for $\langle S, R \rangle$. Let $x \in S$ and let $\{y_i\} = Att(x)$. Then*

1. *If for some y_i, $P(y_i) = 1$ then $P(x) = 0$.*

2. *If for all y_i, $P(y_i) = 0$ then $P(x) = 1$.*

Proof. Let us use Figure 1 where $\{y_i\} = Att(x)$.

Case 1. Assume that $P(y_1) = 1$. We need to show that $P(x) = 0$. We have:

$$P(x) = P(\bigwedge_i \neg y_i) \tag{E3}$$

We also have

$$P(A) = \sum_{\varepsilon \vdash A} P(\varepsilon) \tag{P1}$$

Therefore

$$P(x) = \sum_{\varepsilon \vdash \bigwedge_i \neg y_i} P(\varepsilon)$$

$$P(x) = \sum_{\varepsilon \Vdash \neg y_1 \wedge \bigwedge_{j=1}^{n} \neg y_j} P(\varepsilon) \qquad (i)$$

but

$$P(y_1) = \sum_{\varepsilon \Vdash y_1} P(\varepsilon) = 1$$

Therefore we have

$$\sum_{\varepsilon \Vdash \neg y_1} P(\varepsilon) = 0 \qquad (ii)$$

From (i) and (ii) we get that $P(x) = 0$.

Case 2. We assume that for all i, $P(y_i) = 0$ and we need to show that $P(x) = 1$.

We have

$$P(x) = P(\bigwedge \neg y_i)$$

$$P(x) = 1 - P(\bigvee y_i) \qquad (iii)$$

We also have

$$P(\bigvee y_i) = \Sigma_{\varepsilon \Vdash \bigvee y_i} P(\varepsilon) \qquad (iv)$$

Suppose for some ε' such that $\varepsilon' \Vdash \bigvee y_i$ we have $P(\varepsilon') > 0$. But $\varepsilon' \Vdash \bigvee y_i$ implies $\varepsilon' \Vdash y_i$, for some i.

Say $i = 1$. Thus we have $\varepsilon' \Vdash y_1$ and $P(y_1) = 0$ and $P(\varepsilon') > 0$. This is impossible since

$$P(y_1) = \sum_{\varepsilon \Vdash y_1} P(\varepsilon) \qquad (v)$$

Therefore for all ε such that $\varepsilon \Vdash \bigvee y_i$ we have that $P(\varepsilon) = 0$. Therefore by (iii) and (iv) we get

$$P(x) = 1.$$

$\square$

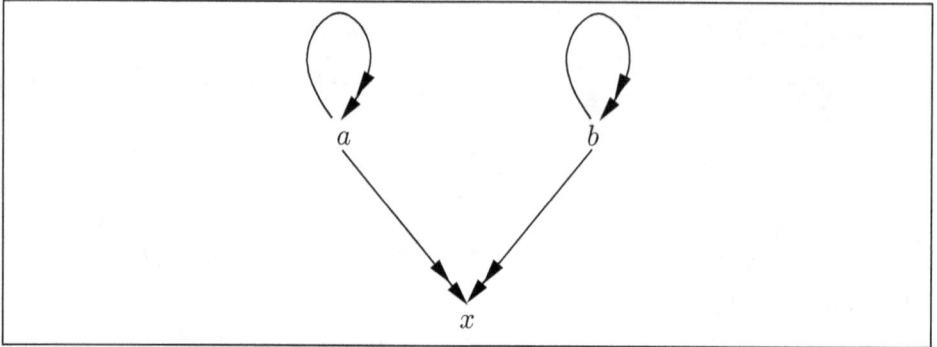

Figure 7: A network with Method 1 and Method 2 probabilities.

Remark 3.8. *Let $\langle S, R \rangle$ be a network and let P be a semantic probability for $\langle S, R \rangle$ (Method 2).*

Let λ be defined as follows, for $x \in S$.

$$\lambda(x) = \begin{cases} \mathbf{in}, & \text{if } P(x) = 1 \\ \mathbf{out}, & \text{if } P(x) = 0 \\ \mathbf{und}, & \text{if } 0 < P(x) < 1 \end{cases}$$

The perceptive reader might expect us to say that λ is a legitimate Caminada labelling, especially in view of Lemma 3.7. This is not the case as Example 3.9 shows.

Example 3.9. *This example shows that in the probabilistic semantics it is possible to have $P(x) = 0$, while for all attackers y of x we have $0 < P(y) < 1$. Thus the nature of the probabilistic attack is different from the traditional Dung one. If $Att(x)$ is the set of all attackers of x and $P(\bigvee_{y \in Att(x)} y) = 1$, then, and only then $P(x) = 0$.*

Thus the attackers of x can attack with joint probability.

The example we give is the network of Figure 7.

This has a Method 1 probability of $P_1(a) = \frac{1}{2}, P_1(b) = \frac{1}{4}$ and $P_1(x) = \frac{1}{4}$.

Thus for any model $\mathbf{m} = \pm a \wedge \pm b \wedge x$ we have

$$P_1(\mathbf{m}) = \frac{1}{2} \times \frac{1}{2} \times \frac{1}{4} = \frac{1}{16}$$

and for any model

$$\mathbf{m}' = \pm a \wedge \pm \wedge \neg x$$

we have

$$P_1(\mathbf{m}) = \frac{1}{2} \times \frac{1}{2} \times \frac{3}{4} = \frac{3}{16}.$$

Figure 7 also has a Method 2 probability model. We can have

$$P_2(a) = P_2(b) = \tfrac{1}{2}$$
$$P_2(x) = 0.$$

Let us check what values to give to the models. The models are:

$$m_1 = x \wedge a \wedge b$$
$$m_2 = x \wedge a \wedge \neg b$$
$$m_3 = x \wedge \neg a \wedge b$$
$$m_4 = x \wedge \neg a \wedge \neg b$$
$$m_5 = \neg x \wedge a \wedge b$$
$$m_6 = \neg x \wedge a \wedge \neg b$$
$$m_7 = \neg x \wedge \neg a \wedge b$$
$$m_8 = \neg x \wedge \neg a \wedge \neg b.$$

We want the following equations to be satisfied.

1. $P_2(x) = 0$. *This means we need to let*

$$P_2(m_i) = 0, i = 1, \ldots, 4.$$

2. $P_2(a) = \tfrac{1}{2}$. *This means we need to let*

$$P_2(m_5) + P_2(m_6) = \tfrac{1}{2}$$
$$P_2(m_7) + P_2(m_8) = \tfrac{1}{2}.$$

3. $P_2(b) = \tfrac{1}{2}$, *yields the equations*

$$P_2(m_5) + P_2(m_7) = \tfrac{1}{2}$$
$$P_2(m_6) + P_2(m_8) = \tfrac{1}{2}.$$

4. *We also need to have the equation*

$$0 = P_2(x) = P_2(\neg a \wedge \neg b)$$

Therefore $P_2(m_8) = 0$.

We thus have the following equations left

(a) $P_2(m_5) + P_2(m_6) = \frac{1}{2}$

(b) $P_2(m_7) = \frac{1}{2}$

(c) $P_2(m_5) + P_2(m_7) = \frac{1}{2}$

(d) $P_2(m_6) = \frac{1}{2}$.

From (b) and (c) we get $P_2(m_5) = 0$. This makes $P_2(m_6) = \frac{1}{2}$. Thus we get the following solution:

$$P_2(m_i) = 0, \ for \ i = 1, 2, 3, 4, 5, 8$$
$$P_2(m_6) = P_2(m_7) = \frac{1}{2}.$$

Note that the equations (E3) hold for P_1 and P_2:

$$P(a) = 1 - P(a)$$
$$P(b) = 1 - P(b)$$

hold of both P_1 and P_2. As for $P(x) = P(\neg a \wedge \neg b)$ we check

$$
\begin{aligned}
\tfrac{1}{4} = P_1(x) &= P_1(\neg a \wedge \neg b) \\
&= P_1(\neg a) \times P_1(\neg b) = \tfrac{1}{4}.
\end{aligned}
$$

For P_2 we have

$$0 = P_2(x) = P_2(\neg a \wedge \neg b)$$
$$P_2(\neg(a \vee b) = 1 - P_2(a \vee b)$$
$$P_2(a \vee b) = P_2(m_1) + P_2(m_2)$$
$$+ P_2(m_3) + P_2(m_5) + P_2(m_6)$$
$$+ P_2(m_7) = 0 + 0 + 0 + \tfrac{1}{2} + \tfrac{1}{2} = 1.$$

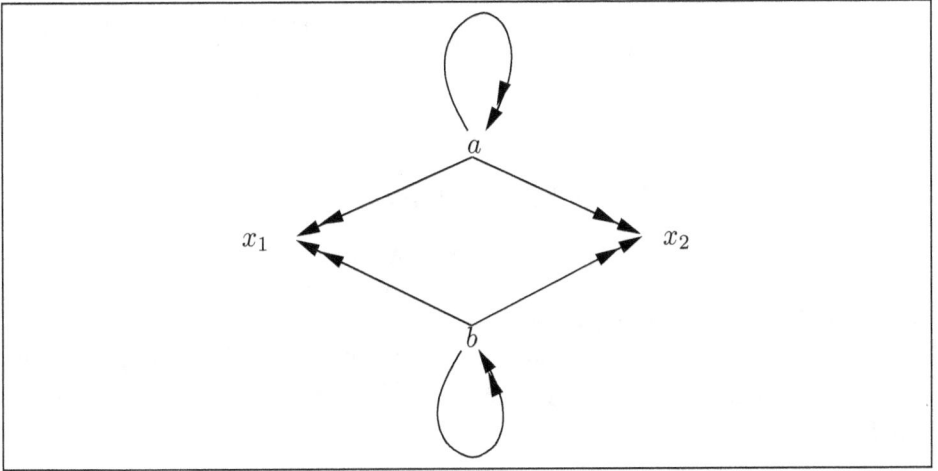

Figure 8: Mirrored network of Figure 7.

Thus $P_2(\neg a \wedge \neg b) = 0$.

So P_1 and P_2 are legitimate probabilities on Figure 7. P_1 is a Method 1 probability and P_2 is a Method 2 probability.

Definition 3.10. *We now define the Gabbay–Rodrigues Probabilistic Labelling Π on a network $\langle S, R \rangle$. Π is a {in, out, und}-labelling satisfying the following.*

There exists a semantic probability P on $\langle S, R \rangle$ such that for all $x \in S$

*1. $\Pi(x) = $ **in**, if $P(\bigvee Att(x)) = 0$*

*2. $\Pi(x) = $ **out**, if $P(\bigvee Att(x)) = 1$*

*3. $\Pi(x) = $ **und**, if $0 < P(\bigvee Att(x)) < 1$*

Example 3.11. *This example is due to M. Thimm, oral communication, 24th October 2014. Consider Figure 8.*

This figure contains Figure 7 and its mirror image. We saw that in Figure 7 (as well as in this Figure 8) any probability on the figures must yield

$$P(a) = P(b) = \frac{1}{2}.$$

*Figure 7 allowed for two possibilities for x. $P_1(x) = \frac{1}{4}$ and $P_2(x) = 0$.
Let us try P for our Figure 8 with*

$$P(x_1) = \frac{1}{4} \text{ and } P(x_2) = 0.$$

This is not possible because we must have

$$P(x_i) = P(\neg a \wedge \neg b).$$

So $P(x_1)$ must be equal to $P(x_2)$.

This example will show in the comparison with the literature section that our probability semantics is different from that of M. Thimm in [13].

See also Example 3.5.

Theorem 3.12. *Let $\langle S, R \rangle$ be a network and let λ be a legitimate Caminada labelling on S, giving rise to a complete extension. Then there exists a probability P_λ on the models (Method 2 probabilistic semantics) such that for all $x \in S$:*

- $P_\lambda(x) = 1$, *if* $\lambda(x) = $ **in**

- $P_\lambda(x) = 0$, *if* $\lambda(x) = $ **out**

- $P_\lambda(x) = \frac{1}{2}$, *if* $\lambda(x) = $ **und**.

Proof. (We use an idea from M. Thimm [13].)

Let $S = \{s_1, \ldots, s_k\}$. Then when we regard the elements of S as atomic propositions in classical propositional logic, there are 2^k models based on S. Each of these models gives values 0 (false) or 1 (true) to each atomic proposition. Each such a model can be represented by a conjunction of the form $\alpha = \bigwedge_i \pm s_i$. α represents the model which gives value 1 to s_i if $+s_i$ appears in α and gives value 0 to s_i if $-s_i$ appears in α. Given a model we can construct the respective α for it. Let

$$\alpha_1 = \bigwedge_{\lambda(s)=\textbf{in}} s; \quad \alpha_0 = \bigwedge_{\lambda(s)=\textbf{out}} \neg s;$$

$$\alpha_{\frac{1}{2}} = \bigwedge_{\lambda(s)=\textbf{und}} s; \quad \text{and } \beta_{\frac{1}{2}} = \bigwedge_{\lambda(s)=\textbf{und}} \neg s.$$

We now define a Method 2 probability P_λ on the models.

1. $P_\lambda(\alpha_1 \wedge \alpha_0 \wedge \alpha_{\frac{1}{2}}) = \frac{1}{2}$

2. $P_\lambda(\alpha_1 \wedge \alpha_0 \wedge \beta_{\frac{1}{2}}) = \frac{1}{2}$

3. $P_\lambda(m) = 0$, for any other model, m different from the above.

Clearly P_λ is a probability. We examine its properties

(i) Let x be such that $\lambda(x) = \textbf{in}$.

Then
$$P_\lambda(x) = \sum_{m \Vdash x} P_\lambda(m).$$

Only (1) and (2) can contribute to $P_\lambda(x)$, so the value is 1.

(ii) Let $\lambda(x) = \textbf{out}$.

The only two models that can contribute to $P_\lambda(x)$ are in (1) and (2) above, but they prove $\neg x$. So $P_\lambda(x) = 0$.

(iii) Let $P_\lambda(x) = \textbf{und}$.

Then clearly $P_\lambda(x)$ gets a contribution from (1) only. We get $P_\lambda(x) = \frac{1}{2}$.

We now need to verify that P_λ actually satisfies the equations of (E3).

Let $x \in S$ and let y_i be its attackers. We want to show that

$$P_\lambda(x) = P_\lambda(\bigwedge_i \neg y_i)$$

or

$$P_\lambda(x) = 1 - P_\lambda(\bigvee_i y_i).$$

(iv) Assume $P_\lambda(x) = 1$. Then $P_\lambda(x)$ gets contributions from both (1) and (2). The only option is that then $\lambda(x) = \textbf{in}$, and so all attackers of y_i of x are out, so $\alpha_0 \Vdash \bigwedge \neg y_i$ and so $P_\lambda(\bigwedge_i \neg y_i) = 1$, because it gets contributions from both (1) and (2).

(v) Assume $P_\lambda(x) = 0$.

Thus neither (1) nor (2) contribute to $P_\lambda(x)$. Therefore $\alpha_0 \Vdash x$ and so $\lambda(x) = $ out and so for some attacker y_i, $\lambda(y_i) = $ in and so $\alpha_1 \Vdash y_i$ and so $P_\lambda(\bigwedge_i \neg y_i)$ cannot get any contribution either from (1) or from (2) and so $P_\lambda(\bigwedge_i \neg y_i) = 0$.

(vi) Assume that $P_\lambda(x) = \frac{1}{2}$.

So $P_\lambda(x)$ can get a contribution either from (1) or from (2), but not from both. So $\lambda(x)$ must be undecided.

So the attackers y_i of x are either **out** (with $P_\lambda(y_i) = 0$)) or **und** (with $P_\lambda(y_i) = \frac{1}{2}$), and we have that at least one attacker y of x is **und**.

Let y_i^0 be the attackers that are out and let $y_j^{\frac{1}{2}}$ be the undecided attackers. Consider

$$e = \bigwedge_i \neg y_i^0 \wedge \bigwedge_j \neg y_j^{\frac{1}{2}}.$$

The only model which can both contribute to $P_\lambda(e)$ is $\alpha_1 \wedge \alpha_0 \wedge \beta_{\frac{1}{2}}$ and thus $P_\lambda(e) = \frac{1}{2}$.

Thus from (iv), (v) and (vi) we get that (E3) holds for P_λ. $\qquad\square$

Remark 3.13. *Note that the P_λ of Theorem 3.12 is strictly Method 2 probability. For example we saw that the network of Figure 2.3 with $a = b = $ und cannot solve Method 1 probability. The next section will see how far we can go with Method 1 probability.*

Summary of the results so far for the semantical probabilistic Method 2. We saw that Dung's traditional complete extensions strictly contain the probabilistic Method 1 extensions and is strictly contained in the probabilistic Method 2 extensions.

4 Approximating the semantic probability by syntactic probability

We have seen in Theorem 3.12 that the Method 2 probabilistic semantics can give us all the traditional Dung complete extensions. This result, together with the probabilistic semantics P_2 of Example 3.9 would show that Method 2 semantics is stronger than traditional Dung complete extensions semantics.

This section examines how far we can stretch the applicability of the syntactical probability approach (Method 1). We know from the "all-undecided" extension for the network in Figure 2 that there are cases where we cannot give Method 1 probability. We ask in this section, can we approximate such extensions by Method 1 probabilities?

We find that the answer is yes.

Let $\langle S, R \rangle$ be a network. Let λ be a legitimate Caminada labelling giving rise to a complete extension $E = E_\lambda$. If the extension is a preferred extension, then there exists a solution f to the $\mathrm{Eq_{inv}}$ equations which yield λ and f is actually a Method 1 (and here also a Method 2) probabilistic semantics for $\langle S, R \rangle$. The question remains as to what happens in the case where λ is not a preferred extension. In this case we are not sure whether λ can be realised by a solution f of the

$\mathrm{Eq_{inv}}$ equations. In fact there are examples of networks where no such f exists. We know from Theorem 3.12 that there exists a probability function P_λ on models that would yield λ according to Definition 3.10. We seek an $\mathrm{Eq_{inv}}$ function which approximates this probability.

We shall use the ideas of Example 2.5.

Remark 4.1. *We need to use some special networks.*

1. *Consider Figure 9, which we shall call U_n. $n = 1, 2, 3, \ldots$.*

   The $\mathrm{Eq_{inv}}$ equations solve for this figure as $u_i = \frac{1}{2}, i = 1, \ldots, n$.

$$u = \frac{1}{2^n}$$

   Thus if u attacks any node x, its "impact" on x is the multiplicative value $1 - \frac{1}{2^n}$. For n very large, the attack is almost

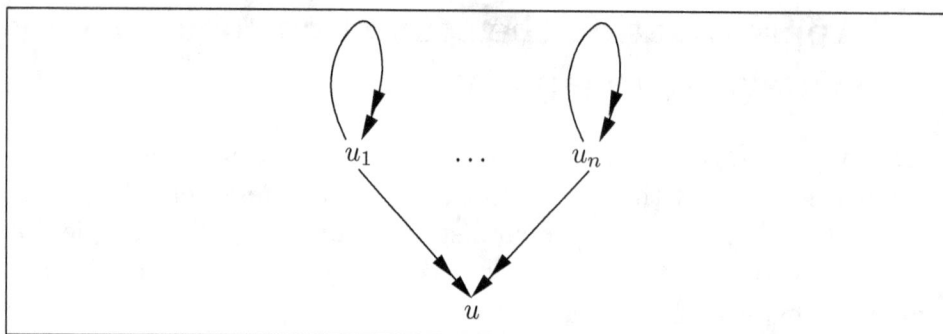

Figure 9: Multiple attacks by undecided nodes.

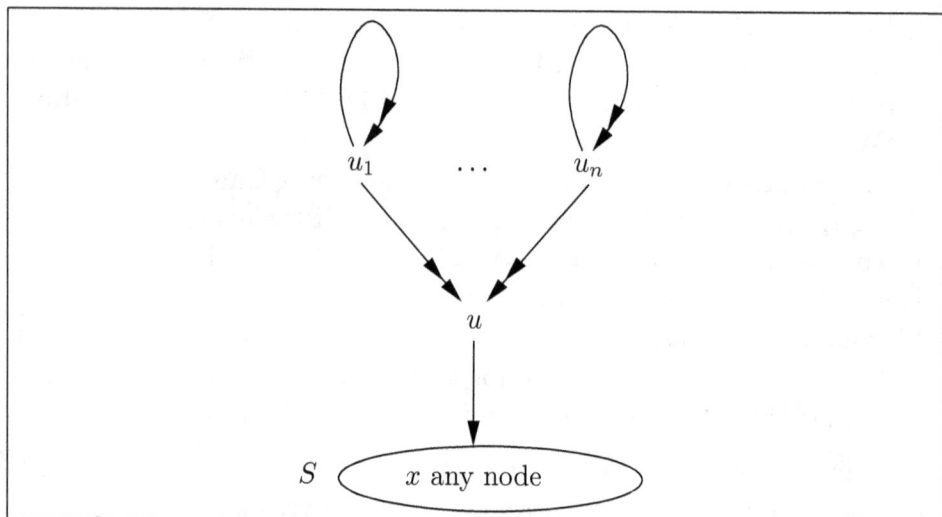

Figure 10: Scenario depicted in Remark 4.1.

negligible.

2. *Let $\langle S, R \rangle$ be any network. Let u be a node not in S. If we add u to S and let it attack all elements of S, we can assume in view of (1) above that the Eq_{inv} value of u is $\frac{1}{2^n}$. Figure 10 depicts this scenario.*

   We suppress $\{u_1, \ldots, u_n\}$ and just record that $u = \frac{1}{2^n}$.

Construction 4.2. *Let $\langle S, R \rangle$ be given and let λ be a legitimate Caminada labelling giving rise to a non-preferred extension.*

Let $u \notin S$ be a new point and assume in view of Remark 4.1 that the value of u is very very small. Let

$$S' = S \cup \{u\}$$

and let

$$R' = R \cup \{(u, v) | \lambda(v) = \mathbf{und}\}.$$

Let $\lambda' = \lambda \cup \{(u, \mathbf{und})\}$.

Let $Att(x)$ be the set of all attackers of x in $\langle S, R \rangle$ and let $Att'(x)$ be the set of all attackers of x in $\langle S', R' \rangle$.

We have if $\lambda'(x) \in \{\mathbf{in}, \mathbf{out}\}$, then $u \notin Att'(x)$.

If $\lambda'(x) = \mathbf{und}$, then $y \in Att'(u)$.

Consider the following set of equations on $\langle S', R' \rangle$.

$$x = 1, \ \text{if } \lambda'(x) = \mathbf{in} \tag{EQ1}$$

$$x = 0, \ \text{if } \lambda'(x) = \mathbf{out} \tag{EQ0}$$

$$x = \Pi(1 - y)_{y \in Att'(x) in \ \langle S', R' \rangle}, \ \text{if } \lambda'(x) = \mathbf{und} \tag{EQU}$$

This set of equations has a solution $\mathbf{f}$.

We claim the following

1. *$\lambda(f)$ is a complete extension*

2. *$\lambda(f) = \lambda'$*

It is clear that $\lambda(f)(x) = \lambda'(x)$, for $\lambda'(x) \in \{\mathbf{in}, \mathbf{out}\}$. Does $\lambda(f)$ agree with λ' on undecided points of λ'? The answer is that it must be so, because λ' is a preferred extension. So $\lambda(f)$ cannot be an extension with more zeros and ones than λ'.

Remark 4.3. *The perceptive reader might ask why do we use those particular equations in Construction 4.2 above? The answer can be seen from Figure 11.*

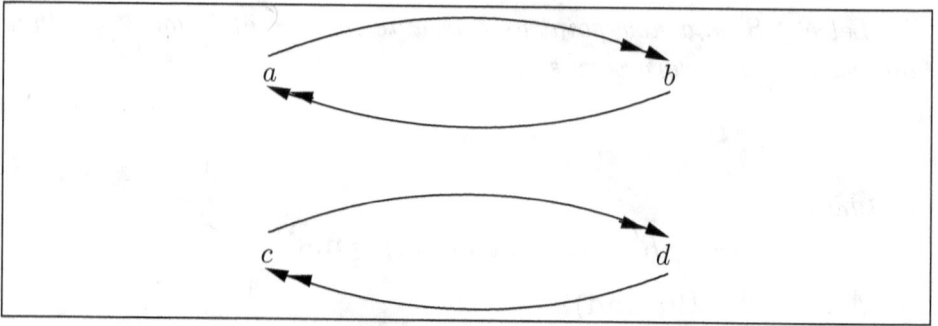

Figure 11: A network with two cycles.

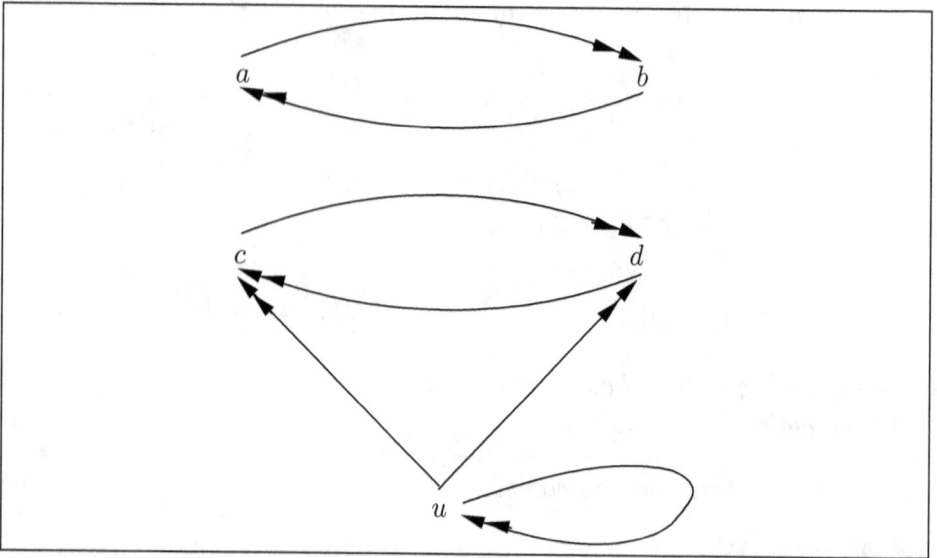

Figure 12: A self-attacking node attacking one of the cycles in the network of Figure 11.

*Consider $\lambda(a) = $ **in**, $\lambda(b) = $ **out**, $\lambda(c) = \lambda(d) = $ **und**.*

We create Figure 12.

We take the equation

$$a = 1, b = 0$$
$$c = (1 - d)(1 - u)$$
$$d = (1 - c)(1 - u)$$
$$u = 1 - u.$$

The solution for the equations for c, d and u are

$$u = \tfrac{1}{2}$$
$$c = d = \tfrac{1}{3}$$

We have to insist on $a = 1, b = 0$. If we do not insist and write the usual equations

$$a = 1 - b$$
$$b = 1 - a,$$

we might get a different solution, e.g.

$$b = 1, a = 0.$$

This not the original λ.

Remark 4.4. *This remark motivates and proves the next Theorem 4.5. We need some notation. Let Q be a set of atoms. By the models of Q (based on Q) we mean all conjunction normal forms of atoms from Q or their negations. So, for example, if $Q = \{a, b, c\}$, we get 8 models, namely*

$$m_1 = a \wedge b \wedge c$$

$$\vdots$$

$$m_8 = \neg a \wedge \neg b \wedge \neg c.$$

If we have atoms

$$Q_1 = \{a_i\}, Q_2 = \{b_j\}, Q_3 = \{c_k\}$$

where Q_i are pairwise disjoint we can write the models of $Q_1 \cup Q_2 \cup Q_3$ in the form

$$\alpha \wedge \beta \wedge \gamma$$

where α is a model of Q_1, β of Q_2 and γ of Q_3.
For example

$$\alpha_1 \wedge \beta_1 \wedge \gamma_1 = (a_1 \wedge a_2 \wedge \ldots) \wedge (\neg b_1 \wedge b_2 \wedge \ldots) \wedge (c_2 \wedge \ldots).$$

Now let $\langle S, R \rangle$ and λ be as in Construction 4.2. Remember we assume that the value of u is very very small, and so the attack value $(1 - u)$ is very close to 1. Consider λ' and f and $\lambda(f)$ again as in Construction 4.2. f is a solution of Eq_{inv} equations (EQ1), (EQ0) and (EQU). Therefore any model of S', say $\alpha = \pm s_1 \wedge \pm s_2 \wedge \pm \ldots \wedge \pm s_k \wedge \pm u$ where $S = \{s_1, \ldots, s_k\}$ will have its probability semantics as

$$P_f(\alpha = \Pi_{i=1}^k f(\pm s_k))) \times f(\pm u) \qquad (*)$$

where

$$f(+s) = f(s)$$
$$f(-s) = 1 - f(s).$$

In particular, we have the following:

1. *Let $E^+ = \{e_1^+, \ldots\}$ be the subset of S such that $\lambda(e_i^+) = $ **in**. Let $E^- = \{e_j^-\}$ be the subset of S such that $\lambda(e_j^-) = $ **out**. Let $E_{\text{und}} = \{b_k\}$ be the set of all nodes in S such that $\lambda(b_k) = $ **und**.*

   We therefore have that any model δ of S' has the form

$$\begin{aligned} \delta &= \bigwedge_i \pm e_i^+ \wedge \bigwedge_i \pm e_j^- \wedge \bigwedge_k \pm b_k \wedge \pm u \\ &= \alpha \wedge \beta \pm u \end{aligned}$$

   where α is a model of $E^+ \cup E^-$ and β is a model of E_{und}.

   Let $\alpha_{1,0}$ be the particular conjunction

$$\alpha_{1,0} = \bigwedge_i e_i^+ \wedge \bigwedge_j \neg e_j^-.$$

Let β be any model of E_{und}. Consider $P_f(\delta), \delta = \alpha \wedge \beta \wedge \pm u$. Then by (*) we have that

$$P_f(\delta) = 0, \ if \ \alpha \neq \alpha_{1,0}. \qquad (**)$$

Since P_f is a probability, we have for any $s \in S'$

$$P_f(s) = P_f(\bigwedge_{y \in Att'(s)} \neg y).$$

Note that for $s \in S, s \neq u$ such that $\lambda(s) \in \{\textbf{in}, \textbf{out}\}$, u does not attack s, and so we have

$$\begin{aligned} P_f(s) &= P(f)(\bigwedge_{y \in Att(s)} \neg y) \\ &= \Pi_{y \in Att(s)}(1 - f(y)) \end{aligned} \qquad (\sharp 1)$$

For u we have that u is very small and so $P_f(u) = \frac{1}{2^n}$.

For $s \in S$ such that $\lambda(s) = \textbf{und}$, we have that u attacks s and so

$$\begin{aligned} P_f(s) &= P_f(\bigwedge_{y \in Att'(s)} \neg y) \\ &= (\Pi_{y \in Att(s)}(1 - f(y))) \times (1 - \frac{1}{2^n}) \end{aligned} \qquad (\sharp 2)$$

The $(1 - \frac{1}{2^n})$ is the attack of u.

We ask what are the attackers of $s \in E_{\text{und}}$? They cannot be nodes x such that $\lambda(x) = \textbf{in}$, because then s would be out. So the value of $f(y)$, (for $y \in Att(s)$) is either 0 or a value in $(0, 1)$.
  So we can continue and write

$$P_f(s) = (1 - \frac{1}{2^n})\Pi_{\substack{y \in Att(s) \\ \lambda(y) = \\ \textbf{und}}} (1 - f(y)) \qquad (\sharp 3)$$

Note that $0 < P_f(s) < 1$, because all the $f(y)$, for $\lambda(y) = \textbf{und}$, satisfy $0 < f(y) < 1$.
  We also have

$$\sum_{\text{all models } m} P_f(m) = 1. \qquad (\sharp 4)$$

Since(**) holds, we need consider only models m of the form $\alpha_{1,0} \wedge \beta \wedge \pm u$.

We can write

$$1 = \sum_{\beta \wedge \pm u} P_f(\alpha_{1,0} \wedge \beta \wedge \pm u) \tag{$\sharp$5}$$

where β is a model of E_{und}. Let us analyse ($\sharp$5) a bit more.
Assume $\beta = \bigwedge_k \pm b_k$.
So

$$P_f(\alpha_{0,1} \wedge \beta \wedge u) + P_f(\alpha_{0,1} \wedge \beta \wedge \neg u) = \Pi_k f(\pm b_k). \tag{$\sharp$6}$$

We thus get that:

$$\sum_{\beta} \Pi_k f(\pm b_k) = 1. \tag{$\sharp$7}$$

($\sharp$7) says something very interesting. It says that f restricted to E_{und} gives a proper probability distribution on the models of E_{und}.
This combined with ($\sharp$3) gives us the following result.
Consider $(E_{\text{und}}, R_{\text{und}})$ where $R_{\text{und}} = R \upharpoonright E_{\text{und}}$. Then $f \upharpoonright E_{\text{und}}$ is a proper probability distribution on $(E_{\text{und}}, R_{\text{und}})$.
Does it satisfy the proper equations?
Let $s \in E_{\text{und}}$. Do we have

$$P_{\text{und}}(s) \stackrel{?}{=} P_{\text{und}}(\bigwedge_{\substack{y \in E_{\text{und}} \\ yRx}} \neg y)$$

Let us check.
The real equation is

$$P_{\text{und}}(s) = P_{\text{und}}(\bigwedge_{\substack{y' \in E_{\text{und}} \\ yRx}} \neg y) \times (1 - u) \tag{$\sharp$8}$$

Since u is very small, we have a very good approximation.[5]

We can now define a probability P on $\langle S, R \rangle$. Let $m = \alpha \wedge \beta$ be a model, where α is a model for $E^+ \cup E^-$ and β is a model for E_{und}.

Then define P as follows

$$P(\alpha \wedge \beta) = 0, \ \text{if } \alpha = \neg\alpha_{1,0}$$

$$P(\alpha \wedge \beta) = P_{und}(\beta), \ \text{if } \alpha = \alpha_{1,0}$$

We need to show that approximately

$$P(s) = P(\bigwedge_{y \in Att(s)} \neg y)$$

If $s \in E^+ \cup E^-$ this follows from ($\sharp$1).

If $s \in E_{und}$, this follows from ($\sharp$3) and ($\sharp$8).

Note that since the f involved came from Eq_{inv} equations, P satisfies the following on $\langle S, R \rangle$.

$$P(s) = 0, \ \text{if some } y \in Att(s) P(y) = 1$$

$$P(s) = 1, \ \text{if for all } y \in Att(s), P(y) = 0 \qquad (\sharp 9)$$

$$P(s) = \ \text{undecided, otherwise.}$$

Theorem 4.5.

1. *Let $\langle S, R \rangle$ be a network and let λ be a legitimate Caminada labelling on S. Then there exists a Method 1 probability distribution P_λ, which almost satisfies equation (E3), namely for every ε, there exists a Method 1 probability P_λ depending on ε, such that for every x and its attackers y_i, we have $|P_\lambda(x) -$*

---

[5]The perceptive reader might ask what happens if we let u converge to 0? The answer is that we get a proper Eq_{inv} extension. However, this may be an all undecided extension (which is what we do want), or it may be a complete extension properly containing all the undecided extensions (which is not what we want!).

We may decide to do what physicists do to their equations. Write the equations in full and simply neglect any item containing higher order u, i.e., u^2, u^3, etc. This is reasonable when the value of each node is small.

$P_\lambda(\wedge \neg y_i)| < \varepsilon$, *such that*

$$\lambda(x) = \mathbf{in}, \ \textit{if } P_\lambda(x) = 1$$
$$\lambda(x) = \mathbf{out}, \ \textit{if } P_\lambda(x) = 0$$
$$\lambda(x) = \mathbf{und}, \ \textit{if } 0 < P_\lambda(x) < 1.$$

2. *P is obtained as follows*

Case 1. λ *is a preferred extension. Then let f be a solution of Eq_{inv} for $\langle S, R \rangle$. Let $P_\lambda = f$.*

Case 2. λ *is not a preferred extension.*

Let $E_{\mathrm{und}}^\lambda = \{x | \lambda(x) = \mathrm{und}\}$. Consider $\langle S', R' \rangle$, where $S' = E_{\mathrm{und}}^\lambda \cup \{u\}$, where u is a new point not in S with value almost 0.

$$R' = R \upharpoonright E_{\mathrm{und}}^\lambda \cup \{u\} \times E_{\mathrm{und}}^\lambda.$$

Then $\langle S', R' \rangle$ has only one extension (all undecided). Let f' be a solution to Eq_{inv} on $\langle S', R' \rangle$. We now define P_λ on $\langle S, R \rangle$.

Let $\alpha_{1,0} = \bigwedge_{\lambda(x)= \text{ in}} x \wedge \bigwedge_{\lambda(y)= \text{ out}} \neg y$.

Let $m = \alpha \wedge \beta$ be an arbitrary model of S, where α is a model of $\{x | \lambda(x) \in \{\mathbf{in}, \mathbf{out}\}$ and β is a model of E_{und}^λ. Define $P_\lambda(\alpha \wedge \beta)$ to be

$$P_\lambda(\alpha \wedge \beta) = 0 \ \textit{if } \alpha \neq \alpha_{1,0}$$
$$P_\lambda(\alpha_{1,0} \wedge \beta) = f'(\beta)$$
$$\textit{where } \beta = \bigwedge_{s \in E_{\mathrm{und}}^\lambda} \pm s$$
$$\textit{and } f(\beta) = \Pi_{\pm s \text{ in } \beta} f(\pm s).$$

Proof. Follows from the considerations of Remark 4.4. $\qquad\square$

Example 4.6. *Let us show how Theorem 4.5 works by doing a few examples.*

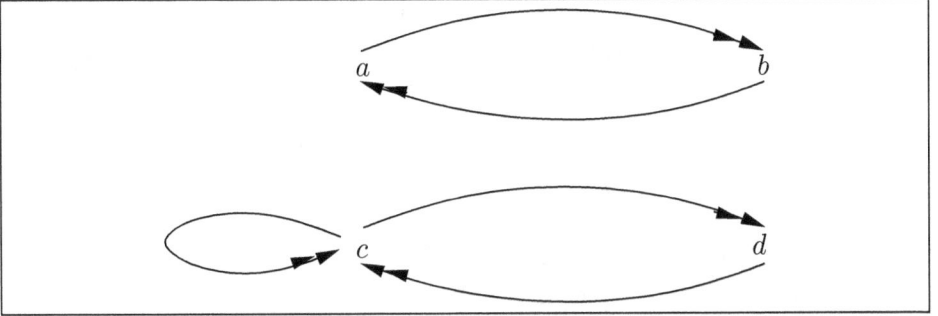

Figure 13: Augmented network of Figure 2 with node a as c and b as d and an extra cycle.

1. *Consider the network of Figure 11 and the extension λ mentioned there, namely $\lambda(a) = \mathbf{in}$, $\lambda(b) = \mathbf{out}$, $\lambda(c) = \lambda(d) = \mathbf{und}$.*

   Following our algorithms we look at the $\{c, d, u\}$ part of Figure 12 and solve the equations. We get $u = \frac{1}{2}, c = d = \frac{1}{3}$.

   The probability P_λ will be as follows:

   $$P_\lambda(\alpha \wedge \beta) = 0$$
   $$if\ \alpha \neq a \wedge \neg b.$$

   Now look at

   $$P_\lambda(a \wedge \neg b \wedge c \wedge d) = \frac{1}{3} \times \frac{1}{3} = \frac{1}{9}$$
   $$P_\lambda(a \wedge \neg b \wedge c \wedge \neg d) = \frac{1}{3} \times \frac{2}{3} = \frac{2}{9}$$
   $$P_\lambda(a \wedge \neg b \wedge \neg c \wedge d) = \frac{2}{3} \times \frac{1}{3} = \frac{2}{9}$$
   $$P_\lambda(a \wedge \neg b \wedge \neg c \wedge \neg d) = \frac{2}{3} \times \frac{2}{3} = \frac{4}{9}.$$

2. *Let us look at Figure 13.*

   With $\lambda(a) = \mathbf{in}$, $\lambda(b) = \mathbf{out}$, $\lambda(c) = \lambda(d) = \mathbf{und}$.

   The $\{c, d\}$ part is Figure 2. Here we solve the equations on the $\{c, d, u\}$ part associated with $\{c, d\}$, which is the same as Figure 3. The solution is found in Example 2.5, with $u = \frac{1}{2}$.

*We get $u = \frac{1}{2}; c = 0.36, 1 - c = 0.764, d = 0.382, 1 - d = 0.618$.
The probability P_λ of this case is $P_\lambda(\alpha \wedge \beta) = 0$, if $\alpha \neq a \wedge \neg b$.*

$$P_\lambda(a \wedge \neg b \wedge c \wedge d) = 0.236 \times 0.382 = 0.09$$
$$P_\lambda(a \wedge \neg b \wedge c \wedge \neg d) = 0.236 \times 0618 = 0.146$$
$$P_\lambda(a \wedge \neg b \wedge \neg c \wedge d) = 0.764 \times 0.382 = 0.292$$
$$P_\lambda(a \wedge \neg b \wedge \neg c \wedge \neg d) = 0.764 \times 0.618 = 0.472.$$

Indeed
$$0.09 + 0.146 + 0.292 + 0.472 = 1.000.$$

We now discuss imposing probability on instantiated networks such as ASPIC+. We begin with simple instantiations into classical propositional logic.

Definition 4.7. *1. An abstract instantiated network (into classical propositional logic) has the form $\mathcal{A} = \langle S, R, I \rangle$, where $\langle S, R \rangle$ is an abstract argumentation network and I is a mapping associating with each $x \in S$, a well-formed formula $I(x) = \varphi_x$ of classical propositional logic.*

2. For any $\mathcal{A}$ as in 1, we associate the theory $\Delta_\mathcal{A} = \{\varphi_x \leftrightarrow \bigwedge_{(y,x) \in R} \neg \varphi_y \mid x \in S\}$.

3. A semantic probability model P on $\mathcal{A}$ is a probability distribution on the models based on S such that for all $x \in S$, we have:

$$P(\varphi_x) = P(\bigwedge_{(y,x) \in R} \neg \varphi_y)$$

Example 4.8. *Consider Figure 14 where part (b) is an instatiation of part (a) with $I(x) = a_1 \vee a_2$ and $I(a_3) = a_3$. The equations any probability assignment needs to satisfy are*

$$\begin{aligned} P(a_1 \vee a_2) &= 1 \\ P(a_3) &= P(\neg(a_1 \vee a_2)) \\ &= P(\neg a_1 \wedge \neg a_2) \\ &= 0. \end{aligned}$$

If we let $P(a_1) = x$, $P(a_2) = 1 - x$, $P(a_3) = 0$, with $x \in [0, 1]$, then P satisfies the equations. Compare with Example 3.4.

Figure 14: (a) A network and (b) one of its instantiations with $x = a_1 \lor a_2$

5 Comparison with the literature

There are several probabilistic argumentation papers around. This is a hot topic in 2014. We highlight two main points of view. The external and the internal views.

Let $\langle S, R \rangle$ be a network and let $\mathbf{f}$ be a function from S to $[0, 1]$. We can regard $\mathbf{f}$ as giving a probability number to each $x \in S$. The internal probability is where the above numbers signify the value of the argument. Its truth, its reliability, its probability of being effective, etc., or whatever measure we attach to it as an argument. Figure 15 represents in this case the Eq$_{\mathrm{inv}}$ solution (and hence probability) of the network of Figures 2 and 3. The external view is to think of $\mathbf{f}(x)$ as the probability of the predicate "$x \in S$". That is, the probability that the argument x is present in S. Consider again Figure 15.

The probability that a is in the network is 0.236 and the probability that b is in the network is 0.382. Therefore, the probability that the network contains both $\{a, b\}$ is $0.236 \times 0.388 = 0.09$. The probability that the network contains only a is $0.236 \times (1 - 0.382) = 0.1458$. The probability that the network contains only b

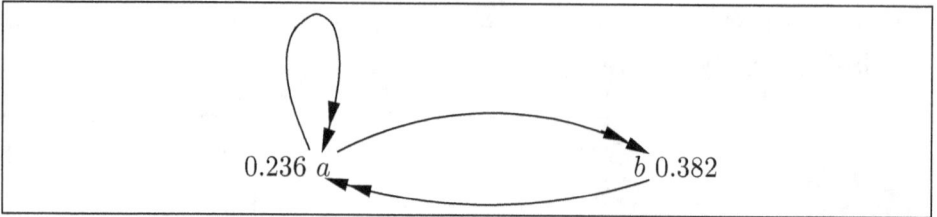

Figure 15: The Eq$_{inv}$ solution to the networks of Figures 2 and 3

is $0.382 \times (1 - 0.236) = 0.292$ and the probability that the network is empty is $(1 - 0.236) \times (1 - 0.382) = 0.472$. It is clear why we are calling this view an external probability view. It imposes probability externally expressing uncertainty on what the network graph is. This is done either by giving the probability to points or more generally by giving probability directly to subsets G of S, expressing the probability that the graph is really that subset of S with R restricted to G. This external view has value in dialogue argumentation or negotiation when we try to estimate what network our opponent is reasoning with. The problem with this external view is how to connect with the attack relation. Note that mathematically in the external view we have probabilities on points in S or probabilities on subsets of S, which are the same options as in our internal view, but the understanding of them is different. We in the internal view considered the subset as a classical model, while the external view considers it as a subnetwork. When we use the internal view, we can connect it with the attack relation via the equational approach (Equation (E3)), but how would the external view connect with the attack relation? We can ask, for example, how to get a value for a single point to be "in" an extension? Intuitively, looking back at Figure 15, we can say the point a for example is "in" in case the network is $\{a\}$ and is also "in" in one of the three extensions in case the network is $\{a, b\}$. So we might take the "in" value to be $0.1458 + 0.09/3 = 0.1458 + 0.03 = 0.1758$. The connection with the attack relation can be done perhaps through the probabilities for admissible sets, since being admissible is connected with the attack relation. There are problems, however, with this approach.

Hunter [7] was trying to lay some foundations for this view, following the papers [3, 9]. See also a good summary in Hunter[8]. Hunter was trying to find a connection between the external probability view and some reasonable values we can give to admissible subsets. He proposes restrictions on the probability function on S. We are not going to discuss or reproduce Hunter's arguments here. It suffices to say that possibly a subsequent paper of ours will critically examine the external view and compare with the internal view.

Let us now compare our work with that of M. Thimm, [13], whose approach is also internal. We quote from [13]:

"In this paper we use another interpretation for probability, that of *subjective probability* [11]. There, a probability $P(X)$ for some $X \in \mathcal{X}$ denotes the *degree of belief* we put into X. Then a probability function P can be seen as an epistemic state of some agent that has uncertain beliefs with respect to $\mathcal{X}$. In probabilistic reasoning [11, 12], this interpretation of probability is widely used to model uncertain knowledge representation and reasoning.

In the following, we consider probability functions on sets of arguments of an abstract argumentation frameworks. Let $\mathsf{AF} = (\mathsf{Arg}, \rightarrow)$ be some fixed abstract argumentation framework and let $\mathcal{E} = 2^{\mathsf{Arg}}$ be the set of all sets of arguments. Let now $\mathcal{P}_{\mathsf{AF}}$ be the set of probability functions of the form $P : 2^{\mathcal{E}} \rightarrow [0, 1]$. A probability function $P \in \mathcal{P}_{\mathsf{AF}}$ assigns to each set of possible extensions of AF a probability, i.e. $P(e)$ for $e \in \mathcal{E}$ is the probability that e is an extension and $P(E)$ for $E \subseteq \mathcal{E}$ is the probability that any of the sets in E is an extension. In particular, note the difference between e.g. $P(\{\mathcal{A}, \mathcal{B}\}) = P(\{\{\mathcal{A}, \mathcal{B}\}\})$ and $P(\{\{\mathcal{A}\}, \{\mathcal{B}\}\})$ for arguments $\mathcal{A}, \mathcal{B}$. While the former denotes the probability that $\{\mathcal{A}, \mathcal{B}\}$ is an extension the latter denotes the probability that $\{\mathcal{A}\}$ or $\{\mathcal{B}\}$ is an extension. In general, it holds $P(\{\mathcal{A}, \mathcal{B}\}) \neq P(\{\{\mathcal{A}\}, \{\mathcal{B}\}\})$.

For $P \in \mathcal{P}_{\mathsf{AF}}$ and $\mathcal{A} \in \mathsf{Arg}$ we abbreviate

$$P(\mathcal{A}) = \sum_{\mathcal{A} \in e \subseteq \mathsf{Arg}} P(e).$$

Given some probability function P, the probability $P(\mathcal{A})$ represents the degree of belief that $\mathcal{A}$ is in an extension (according to P), i.e. $P(\mathcal{A})$ is the sum of the probabilities of all possible extensions that contain $\mathcal{A}$. The set $\mathcal{P}_{\mathsf{AF}}$ contains all possible views one can take on the arguments of an abstract argumentation framework AF.

Example 4. We continue Ex. 1. (Comment by Gabbay and Rodrigues: This is the network of our Figure 4.) Consider the function $P \in \mathcal{P}_{\mathsf{AF}}$ defined via $P(\{\mathcal{A}_1, \mathcal{A}_3, \mathcal{A}_5\}) = 0.3, P(\{\mathcal{A}_1, \mathcal{A}_4\}) = 0.45, P(\{\mathcal{A}_5, \mathcal{A}_2\}) = 0.1, P(\{\mathcal{A}_2, \mathcal{A}_4\}) = 0.15$, and $P(3) = 0$ for all remaining $e \in \mathcal{E}$. Due to Prop. 1 the function P is well-defined as e.g.

$$
\begin{aligned}
P(\{\{\mathcal{A}_5, \mathcal{A}_2\}, \{\mathcal{A}_2, \mathcal{A}_4\}, \{\mathcal{A}_3\}\}) \\
= P(\{\mathcal{A}_5, \mathcal{A}_2\}) + P(\{\mathcal{A}_2, \mathcal{A}_4\}) + P(\{\mathcal{A}_3\}) \\
= 0.1 + 0.15 + 0 = 0.25.
\end{aligned}
$$

Therefore, P is a probability function according to Def. 3. According to P the probabilities to reach argument of AF compute to $P(\mathcal{A}_1) = 0.75, P(\mathcal{A}_2) = 0.25, P(\mathcal{A}_3) = 0.3, P(\mathcal{A}_4) = 0.6$, and $P(\mathcal{A}_5) = 0.4$.

In the following, we are only interested in those probability functions of $\mathcal{P}_{\mathsf{AF}}$ that agree with our intuition on the interrelationships of arguments and attack. For example, if an argument $\mathcal{A}$ is not attacked we should completely believe in its validity if no further information is available. We propose the following notion of *justifiability* to describe this intuition.

Definition 4. A probability function $P \in \mathcal{P}_{\mathsf{AF}}$ is called *p-justifiable* wrt. AF, denoted by $P \Vdash_{\mathcal{J}} \mathsf{AF}$, if it satisfies for all $\mathcal{A} \in \mathsf{Arg}$.

1. $P(\mathcal{A}) \leq 1 - P(\mathcal{B})$ for all $\mathcal{B}, \in \mathsf{Arg}$ with $\mathcal{B} \to \mathcal{A}$ and

2. $P(\mathcal{A}) \geq 1 - \sum_{\mathcal{B} \in \mathcal{F}} P(\mathcal{B})$ where $\mathcal{F} = \{\mathcal{B} | \mathcal{B} \to \mathcal{A}\}$.

Let $P_{\mathsf{AF}}^{\mathcal{J}}$ be the set of all p-justifiable probability functions wrt. AF.

The notion of p-justifiability generalizes the concept of complete semantics to the probabilistic setting. Property 1.) says that the degree of belief we assign to an argument $\mathcal{A}$ is bounded from above by the complement to 1 of the degrees of belief we put into the attackers of $\mathcal{A}$. As a special case, note that if we completely believe in an attacker of $\mathcal{A}$, i.e., $P(\mathcal{B}) = 1$ for some $\mathcal{B}$ with $\mathcal{B} \to \mathcal{A}$, then it follows $P(\mathcal{A}) = 0$. This corresponds to property 1.) of a complete labelling (see Section 2). Property 2.) of Def. 4 says that the degree of belief we assign to an argument $\mathcal{A}$ is bounded from below by the inverse of the sum of the degrees of belief we put into the attacks of $\mathcal{A}$. As a special case, note that if we completely disbelieve in all attackers of $\mathcal{A}$, i.e. $P(\mathcal{B}) = 0$ for all $\mathcal{B}$ with $\mathcal{B} \to \mathcal{A}$, then it follows $P(\mathcal{A}) = 1$. This corresponds to property 2.) of a complete labeling, see Section 2. The following proposition establishes the probabilistic analogue of the third property of a complete labelling.

Proposition 2. Let P be p-justifiable and $\mathcal{A} \in \mathsf{Arg}$. If $P(\mathcal{A}) \in (0, 1)$ then

1. there is no $\mathcal{B} \in \mathsf{Arg}$ with $\mathcal{B} \to \mathcal{A}$ and $P(\mathcal{B}) = 1$ and
2. there is a $\mathcal{B}' \in \mathsf{Arg}$ with $\mathcal{B}' \to \mathcal{A}$ and $P(\mathcal{B}') > 0$.

From our point of view, Thimm's approach is a variant of our semantic Method 2 approach without the strong equation (E3) but the weaker Definition 4 of Thimm. Thus Thimm will allow for different values for nodes x_1 and x_2 in our Figure 8, while we would not (see Example 3.5).

Although Thimm's approach is mathematically close to us, conceptually we are far apart. Thimm motivates his approach as a degree of belief in a subset $E \subseteq S$, considering E as an extension. We consider E as representing a classical model m of the classical propositional logic with atoms S

$$m = \bigwedge_{s \in E} s \wedge \bigwedge_{s \notin E} \neg s$$

and assign probability to it and then we export this probability to argumentation via the equational approach, equation (E3).

This is an instance of our methodology of "Logic by Translation", From our point of view, equations (E3) are essential, conceptual and non-technical. For Thimm, the inequalities of his Definition 4 appear to be technical to enable the probabilities to work of ground extension.

Our point of view also leads us to the Eq_{inv} Method 1 probabilities and to the approximation results of Section 4.

In Thimm's conceptual approach, this way of thinking does not even arise.

To summarise, this paper presented an internal view of probabilistic argumentation. There is a need for two subsequent research papers

1. The external view done coherently and its connection to the internal view

2. A conditional probability view and its connection with Bayesian Networks views as Argumentation Networks

6   Conclusions

This section explains and sets our approach in a general generic context.

Suppose we are given a system $\mathbb{S}$ such as an argumentation system $\langle S, R \rangle$ and we want to add to it some aspect $\mathbb{A}$.

There is a generic way to add any new feature to a system. It involves 1) identifying the basic units which build up the system and 2) introducing the new feature to each of these basic units. In the case where the system is argumentation and the feature is probabilistic we have the following: the basic units are **a.** the nature of the arguments involved; **b.** the membership relation in the set S of arguments;[6] **c.** the attack relation; and **d.** the choice of extensions.

---

[6]Note that the set S itself may not be fully or accurately known, especially modelling an opponent in dialogue systems.

Generically to add a new aspect (probabilistic, or fuzzy, or temporal, etc) to an argumentation network $\langle S, R \rangle$ can be done by adding this feature to each component. **a.** We make the effective strength of the argument probabilistic; **b.** we give probability to whether an argument is included in S;[7] **c.** we make the attack relation probabilistic; and **d.** we put probability on the extensions.

These features interact and need to be chosen with care and coordination. We need a methodological approach to make our choices. One such methodology is what we called "logic by translation".

We meaningfully translate the argumentation system into classical logic which does have probabilistic models and then let probabilistic classical logic endow the probability on the argumentation system. As we mentioned, this of course depends on how we translate.

We gave in this Chapter an object-level translation. The arguments of S became atoms of classical propositional logic, we then used probability on the models of classical logic and used the attack relation R to express equational restrictions on the probabilities. In this kind of translation, the attack relation did not become probabilistic.

We could have used a meta-level translation into classical predicate logic, using a binary relation R for expressing in classical logic the attack relation and using unary predicates to express that an argument x is "in", x is "out", etc., with suitable coordinating axioms. In this case all predicates would have become probabilistic including the attack relation R. As far as we know nobody has done this to R.

In this context of possible options what we have done is one systematic approach and we compared it with other approaches. It should be noted that we could have followed the same steps to get fuzzy argumentation networks; temporal argumentation networks; or indeed any other feature available for classical propositional logic.

---

[7] **a.** and **b.** are distinct, because **a.** represents how effective an argument is, whereas **b.** is the decision of whether or not to include an argument for consideration. An argument may be deemed very effective but not included for consideration for completely different reasons.

References

[1] M. Caminada and D. Gabbay. A logical account of formal argumentation. *Studia Logica*, pages 109–145, 2012.

[2] M. Caminada and G. Pigozzi. On judgment aggregation in abstract argumentation. *Autonomous Agents and Multi-Agent Systems*, 22(1):64–102, 2011.

[3] P. M. Dung and P. Thang. Towards (probabilistic) argumentation for jury-based depute resolution. In B. Verheij, S. Szeider, and S. Woltran, editors, *Proceedings of COMMA III*, Frontiers in Artificial Intelligence and Applications, pages 171–182. IOS Press, 2012.

[4] D. Gabbay. *Logics for Artificial Intelligence and Information Technology*. College Publications, 2007.

[5] D. Gabbay. Equational approach to argumentation networks. *Argument and Computation*, pages 87–142, 2012.

[6] D. M. Gabbay and O. Rodrigues. A self-correcting iteration schema for argumentation networks. In S. Parsons, N. Oren, C. Reed, and F. Cerutti, editors, *Proceedings of COMMA V*, Frontiers in Artificial Intelligence and Applications, pages 377 – 384. IOS Press, 2014. DOI: 10.3233/978-1-61499-436-7-377.

[7] A. Hunter. Some foundations for probabilistic abstract argumentation. In B. Verheij, S. Szeider, and S. Woltran, editors, *Proceedings of COMMA IV*, Frontiers in Artificial Intelligence and Applications, pages 117–128. IOS Press, 2012.

[8] A. Hunter. A probabilistic approach to modelling uncertain logical arguments. *International Journal of Approximate Reasoning*, 54:47–81, 2013.

[9] H. Li, N. Oren, and T. Norman. Probabilistic argumentation frameworks. In *Proceedings of the First International Workshop on the Theory and Applications of Formal Argumentation (TAFA'11)*, volume 7132 of *Lecture Notes in Computer Science*. Springer, 2012.

[10] S. Modgil and H. Prakken. the ASPIC+ framework for structured argumentation: A tutorial. *Argument & Computation*, 5(1):31–62, 2014.

[11] J. B. Paris. *The Uncertain Reasoner's Companion. A Mathematical Perspective*. Cambridge University Press, 2006.

[12] J. Pearl. *Probabilistic Reasoning in Intelligent Systems. Networks of Plausible Inference*. Morgan Kaufmann, 1998.

[13] M. Thimm. A probabilistic semantics for abstract argumentation. In

Proceedings of the 20th European Conference on Artificial Intelligence (ECAI'12), 2012.

CHAPTER 3
HEAL2100: HUMAN EFFECTIVE ARGUMENTATION AND LOGIC FOR THE 21ST CENTURY.
THE NEXT STEP IN THE EVOLUTION OF LOGIC

1 Logic (up to the year 2016)

Logic began with Aristotle.[1] He realised that in order to write his books he needed logic as a tool (organon) and he wrote his five books on syllogistic logic. Aristotle's logic was refined in later periods and the next significant step came with Peter Abelard who worked in the early 12th century. His treatise the Dialectica [2] contained new ideas such as *de re* and *de dicto* modalities. It became possible to apply logic to language, theology and philosophy. New handbooks of logic appeared in later centuries, by Peter of Spain, Lambert of Auxerre and William of Sherwood. Later logicians were William of Ockham, Jean Buridan, Gregory of Rimini and Albert of Saxony. The best known textbook was by Antoine Arnold and Pierre Nicole The Port Royal Logic [3], J S Mill, *A System of Logic*, [61], 1843 in the 19th century.

Two points to be borne in mind about the development of logic up to the 19th century:

- It was mainly syllogism with extras.

---

[1] The Stoics invented propositional logic in antiquity, and Aristotle himself was the first to systematize dialectic in *Topics* and *On Sophistical Refutations*.

- It dealt with human beings, their language reasoning, *their argumentation*, and their behaviour (as opposed to pure mathematics).

Modern mathematical logic was developed in the late 19th century carrying on until the middle of the 20th century [4, 5]. There were four pillars to mathematical logic: model theory, set theory, proof theory and recursion theory. Emphasis was diverted from the study and application of logic to the humanities, to the study and application of logic to mathematics and its foundations [5].

Dov Gabbay and John Woods [12], called this *The Hundred-Years' Detour*. This has changed with the rise of computer science, artificial intelligence, computational linguistics etc. There was a strong consumer demand for devices using this new technology. *In turn*, there was an urgent need to develop and evolve logic to serve these demands. Emphasis in logic reverted back to the analysis of day-to-day human activity. New logics were developed by diverse non-cooperating non-communicating communities, each driven by the needs of certain types of application or device. The landscape of logic became a *chaos of different methods*. New proposals for what Logic is have been pushed forward by Dov Gabbay and colleagues, such as *New logic with mechanisms* and *New logic with mechanisms and networks*, see Figure 2. For an evolutionary survey of modern logical systems see [23].

The above mentioned 20th century developments (problems and New Logic proposals) turned logic out of its 100 years detour, back to the modelling of the human approach, nevertheless, it still suffers from three limitations.

1. Logic remains a mathematical, formal system which cannot come completely to grips with human reasoning

2. It excludes the study and use of fallacies (see Section 2 below) and so ignores the most effective human use of (fallacious) logic

3. Worse yet, the new developments, though also sometimes applied in the humanities (logic and law, logic and analytic philosophy, logical analysis of language, logic and theology, logic and

argumentation and debate), does not include a unified coherent logical theory nor is there a perception of differences in systems of thought arising from differing cultures such as Western vs. Islamic or Christian vs. Jewish. Frequent misunderstandings arising from such differences are not surprising and very damaging.

There is some realisation among a few of these diverse logical research communities that there needs to be more communication between them and unifying principles are indeed being sought.

The logic with which we are familiar reflects a Western cultural way of thinking and behaving. There are other major cultures which think and behave differently.

The following are strong communities developing the new and old areas of logic:

- The traditional mathematical logic community

- The fuzzy logic community

- The argumentation communities

- The informal logic community

- The researchers dealing with fallacies

- The non-classical logic community and research groups

- The logic and language community

- The probability and Bayesian network community

- The philosophical logic community

- The logic programming community

- The automated reasoning community

- The belief revision community

- The legal reasoning community.

Such communities of course share many common members.

There were several logicians and groups who since the late seventies tried in research, conferences and social administration to encourage unification and communication among the various communities, through publishing many research books, a large number of handbooks and journals and many conferences, workshops and summer schools.

Bringing communities together is not easy. A major obstacle is that the majority of rank and file researchers work in their own restricted area and are concerned with quick publications leading to a promotion. Even when they become established famous and senior, some of them develop a territorial protectiveness and shy away from other communities.

There is another, more scientific difficulty. In most cases to show a connection requires further research and generalisation and this may take time and not be easy to do by a single individual and may not be easily understood.

In our case we want to bring the fallacies and the argumentation communities together by accepting the fallacies as legitimate reasoning schemas (see however Remark 1.1 below). This is both necessary and possible now because of several scientific and social developments:

1. Social media and internet made it clear that fallacious reasoning and patterns of thinking are most effective to the extent that such ways of thought can topple governments, influence elections and support and foster terrorism. The fallacies have been weaponised on a large scale. Counter measure arguments are urgently needed and patterns of reasoning (such as HEAL2100) are urgently required.

   This increased use of fallacies is a result of two trends. The traditional media lost ground to social media and their moderating influence was decreasing. Traditional media wanted to appeal/sell to maximal number of people so they followed a middle reasonable non extreme course. The news/opinion makers on social media was free and abundant, so to compete they adopted extreme views as well as used fallacies and fake news to push these views and improve their ratings.

2. Developments in big data and the internet of things give us the means to develop the new logic. When a fallacy is encountered by a user, then he can use a big data application to find many other examples of the same pattern and responses to it, and thus construct his own response. Currently this is not possible in real time.

3. Good work on argumentation and fallacies is mature enough and detailed enough to enable us to move to the next step.

4. The outstanding technical success and applicability of the developments of the Fuzzy community and Bayesian community is also an enabling factor towards the next step in the evolution of logic.[2]

5. Developments in theories of universal distortions and the use of logic in the sex offenders therapist communities show the effectiveness of countermeasures to fallacies.

6. There is an important social trend that enforces the importance of developing HEAL2100. Traditional media (often known as "legacy media" such as printed newspapers and TV cable and broadcast channels are declining in popularity while their target audiences are ageing.

   These are being replaced by digital media on platforms such as YouTube and Twitter (and their successors) created by individuals or small groups with limited budgets. Tight resources result in one person expounding to a static camera or a couple

---

[2]Note, however, Gilbert Harman's [62], a Discussion of the Relevance of the Theory of Knowledge to the Theory of Induction (with a Digression to the Effect that neither Deductive Logic nor the Probability Calculus has Anything to Do with Inference). He points out that no human reasoner is capable of fulfilling the Bayesian constraints. See also, independently, Woods and Walton in chapter 1 of their *Fallacies: Selected Papers*, [31].

of people discussing an issue at much greater length than could have been allocated by a TV or radio station. Most viewers understand that the content creators will have partisan views or a low commitment to veracity but nevertheless an increasing proportion of the demographic below 40 years old is consuming this new media and is exposed to argumentation and logic, to differing degrees, in a way that was not previously available. There will be many times when consumers watch one discussion which seems reasonable enough until they then open up the next channel and find the arguments in the first debunked as either untruthful or specious.

Modern media has often been criticised as the death of civilisation but in many ways it offers an opportunity for the general public to learn about argumentation in a way that has been available only to the better educated in universities and elite schools but has not been open to the general population since the days of the Athenians meeting on the Pnyx, See Figures 1 and 2.

Remark 1.1. *We said above that we hope to bring together with us the fallacies and argumentation communities. We need to make a quick remark here, which will be developed fully in a later section. These communities regards fallacies in the context of deductive reasoning. The weaponised fallacies we have identified in the media are not only deductive fallacies but also what we now call "Action-Fallacies" (we need to coin this new concept now). This will be defined in a later section but meanwhile let us give a schematic explanation.*

Assume a deductive encounter between a witness and a defense attorney in front of a Jury. Call the deductive encounter level 1 encounter. It is important for the defense attorney to discredit or falsify or argue against the testimony. If the defense attorney fails then his client may be jailed. The attorney may move to a level 2 encounter (meta-level) by arguing that the level 1 encounter should be canceled altogether. For example the attorney may argue (fallaciously or not) that the proceedings is an appeal proceedings and no new witnesses are allowed. There are, however, many action-fallacies which can be used, ranging from the extreme Mafia "fallacy" of assassinating the witness, or the lesser options; intimidate the witness, insult the witness, drug and confuse the witness and so on. These are level 2 non-deductive action - fallacies designed to abort the level 1 proceedings.

A real example of this is reported in Example 4.4 below.

So to summarise, when we talk about integrating the fallacies we mean action-fallacies, which may be deductive fallacies or real actions fallacies used in a higher level to abort a lower level..

Aristotle	Syllogism, 13 fallacies The concept of logic is based on human reasoning		
Middle ages	Studied aspects of various languages	Studied logical rules connected with religion	Classified and stud-ied more fallacies
Mid 19th century	Boole/De Morgan	**Big detour from human based logic towards mathematical logic**	
Mid 20th century	Logic for computer science Deductive human reasoning (problems and New Logic proposals, see Figure 2)		
21st century	Deductive human reasoning + integrating fallacies		

Figure 1: Time-line for the evolution of logic, from Aristotle to the present

2 Fallacies

This Section presents our views leading to the idea of integrating the fallacies into *New logic with mechanisms, networks and fallacies* which we also call HEAL2100. Defining HEAL2100 is research in progress. We do not know yet what form it will take.

We follow several Subsections. Subsection 2.1 presents a short objective survey of the state of affairs up to now. We found [10] very helpful and we are following its presentation. Subsection 2.2 discusses our view/interpretation of the survey in Subsection 2.1

2.1 Historical and current view of fallacies

Aristotle

Aristotle was the first to systematise logical errors into a list and to establish the convention that being able to refute an opponent's thesis is one way of winning an argument [7]. Aristotle's "Sophistical Refutations" (De Sophisticis Elenchis) identifies thirteen fallacies. He divided them up into two major types: linguistic fallacies and non-linguistic fallacies, some depending on language and others that do not. These fallacies are called verbal fallacies and material fallacies, respectively. A material fallacy is an error in what the arguer is talking about, while a verbal fallacy is an error in how the arguer is talking. Verbal fallacies are those in which a conclusion is obtained by improper or ambiguous use of words.[3]

---

[3]Aristotle's 13 fallacies:

I. Fallacies dependent on Language (De Soph Elen 4, 165b24-166b28)
Ambiguity (equivocation or homonymy)
Amphiboly (or ambiguity)
Combination
Division
Accent
Form of expression

II. Fallacies outside of language (De Soph Elen 5, 166b28-168a18)
Accident
The use of words absolutely or in a certain respect

1960	Traditional mathematical logics. Intuitionistic and classical logic. Let us refer to this as **TDL**, Traditional Deductive (modern formal, classical or intuitionistic or other axiomatic) Logic The traditional view of the Fallacies is called **SDF**, Standard Definition of Fallacies, See [18, p. 52]
1960–1990	Intensive development in computer science and AI The rise of many new logics
1980-2000 Dov Gabbay and Many Colleagues Published Multi volume Handbooks of Logic	Systematizing and legtimizing many logics Handbook of Philosophical Logic, Handbook of Logic in Computer Science, Handbook of Logic in AI, and many more.
2000 Dov Gabbay–John Woods See paper [12] proposing the *New logic with mechanisms*.	*New logic with mechanisms* = whatever system is working in the head of a logical agent = traditional deductive logic (**TDL**) + various logical mechanisms (which arose in artificial intelligence, theoretical computer science and study of language during the period 1980-2000)
	...continued over the page

2009 Dov Gabbay Luxembourg Lecture of Dov Gabbay. See [23]	*New logic with mechanisms and networks = New logic with mechansims + Networks + Argumentation + Axioms + Action sequences + A variety of Metal-level Postulates and Algorithms*
Introduced in a 2009–2017	Incredible developments of Smartphones and Social Media: Facebook, YouTube, Twitter, Wikipedia as well as technical developments of the internet and the emergence of the new area known as Big Data
2017 Dov Gabbay–Lydia Rivlin	*New logic with mechansims, networks and fallacies (which we call HEAL2100)= New logic with mechanisms and networks + Integrated fallacies*

Figure 2: Time-line for Logic in the period 1960–2017

Modern times, first wave

Irving Copi in his influential textbook from the mid-twentieth century — defines a fallacy as "a form of argument that seems to be correct but which proves, upon examination, not to be so", see [18]. Copi

Misconception of refutation
Begging the question
Consequent
Non causa pro causa
Complex question.

lists (1961) 18 fallacies, (of which 11 are from Aristotle, also called by John Woods ([17] (1992), "The Gang of 18"[4]). His view is what is known as the traditional view, **SDF**. This view is supported by other distinguished researchers such as Woods [17] (1992), Walton [16, p. 179](2010) (Walton says that a fallacy is an argument that seems to be correct but is not), Salmon [20], and Powers [19].

It was Hamblin [9], who wrote the first book totally devoted to Fallacies, who first criticised **SDF**.[5] He was followed by others. Finocchiaro distinguishes six ways in which arguments can be fallacious. They all have deductive aspects. Finocchiaro [51] observes that it is adequate to classify all the kinds of errors Galileo found in the arguments of the defenders of the geocentric view of the solar system.

Gerald Massey [24], in 1987, has voiced a strong objection to fallacy theory and the teaching of fallacies. He argues that there is no theory of invalidity — no systematic way to show that an argument

---

[4]The Gang of Eighteen fallacies:
ad baculum
ad hominem
ad misericordiam
ad populum
ad verecundiam
affirming the consequent
amphiboly
begging the question
biased statistics
complex question
composition and division
denying the antecedent
equivocation
faulty analogy
gambler's
hasty generalization
ignoratio elenchi
secundum quid.

[5]The term **SDF** was coined by van Eemeren and Grootendorst. It was the name they gave to what Hamblin had said:

SDF: A fallacy is an argument that seems to be valid but is not so.

Hamblin made the historical claim that everyone since Aristotle held this view about what made for a fallacy. Hansen [57] showed that Hamblin was wrong.

is invalid other than to show that it has true premises and a false conclusion [24, p. 164]. By the way this is now available (called refutation system, see [21], 2011). Note that Massey's view/objections to the fallacies is also deductively based, it requires a logical system generating the fallacious arguments as well as the valid arguments.

Johnson and Blair in their textbook *Logical Self-Defence*, first published in 1977, see [22] introduced new ideas for the time; the idea of an argument between two parties, in the presence of audience. Their emphasis is on arming students to defend themselves against fallacies in everyday discourse. In place of a sound deductively valid argument with true premises — Johnson and Blair posit an alternative ideal of a cogent argument, one whose premises are acceptable, relevant to and sufficient for its conclusion. Acceptability replaces truth as a premise requirement, and the validity condition is split in to two different conditions, premise relevance and premise sufficiency. Acceptability is defined relative to audiences — the one's for whom arguments are intended — but the other basic concepts, relevance and sufficiency, although illustrated by examples, remain as intuitive.

We note the importance of the idea of self defence, which is compatible with our view of weaponising and defending against the use of fallacies. We also note that what they call cogent argument, which is not considered logical deductive in traditional logic (**TDL**), is considered logic in our *New logic with mechanisms*, (see Figure 2) because it is an instance of non-monotonic reasoning. The Johnson and Blair defence is just a *New logic with mechanisms* counter argument.

To conclude this subsection it would be useful to give an example,which will illustrate both an instance of a *New logic with mechanisms* system and an opportunity for self defence.

Example 2.1. *Consider the story below.*

The common practice in the 1970's in top North American philosophy departments is to find jobs for their students who just received a PhD. This is wonderful practice to be highly praised. Our story deals with the case of one student who got a PhD, let us call him H, ("the Hippy"), and the following is a departmental staff discussion of whether to spend resources and effort and take responsibility for placing H (finding him a position in another university).

*Professor A (Reverend, Philosophy of Religion): We should aban-
don H. He is wild, looks like a savage, and although his thesis was
strong, he will either fail his interview, or else shame us and be sacked
within 6 months of being appointed.*

*Professor B (Social Choice Theory, H's advisor): We still have
time until the interview. By the time of his interview, he will be
presentable, shave his hair, wear a shirt and a tie and look like a
normative candidate.*

*Any sane person would want a good job and prepare for it and I
am confident H will do the same.*

*Professor A: H is too wild, it will not work. I appreciate your
commitment to your student, but the department should not be involved.*

Possible replies for Professor A.

1. *Argue and give evidence that H will behave.*
   *This is compatible with the idea of Logical self defence. The
   defence would be in* New logic with mechansims, *maybe present a
   detailed plan how to prepare H and evidence that H will comply.*

2. *Attack with a Fallacy.*
   *Reverend, you seem to dislike H, ever since he said that Jesus
   was nothing but a political agitator! You should overcome that!*

We now describe the system of New logic with mechanisms, *needed
to model this argument.*

   i. *We need a language for facts and their negations.*

   ii. *We need a language for actions of clauses of the form:*
   Facts ⇒ Execute new fact and override existing fact.

   iii. *We can have common sense mechanisms which can take a set of
   facts and expand it.*

   iv. *We define a consequence relation between sets of facts S and a
   new fact x to be S| − x iff there exists a sequence of actions and
   mechanisms leading from S to x.*

The argument between the professors is about the sequence of actions proposed.

Note that this logic is practical. If you have a business and go to the bank and ask for a loan, this is how you argue that you can easily pay it back. You present a business plan which is a sequence of actions which can generate and maintain income. We note here that an action fallacy can also be used, for example passing a note to the Reverend saying that unless he immediately concedes the point his adulterous relationship with a student will be immediately revealed.

This subsection is continued in Appendix A.

2.2 Our initial position on fallacies

We first recall our our distinction between "Deductive-Fallacies" and "Action-Fallacies" as intuitively explained in Remark 1.1. The Deductive- Fallacies are what is commonly called Fallacies. We also recall recurring comment in Subsection 2.1, that if we consider the deductive logic against which we measure the fallacies as *New logic with mechanisms and networks*, see [23], then the statement below is still valid: A deductive-fallacy is the use of an invalid or otherwise faulty argument or dialogue move which *appears* valid. It is important to note that an action fallacy in say a weaker logic may become a deductive fallacy in a stronger logic, if the stronger logic incorporates as a legitimate move that kind of actions. We may also have a reverse change, a legitimate action in an earlier logic becomes illegitimate in a later logic. A striking example is the historic **Trial by Combat** rule. (Trial by combat was a method of Germanic law to settle accusations in the absence of witnesses or a confession in which two parties in dispute fought in single combat; the winner of the fight was proclaimed to be right. In essence, it was a judicially sanctioned duel. It remained in use throughout the European Middle Ages, gradually disappearing in the course of the 16th century. See `https://en.wikipedia.org/wiki/Trial_by_combat`, accessed on July 18, 1700 hours UK time)

A fallacious argument or move may be deceptive by appearing to be better than it really is. Some fallacies are committed intentionally to manipulate or persuade by deception, while others are committed unintentionally due to carelessness or ignorance. Lawyers acknowledge that the extent to which an argument is sound or unsound depends on the context in which the argument is made.

Fallacies are among the most effective arguments used by people and are among the most successful in affecting human actions and behaviour in social, political, legal and interpersonal interactions.[6] This being so we still have not been able to model and understand them. To this day, logicians have dismissed them simply as wrong reasoning and their use a sign of ignorance. See however, Woods' *Errors of Reasoning*, [56] and see the discussion in Subsection 2.1.

Fallacies are commonly divided into "formal" and "informal". A formal fallacy can be expressed neatly in a standard system of logic, such as propositional logic, while an informal fallacy originates in an error in reasoning other than an improper logical form, see [9, 16, 17] and Subsection 2.1. Arguments containing informal fallacies may be formally valid, but still fallacious.

Modern argumentation and informal logic identifies, discusses and classifies over a hundred fallacies, [11], listing over a 100 fallacies and [59], listing 137 fallacies, and there are hundreds (at least 500) articles about fallacies (see [60], Hansen and Fioret in Informal Logic, 2016). See also [16, 17].

However, no one in the logic and argumentation community considers fallacies as an effective instrument of reasoning and no one has tried to model them from this point of view, systemise their use, offer counter-fallacies in debates and in general turn them into another pillar of logic and language. This is not a criticism. The tradition since Aristotle has been to regard a fallacy as a failure of reasoning which should be avoided. However, it has become increasingly obvi-

---

[6]We have not conducted a scientific study to back up this claim. However, the second author has been in politics for many years, (even running as a candidate for British Parliament) and has been following debates in the social media. The first author has been following Middle Eastern politics for years. This is the basis for our conclusion of the effectiveness and use of fallacies and that that it is time to integrate fallacies into Logic.

ous to those of us who have been studying the internet that logical fallacies have proved not only effective in argumentation but often more effective than pure logic. We have conducted a wide study of internet arguments, both in videos and in media such as Twitter and the evidence presented forced us to accept fallacies as a form of dialogue, which then prompted us to study how to integrate their use into formal theories of argumentation and logic. Furthermore our involvement with advances in the notion of logical systems, coming from human reasoning modelling in theoretical computer science and artificial intelligence made it possible for us to initiate the first integration of fallacies into such models.

There has been a lot of excellent research studying fallacies among the philosophical logic and informal logic communities, as we have seen in Subsection 2.1, which essentially forms the ground work for such integration. We have no doubt that the fallacies community at large would have reached the same conclusion as ourselves had they been exposed as a whole, as we have been, to internet debates and use of fallacies and to AI modelling of human reasoning. We are going to rely upon and use, as our starting point, the work on fallacies of prominent researchers who devoted their lives and many books to analysing these issues. Note however our comments on action-fallacies in Remark 1.1. We especially note the seminal work of John Woods [56], whose brilliant analysis of deductive fallacies is a good compatible starting point for us. See Woods' EAUI approach [56, p. 136]. Given this most valuable body of work, what we need now is to move to a NEW AREA of *Argumentation, Human Effective Argumentation and Logic (HEAL2100)* — THE NEXT EVOLUTIONARY STEP FOR LOGIC.

See Figures 3, 4 and Figure 5.

	Fallacies
Aristotle	13 fallacies classified into 2 types, rejections and mistakes
1970	The Gang of 18 Fallacies
2008	Over 100 fallacies classified many types. Still rejected as mistakes but analysed and refined by a very strong and vibrant informal logic and argumentation communities.
2008–2017	**Powerful use of fallacies as weapons of reasoning**
2017 Gabbay–Rivlin	Proposal to integrate the fallacies into deductive logic.

Figure 3: Evolution of the view of the logic community on fallacies

3 Big data

The means to model the use of the fallacies come from recent advances in computer science and AI in the area of Big Data (see [52]). The internet allows us access (in real-time of patterns of data, such as the use of fallacies), previously inaccessible and until recently non-existent data repositories, such as:

- Social Media (e.g. Facebook)

- Publicly Available Sources (government , databases, newspapers, online blogs, etc.)

- YouTube videos

- Streaming

- Advertising

- and so on.

Why integrate fallacies?

- The use of Fallacies in interactions between humans is more effective than traditional deductive arguments; it is extensively used.

- Modelling and integrating fallacies into the *New logic with mechanisms, networks and fallacies* can help develop logic on its evolutionary path and will include new models of formal logic, practical reasoning and practical Artificial Intelligence.

- Allows for better understanding of human reasoning and interactions, as it is now (2017 and onwards) extensively used and is here to stay in the social media. This we hope will result in better reasoning awareness among the public, better grasp of reality, normative laws, regulations, persuasion, political culture, etc.

Figure 4

Research activity Goals 2008

Our purpose was to propose how to integrate symbolic logic with network (neural and argumentation) reasoning.

Let us consider the human agent in his daily activity.

We ask: what 'logic' does he have in his head?

Current relevant buzz words circulating in the community are, among others: time, action, knowledge, belief, revision, deduction, learning, context, neural nets, probabilistic nets, argumentation nets, consistency, etc.

We want to understand what kind of integrated logic engine the human uses in his daily activity.

Research activity (work packages) goals 2017

Add and integrate the fallacies.

Figure 5: Change in research activity

The 2000 years of study and classification of the fallacies together with big data and our capabilities to search and mine the extensive use of fallacies in social media now give us the tools to embark on the next phase of our study in modelling a form of self-protection from fallacies as well as their use as a reasoning weapon. Such knowledge will also enable us to model cultural systems of thought — such as the Western European, rule-based system, the Jewish Talmudic system (which played an unacknowledged but fundamental role in the formation of Christian medieval commentaries), the Islamic Quranic and Sharia way of thought and Hindu darsanas, among the major ones.

The Use of Big Data. We have two main uses of big data:

1. To seek find and study the use of the fallacies in the social media. We need this to classify their use and integrate them into logic. So we need to use big data expert to work with us throughout the project

2. The new logic we are building will require a response to a fallacy by another fallacy, as our examples show. So part of the logic must be a big data mining application which given a context and a fallacy will offer candidate fallacies for response. For example, it could be a counter example using theorem prover if the fallacy is logical or a counter threat if the fallacy is a threat.

Also of great importance is the expected rise of the role logic and argumentation in everyday life, as discussed in item 6 of section 1.

4 Case study: The fallacy of ad hominem

Let us start by quoting from one of the most important of Big Data resources, Wikipedia:

> "Ad hominem (Latin for 'to the man' or 'to the person'), short for 'argumentum ad hominem', is a logical fallacy in

which an argument is rebutted by attacking the character, motive, or other attribute of the person making the argument, or persons associated with the argument, rather than attacking the substance of the argument itself.

Fallacious ad hominem reasoning is normally categorised as an informal fallacy, [3, 4, 5] more precisely as a genetic fallacy, a subcategory of fallacies of irrelevance. However, in some cases, ad hominem attacks can be non-fallacious; i.e., if the attack on the character of the person is directly tackling the argument itself. For example, if the truth of the argument relies on the truthfulness of the person making the argument—rather than known facts—then pointing out that the person has previously lied is not a fallacious argument."

This fallacy can be further refined into a different type of subfallacies, depending on the type of the attack. We chose this fallacy to illustrate how we are going to deal with it in the new area of logic HEAL2100.

According to the discussion in Subsection 2.1, when this fallacy is used in a debate or in argument discussion between two people, (such as in Example 4.3 and Example 5.6), it is a violation of correct procedure in the system. This will be agreed by the Pragma-dialectic approach, by the Walton Pragmatic approach and by the Johnson and Blair Self Defence approach,[7] as all three envisage a dialogue between two parties. In fact it will be agreed as a fallacious move by everybody.

We have to be careful here, as the next Remark 4.1 (by John Woods) shows.

Remark 4.1 (Smoking). *When 15 year old Billy says to his Dad, "But why shouldn't I smoke, Dad, given that you suck down 20 cigarettes a day?", does anyone in his right mind really think that, in*

---

[7]Ralph Johnson accepts the dialogue approach but Tony Blair now does not (as of January 2017, as he stated in a CRRAR meeting, http://www1.uwindsor.ca/crrar/crrar-in-the-news).

saying so Billy has committed an inapparent error of reasoning, or has broken an Amsterdam bylaw for "critical discussions"?

It is generally agreed that ad hominem remarks can be very effective modes of persuasion. Even more so, they are entertainments designed to move the already-convinced and tick-off the otherwise-minded. The only reason that they got on the fallacies list is when used as premises of arguments with generally unvoiced conclusions or other missing premises. Let's come back to Billy. Suppose we reconstruct what he said along these lines.

1. *Dad thinks that the anti-smoking thesis is true.*

2. *But Dad himself sucks down 20 cigarettes a day.*

3. *[So Dad's practice discomports with his policy.]*

4. *[Therefore, the anti-smoking thesis is false.]*

Of course, this is a bad argument, but nowhere close to a fallacious one. Its badness is not inapparent, and hardly any ad hominem retort is made with the intent of this argument.

Remark 4.1 above is a good one. There are many other cases like the smoking example, such as the cross examining of an expert witness, where a personal attack on the expert and his qualifications may even be expected. What we have in mind, however, are cases where the ad hominem attack is a weapon in the meta-level to completely destroy the opponent. It may not even be an argument. Consider the following real examples, namely Example 4.2, Example 4.3 and the incredibly illogical but deadly Example 4.4.

Our question is: How do we respond to such a fallacious move? Do we explain to our opponent (the user of the fallacy) the reasons why this is a fallacy in the context of our discussion and ask the opponent politely to make another move?

This is not what we see in Social media practice. The fallacy is legitimately used as a weapon and the only way not to lose the argument is to respond with another fallacy. Thus ad hominem is a good case study for us to illustrate our HEAL2100 point of view.

We start by illustrating how this fallacy can be used as a reasoning weapon.

Example 4.2 ("Milk-snatcher" Thatcher). *We quote from:*
http : // www. telegraph. co. uk/ news/ politics/ 7932963/
How-Margaret-Thatcher-became-known-as-Milk-Snatcher.
html ; (accessed on UK 1130 hours May 06, 2017)

> *"The Conservative government had to find substantial cuts to meet election pledges on tax. Removing free school milk for the over-sevens became the most notorious saving introduced. Edward Short, then Labour education spokesman said scrapping milk was Ôthe meanest and most unworthy thing' he had seen in 20 years. It earned Mrs Thatcher the nickname, "Milk Snatcher" and haunted her throughout her career. In 1985 she was refused an honorary degree from Oxford University because of her education cuts."*

> **After the war under Clement Attlee the 1946 Free Milk Act was passed providing one third of a pint to all children under the age of 18.*

Edward Short's argument was emotional and fallacious. Under traditional, rule-based logic Mr. Short would have been expected to give good reasons why Thatcher's policy was wrong and Thatcher could then have responded giving her reasons for the cuts.

However, the emotional argument and personal attack on Mrs Thatcher as a "milk snatcher" was much more effective. The only defence which would have made any impact on public perceptions of the situation would have been for Mrs. Thatcher to attack the Labour Party—possibly for the devaluation of the pound in 1967 opening the Labour Party to the charge of being called "pick-pockets" for stealing money old people and innocent children whose pensions and pocket money was subsequently worth less.

See: Dynamics of a Non-Decision: the Failure' to Devalue the Pound, 1964–7 TIM BALE 20 Century Br Hist (1999) 10 (2): 192-217. DOI: https: // doi. org/ 10. 1093/ tcbh/ 10. 2. 192
Published: 01 January 1999 (accessed UK 1130 hours May 06, 2017).

Instead, the Conservative government of the time stuck to explaining the economic situation, an argument which cut little ice with the parents at the school gates.

Had Mrs Thatcher been in possession of our intended HEAL2100 logic model, and a big data computer at her disposal at the time and the inclination to respond in kind she could have taken the following steps:

- *Edward Short attacks Mrs Thatcher personally, using a fallacy*

- *Mrs Thatcher identifies the structure of the weaponised fallacy-attack*

- *She uses big data to find similar emotive issues around Labour Party policies[8]*

- *She finds a most similar case, although this is not strictly necessary, it could be anything (see the example of the Starkey-Hassan argument below)*

- *She counterattacks by presenting a case found by a HEAL2100 Big Data search*

Compare the above to the traditional deductive rule base logic behaviour:

- *Edward Short presents logical arguments against the cuts*

- *Mrs Thatcher analyses these arguments using facts and logic*

- *She presents her logical counter arguments*

Example 4.3 (You got my name wrong). *This example is from a televised debate (BBC Question Time) which is now available on YouTube and entitled: "Mehdi (Ahmed) Hasan debates David Starkey on Question Time",* `https: // www. youtube. com/ watch? v= CzYlkGbYG1M`, *(accessed on UK 1130 hours May 06, 2017).*

---

[8]We do not currently have an application which can do that in real time. The projects aims to develop one.

Starkey starts by mistakenly referring to Mehdi by the name Ahmed.

At minute 1.23 of the video Starkey implies that Mehdi is prevaricating by pointing out that what he is saying in the televised debate is not what he said on the same subject when speaking to a group of Moslems in a mosque. Mehdi replies at minute 1.40 that Starkey cannot even get his name right, having called him Ahmed and not Mehdi. When Mehdi makes this point, the audience bursts into loud and enthusiastic applause.

People from Moslem societies frequently use a style of argument which is also increasingly used by politicians and ideologues — of whatever cultural background which we will classify as based on an appeal to emotions. This method of argument has as its goal the winning of the argument but not the discovery of any truth or an arrival at consensus. It is a form far better suited to all expressions of modern mass media in which the object is to get a message across to an audience with a widely diverse level of educational attainment and in many cases an extremely limited level of concentration.

In the same vein, Starkey could have answered along the lines of: "Nice to know you care about your name rather than the starving children of your people (or any other emotive issue)." Again our big data HEAL2100 logic could have offered structural analysis and responses. Starkey would not have needed big data to make this response but maybe there was some other additional useful related information about Mehdi.

We are specifically studying the appeal of emotional argument to a more primitive part of the brain because this sort of argument has extremely important implications for how we relate to electronic media.[9]

---

[9]We are grateful to Doug Walton for pointing out that this relates to what the psychologists call heuristics, short-cuts that appeal to emotions.

The heuristics and biases research program of Tversky and Kahneman [54], (Judgment Under Uncertainty: Heuristics and Biases) [1974] was criticised by Gerd Gigerenzer (see [55] for a survey), and others for being too focused on how heuristics lead to errors. The critics argued that heuristics can be seen as rational in a certain sense (bounded rationality), arguing that they can be good enough for

Let us explain: Suppose you are of Indian descent but have had little contact with either your family or culture for a long time. Then you walk into a house where, as you come into the hall, you can smell curry through the open kitchen door. Immediately you are transported back to your childhood, and are filled with memories of your mother's cooking, family meals, rows with your sister, etc.

The sense of smell is well recognised as being wired into the most primitive parts of the brain and smells are also well recognised as emotional triggers at a far deeper level than any other sense.

Compare this with a scenario in which you see a recipe for curry, recognise it, analyse it and are then reminded of your mother's cooking. The chain of associations is much slower and not nearly so personal.

Example 4.4 (CNN Interview). *See this YouTube video: https://www.youtube.com/watch?v=CBZOC43O7OU (accessed on UK 1130 hours May 06, 2017).*

Katie Hopkins is being interviewed by CNN's Hala Gorani. Hopkins attempts to distract Gorani by referring to her first as "darling" and then, when that elicits no response, a little later, as "honey". At this point Gorani cannot ignore it any more and is forced to protest at the slight, thus diverting her fire from the argument. The question of legitimacy is important here. Hopkins' technique works well for a woman-to-woman argument (that is, contempt between equals) but it would not be legitimate if the interviewee were a man. Had a man called her "honey", Gorani would probably have terminated the conversation straight away and thereby would have "won".[10]

---

some purposes without being too taxing on the limited rationality of the human brain.

[10]As a side matter, it is also interesting to note the degrees of pressure Hopkins employs. The word "darling" can be used legitimately by men and women to talk to friends, especially among media people who wish to create an illusion of friendliness with an interculator whose name they cannot remember. Gorani would therefore find this term of address as slightly unsettling but not entirely outside the bounds of normal usage. However, the sobriquet "honey" is never used among friends, but only romantically or de haut en bas. Gorani could not ignore that. The first "darling", therefore, was a testing shot, preparing the ground for the

Worth watching.

Example 4.5 (Arguing with different logics). *John gives the proof:*

$$
\begin{array}{ll}
1. & \text{Assumption } (c \to a) \to c \\
2. & \text{Assumption } \; c \to a \\
\hline
3. & \text{Conclusion } a
\end{array}
$$

Proof: From 2 and 1 derive c and then from c and 2 derive a, all using modus ponens

Mary objects to the proof. She says: but you have used assumption 2 twice!

John uses say classical logic but Mary uses Resource logic.

This is a simple clear example but if the differences between John and Mary are subtle, how can Mary explain her different point of view to John. Big data can help.

Example 4.6 (The Taxi Driver-Analogy). *This is a real example, that happened in Israel. A passenger logician was returning by Taxi in a journey taking 50 minutes. The taxi driver was an immigrant from Uzbekistan, very right wing and a supporter of Prime Minister Benjamin Netanyahu. The prime Minister was investigated for accepting gifts (not too expensive but still considerable) from a very rich friend. Netanyahu did not report these gifts at the time and some investigative journalist discovered it and and the police were looking into the case. It was not a bribe, but just wrong behaviour. The taxi driver was arguing in favour of and supporting the prime minister.*

His argument was as follows

1. *What is wrong in accepting gifts from a friend?*

   Look at me, I wanted to meet my (male) friend from Uzbekistan, I sent him a ticket to come to Israel, I paid for his hotel, I did everything for his visit. What is wrong with accepting this, it is natural between friends.

---

next, and fatal assault.

The answer to that is that it is OK for your friend but not OK for the prime Minister of a country. He should have declared everything he was receiving.

The problem with this answer was that there was no chance in Hell that the taxi driver would understand it. He came from an ex-communist country which was still totalitarian and the fine aspects of democracy were beyond his conceptual world. The passenger clearly needed a better answer for him to grasp the concept, but he was in a taxi which was about to arrive in 15 minutes and he needed to produce an answer immediately. The passenger logician did not find the answer until the next day. It was really simple.

> *2. Answer. Imagine (the passenger could have said to the taxi driver) that your friend is a woman who in the meantime got married. Had she, without telling her husband, come to Israel on a ticket you had bought, staying in a hotel you paid for, and then her husband had found out. What would he think?*

> *She should have told her husband immediately and asked for his blessing.*

> *When in a democracy the prime minister receives gifts it is similar.*

Now, had the passenger had a Big Data application, he would possibly have used his mobile phone to search for an analogous example using the right key words.

The taxi driver could have said that the women friend case is not the same as the prime minister case. Such a response is quite likely, but it would have offered the opportunity for further discussion. The taxi driver, at least, would have seen where the passenger's attempted counter-argument is coming from. Without such or a similar example, there would have been nothing to discuss.

Example 4.7 (Labelling). *This is a simple technique of attack; label your opponent with a strongly emotional totally negative predicate, for example label as a racist. There are many such labels you could use, which carry such a strong emotional reaction that once an opponent* *is labelled by such a word people will reject anything he says. Here are some examples:*

- *Racism*

- *Apartheid*

- *Contrary to international law*

- *A crime against Humanity,*

etc.

    The label need not be so powerful or even negative. It is enough to create a context which weakens the opponents arguments. If we use fuzzy logic where arguments have numerical strength, we can say something like "of course you would say so, it is to be expected, it is in your interests to say so". This is a generic weakening label, which is not negative, but which will weaken the strength of the opponent's argument. In general there is no good answer to such generic argument, but there are exceptions.

    Recently there was (June 2017) on television an interview with a politician. He was a minister and resigned because of ideological disagreement with the prime minister (no scandal or mis-behaviour), see [15]. He started his own party. In the interview he was accused of trying to build himself a political career and possibly aim for a government position. This is a generic labelling attack on any politician. He replied to the television interviewer, "What are you talking about? I was a minister already"!

Example 4.8 (Prime example of labelling and counter-labelling). *From an interview on the BBC radio programme "Midweek", broadcast on October 9th, 2005:* https://www.youtube.com/watch?v=Hy-Ap4LQB-4.

    Labelling is very often not direct but by implication, which is an especially deadly form of attack. When the accusation is delivered in an oblique way it is far more difficult to refute because before a rebuttal can be made the accused person has to put into words the full meaning of the implication only half-suggested by the accuser.

Darcus Howe is a master of this technique and, in the opinion of the authors, he made his living by wrong-footing the sort of people who are terrified of having even a hint that they might harbour politically incorrect opinions.

Howe starts the interview of a famous American comedienne, Joan Rivers, with his usual procedure of implying that Rivers has unholy attitudes — i.e. labelling her as being at the very least a sort of passive racist — but he frames it in such a way that the meaning is ambiguous ("since black offends you"), this leaves him an escape route which he takes when Rivers flies into a fury. He then posits that the "use of the term 'black' offends you".

In other circumstances this would have been effective. The accused person would seize on the opportunity to have a conversation and would accept the lesser label of being uncomfortable with the word "black". Howe would retain the upper hand and anything the accused person would say thereafter would be slightly tainted. Rivers does not accept this. Instead she then starts labelling him, first by taking offence at his implication she is a racist (i.e. using the argument of offence) and then saying that he has a chip on his shoulder (that is, labelling him as unreliable because he has an unworthy agenda). Then again she repeats that he called her a racist ("don't you DARE call me a racist") driving home to the listener both that she is outraged by such a suggestion (self-labelling of innocence) at the same time as reminding the audience that Howe is not only being unfair but doing this with a dishonest agenda (labelling him as unscrupulous).

Howe then suggests that it is a "language problem" which is his attempt to redirect the discussion. This is the argument of redirection and at the very least, labelling Rivers as being uneducated or not that bright. If Rivers had been playing the game she would have agreed that perhaps it was a language problem. She would then have been labelled as someone who does not understand Howe's mode of expression. This would have got Howe out of trouble without an apology or explanation being necessary and Rivers would have been weakened by the implicit racism of not understanding Howe sufficiently well.

Again, she refuses to accept the offered compromise. She labels him "stupid", by defining his first statement as stupid. It is interesting that Howe does not react to that, most people would. But if he does react, he would then have to say something like "I am NOT stupid". This is exactly the sort of reaction he was trying to force out of Rivers at the outset and he knows the rules. He is also aware by this time that Rivers not only knows the rules, too, but is a superior exponent of them. He remains silent. Rivers then launches into an attack on his parenting responsibilities (Howe deserted his family in the West Indies). Again, Howe does not answer this for the same reasons that he has not reacted to the accusation of stupidity. He appeals to the interviewer to put the conversation back on to the original track. Rivers presses home her advantage and then accuses Howe of racism, turning the label 180 to his direction. By that time Howe has to concede that Rivers is not a racist in order to prevent any further attacks. Rivers finishes the exchange by stating that she would not choose to meet Howe in any other circumstances (an unpleasant person label).

It is obvious to the authors that Rivers had done some research on Howe before the interview. She might have looked at his work online or spoken to someone about what he does and was therefore ready for any reference to racism he might make. She also knew about his deserted family. We are sure she was awaiting the opportunity to take towering offence at the smallest provocation, providing her with the excuse to return to him a whole list of labels. She accomplished her aim of defending herself against Howe without using any backing to support her arguments (e.g. I am not a racist because I have worked with black people, I have supported black artists, etc). In this sort of an exchange, proving innocence is a weak and ineffective defence.

5   Structure of possible future research

This section gives more details about the program of work of what do we need to do to get HEAL2100 to be accepted/adopted by the community of logicians.

5.1 Orientation: New logic mechansims and networks

Traditional logic **TDL** is based on rules. Even the various components of the *New logic with mechanisms and networks* (see [23]) such as nonmonotonic logic is based on rules with exceptions and priorities. Argumentation logic and dialogues logic are all based on all kinds of procedures, algorithms and conventions. The semantics for such logics is defined mathematically and is precise and clear

- Different choices of rules, algorithms and semantics gives different logics and these can be rejected or can be agreed to and adopted and applied to a variety of application areas.

- The connections between different logics and their properties can be studied mathematically and much of the activity of the logic community is devoted to such study.

What is happening in current modern logic (up to and including *New logic with mechanisms and networks*) is basically the same as what is happening in mathematics.

Some researchers define and invent new logics, other researchers investigate their properties, some logic communities adopt, apply and possibly even modify chosen logics which suit their needs, giving rise to new logics. And so the cycle continues.

In many ways this cycle is just like the development of major areas of applied mathematics: e.g. fluid dynamics. mathematical biology and other exact science modelling,

The operative consequences of this entire traditional modern logic activity up to the *New logic with mechanisms and networks* in contrast with our proposed New logic with mechanisms, networks and fallacies "HEAL2100" is that for any new candidate for a logic, or for any sequence of of arguments and counter-arguments, which can be put forward in the context of *"New logic with mechanisms and networks"* we can decide on the following questions:

1. Is this candidate a *New logic with mechanisms and networks* acceptable system at all?

2. What is its relationship to other known systems of *New logic with mechanisms and networks*?

3. What are its mathematical properties?

4. What application it is supposed to model?

5. What constitute a fallacy in the system?

and so on.

There are many more traditional questions, (can the logic be axiomatised, what is its complexity, its semantics, proof theory, automated deduction, etc).

To give an example, imagine that we have a program on a computer implementing some known Artificial Intelligence *New logic with mechanisms and networks*. Assume the program is corrupted by a virus and starts behaving in a new way. We can then ask if the corrupt program is or is not a logic and we answer this question, using the mathematical tools of *New logic with mechanisms and networks* to test it and see what it does.

5.2   Our new logic HEAL2100

Let us now examine the challenges facing us in our *New logic with mechanisms, networks and fallacies* = HEAL2100.

We are trying to discover what legitimises a fallacy as a method of argumentation. This means we no longer say they are mistakes and put them aside but accept them as instruments of reasoning. Therefore we need to explain and define when, in HEAL2100, such uses of the fallacies are legitimate — as opposed to **TDL** where it is a given that the use of the fallacies is not legitimate, so there is nothing else to say. We in HEAL2100 have a lot to say. Therefore let us use the term "2100-legitimate" for correct uses of the fallacies.[11]

---

[11]We need to decide how far is a fallacy's legitimacy conferred by community acceptance? Wouldn't this make for very different fallacies in Berlin and Beirut? What about sub-communities — e.g. the South Side of Chicago compared

The objectives are clear, namely to integrate the fallacies into the current state of logic, as described in the previous background section. The methodology of work is described by way of listing work packages groups, Group A, D, B, F and I

- Group A is a work package developing a big data application. Given a key word, say insult "you are a liar and a cheat", the application will find in real time some examples of that.

- Group D is a theoretical research and consultation work package with the mainly Canadian fallacies research community, trying to understand how fallacies work, in order to model them.

- Group B restructures/redesigns existing *New logic with mechanisms and networks* in a way that it can accept/integrate the fallacies.

- Group F classifies the fallacies, understands them and gets them ready for insertion into the restructured logic of Group B. This classification is motivated by the way the fallacies are being used in social media and is likely to be different from any traditional classification

- Group I integrates the fallacies into the restructured logic of group B to form the *New logic with mechanisms, networks and fallacies.*

Note that the above is an iterated process, which we can call the ADFBI process: We iteratively try to develop the groups: A, D, F, B, I, A, D, F, B, I,...
We now describe the work packages for the research Groups:

Work package for Group A.   Task A1: Develop a real time search engine for certain search phrases arising from fallacious arguments. Task A2: Develop guidelines of how to query the application of Task

---

to metropolitan Chicago? How extensively are we prepared to press the fallacies relativity line?

A1 for different fallacious arguments.

Task A3: Map the limitations of the use of Big Data. Preliminary searches (without Tasks A1 and A2) were not promising. It was not like searching the web for the meaning of a foreign word , which one can get and use instantly in a conversation.

Work package for Group D. Tasks D1-D18: Discuss the nature of the Gang of 18 fallacies, respectively each fallacy a separate respective task. This requires careful study of the uses of each fallacy. To get an idea how it works , see our starting preliminary study of the ad hominem fallacy in this paper.

Work package for Group B: Background work. This is the hard work of defining a generic New logic with mechanisms and networks system and showing how the traditional views of the fallacies, as described in Subsection 2.1 and as further put forward by other major researchers in the fallacy community (in work package D), can be embedded/integrated into our generic system. Doing this requires ingenuity, imagination and technical skill and it will take many man months to do. We can do it, using ideas and methods from [48, 49, 50].

Work package for Group F: Classification of the fallacies. When is a Fallacy 2100 legitimate? To see the difficulty of classification let us look at some real examples.

We will start with the fallacy of attacking your opponent (*argumentum ad hominem*).

Example 5.1. *A true case of two university professors arguing:*

> *A1 says to A2:* you are a habitual liar
> *A2 retorts to A1:* You are an adulterer and a drug addict

Example 5.2. *From an Al- Jazira debate.*

> *B1 says to B2:* I say you are a liar and a traitor
> *B2 to B1:* B2 takes off his shoe and throws it at B1.

(This method of argument is all too common on Al Jazira, throughout the Middle East generally and in parts of the Eastern Mediterranean as well as in some African and Far Eastern countries.)

Example 5.3 (Two cars collide on the road). *The drivers are rolling on the asphalt trying to strangle each other. This is an incident witnessed by one of the authors of this paper in Jerusalem 60 years ago. NO WORDS ARE SPOKEN.*

Question: which of the above uses the *ad hominem* fallacy, Examples 5.1, 5.2 and 5.3, we consider as 2100-legitimate?

More generally, when is a use of a fallacy legitimate and when can we consider it a step in some argument sequence? Let us be systematic in trying to answer this question. First of all, we need to collect data. We already have list of fallacies grouped into types. Aristotle listed 13, nowadays we list over 100. Let us write some steps. We rely on our results from Research Group B, because we need several candidates for our good generic system from Group B, to inject and integrate fallacies into them.

Task F1: Collect and classify known lists of fallacies and their fine tuning variations. Such lists exist in the literature but they are viewed and classified from the point of view of fallacies being illegitimate and to be discarded. HEAL2100 views them as weapons of reasoning being put to effective practical use. Let us call this our starting list.

Task F2: We need to use the Internet to collect many instances where fallacies are used, assess their success and reclassify them accordingly. Our research will initially classify them as theoretically 2100-legitimate in principle, with a view to deciding what is 2100-legitimate, pending a closer examination of how the community reacts to such fallacies. We can access big data to collect examples and see if these can help define legitimate use.

Task F3: try to identify what cases are considered illegitimate. We will seek key properties for 2100-illegitimate use.

Task F4: reclassify and possibly identify more fallacies in view if our findings in Task F1–3. We will call the new list our modified start list.

Task F5 : We iterate the process of Task F1–4 several times, using the modified fallacies collected at the previous iteration (see Task F4).

Note that this is a completely new type of work package and may take 18 months to execute.

Work package Group I: Interaction with rule based logics of *New logic with mechanisms and networks.*

Many fallacies are deductive. They can be remedied within *New logic with mechanisms and networks* or they can be remedied within HEAL2100. How do the two possibilities reconcile with one another? In practice correct reasoning can be combined with fallacies. How do we view this and integrate it smoothly? How does the interaction go? For example do we structure the argument interaction network into a network of meta-levels (i.e. a network of networks) and the fallacies move us from one meta-level to another?

Task I-generic. Develop a generic integrated system with several higher levels of reasoning and actions.

TaskI1–I10. Develop ten typical major integrated systems (we do not believe we can have one comprehensive system, in the same way that there is no single one major logical system).
This research can take up to 18 months

Let us give some examples:

Example 5.4 (The jump approach). *This approach to be examined is where we reason logically and then insert a step which is a fallacy and then continue to reason logically. The simplest example of this is what is now known as "alternative facts", in which there is an introduction of fabricated facts into an argument.*

For example Soviet history books contain many fabricated or semi-fabricated Russian innovations, such as the discovery of America, the steam engine, radio and the helicopter, amongst others. Most of these claims are hyperbolic at best but having been established as fact, whatever argument they supported started from this point.

Example 5.5 (YouTube alternative facts ; start at 2.44m). *Recently we found a YouTube video in which it was claimed that Arabic was "the first language", and that all the characters in the Bible (and in the surrounding non-Biblical civilisations) spoke Arabic. The very disturbing aspect of this particular piece is that the speaker is a University professor and obviously intelligent. We might find such concocted history amusing but it is precisely this admixture of fact and fantasy that is killing hundreds of thousands in the Middle East at this time of writing.*
`https://www.youtube.com/watch?v=i_1wZSXEofE`
"Palestinians: Where does the name 'Palestine' come from?'
Corey Gil-Shuster
Published on 26 Oct 2016.
    See also wikipedia article about alternative facts.
`https://en.wikipedia.org/wiki/Alternative_facts`.
    There are many more examples and we need to study how this is done, whether or not it is 2100-legitimate and possibly more importantly — how to deal with it.

Example 5.6 (An example of integrating a fallacy). *We have a single mother who is a top executive in a successful international corporation. Although she is busy she is still deeply devoted to her teenage daughter. The following happens one morning:*
    Mother goes into her teenage daughter's bedroom. Her instant observation is that it is a big mess. There is stuff scattered everywhere.
    Mother's impression is that it is not characteristic of the girl to be like this.
What has happened?
Conjecture: *The girl has boyfriend problems.*
Further Analysis: *Mother notices a collapsed shelf. Did the girl smash*

it? Upon further inspection, mother notices that the pattern of chaos shows that a shelf has collapsed because of excessive weight and scattered everything around, giving the impression of a mess. But, actually, it is not a mess, it does make some (gravitational) sense.
There are several modes of reasoning:

1. *Neural nets type of reasoning.*
   She recognises the mess instantly, like we recognise a face.

2. *Nonmonotonic deduction.*
   Mother reasons from context and her knowledge of her daughter is that the girl is not disorganised like this. She asks 'what happened?'.

3. *Abduction/conjecture.*
   She offers a reasonable explanation that the girl has boyfriend problems, since this is common to that age.

4. *She then applies a database AI deduction and recognises that the mess is due to gravity. This deduction is no longer a neural net impression. It is a careful calculation.*

5. *It could have been a neural net impression.*
   For example, a man who sees many shelf-collapsing cases may recognise the pattern as if it were a face.

Mother story, continued

- *Mother to daughter: why do you leave your room in such a mess?! You should have fixed this before going to bed last night.*
  Daughter's possible logical replies.

  1. *I was too tired*
  2. *I had pressing homework yesterday*
  3. *I am in shock*

  etc.

  Instead the daughter responds with an emotional fallacy.

- *Daughter to Mother: What do you care, you are always at work, you hardly talk to me, you don't care about me, all you care about is your corporate career, you have no right to criticise me!*

Given this emotional action-fallacy the mother cannot continue with any rational deductive argument. If we consider the previous mother-daughter reasoning interaction as level 1, the object level matter of fact reasoning explaining and discussing the mess in the room, the daughter's outburst argument is moving to level 2, a meta-level seeking to abort any such discussion. Nothing will be effective to move back to level 1 except a counter fallacy. Once the counter action-fallacy is successful in level 2 a rational discussion about the mess can continue in level 1.

The authors' recommendation to Mother:

1. *Look sad, tell the girl sorrowfully how hard you work to support her. Remind her of past emotional family scenes. Tell her how much she is hurting you, (you might even try a tear or two).*

Other Options:

2. *Act insulted clear up the mess yourself, then complain it has hurt your back and you can't go to work and blame it on her, hoping you can then talk sense to her.*

3. *(Not recommended) Go into an angry fit and throw the books at her or beat the hell out of her.*

Once the chosen counter fallacy is completed, rational discussion might resume (although in case 3 we rather doubt it).[12]

---

[12]The following is another version (by John Woods) of this example:

Mom remonstrates with daughter and daughter loses her cool and picks a fight. The fight has some factual basis. Senior people in the corporate world have limited time for children, and children often (but not always) resent it. This puts Mom at a clear disadvantage. Given that neither party seeks for a permanent and irreparable alienation, the sooner this is over the better. Dispassionate disquisition about the impact of the modern life on families isn't going to achieve anything quickly. Better, then, for Mom to counterattack, and the sooner she does, the sooner they'll

5.3 Intermediate summary and time scale of proposed research

Note that our HEAL2100 logic differs from traditional rule-based logic on two counts:

- It is not just a set of axioms and rules (whether monotonic, nonmonotonic or any other traditional system) but a program of gathering information, classification and correlation of this information. It is an argumentation system of attacks and counterattack where each move and countermove is justified not by a deductive base logic but by human behaviour pattern discovered and mined by big data.

  So a logical reasoning unit is a structure of data, put together with a view to attack. It is a weaponised structured argument unit.

  The logic is what the program relying on big data tells us to respond to, sequentially.

  As the big data changes, the logic changes!

- We accept the fallacies as effective reasoning structures. We fine tune adjust them by refining them to further reasoning substructures. We use big data to do that as well as finding further reasoning to such instances of fallacies from big data.The calibration of the effective counterattack to such fallacies will be fine-tuned and enriched over time by continually maintained big data programs.

- So logic becomes time dependent as human behaviour changes.

---

make up. After that, as you say, a reasoned discussion might be resumed. (But probably not right then!)

You speak of these outbursts as emotional fallacies, but they aren't that in the traditional sense. In the traditional sense, an argument commits an emotive fallacy when conducted in such a way as to stir the emotions of those to which it is addressed. The *argumentum ad misericordium* is a typical example, as when a defence counsel asks a jury for mercy. But in the Mother Story, nothing like this is going on. Rather what we have there are emotional outbursts.

- We may end up in the unfortunate and uncivilised situation of reasoning only irrationally by shooting fallacies at each other. (We do not believe so. Some fallacies do not work in the wrong context. If I claim I can prove the famous problem $P = NP$, and you ask to see the proof, it is no use me shouting "ARE YOU CALLING ME A LIAR?"

- It can take 4-5 years of research to do this properly

5.4 Expected benefit

- Make people more aware/critical of false news, false arguments, etc. and thus protect our democratic processes. Now with the new media available any small group of people can cause serious problems.

- The success of terrorist arguments to recruit ordinary people in the west can be defended against using the same type of HEAL2100 appropriate counter-arguments.

HEAL2100 can be applied to all CURRENT consumer areas of logic where human behaviour is concerned.

Acknowledgments

We are grateful to Michal Chalamish, Hans Hansen, Douglas Walton and John Woods for penetrating and valuable comments.

References

[1] Douglas N. Walton, Dialog Theory for Critical Argumentation, Amsterdam: John Benjamins Publishing Company, 2007;

[2] Dialectica. Edited by L. M. De Rijk in Petrus Abaelardus: Dialectica, Assen: Van Gorcum 1970 (second edition).

[3] Arnauld, Antoine, 1612-1694; Nicole, Pierre, 1625-1695, Logic, or, The art of thinking : being the Port-Royal logic , Edinburgh : Sutherland and Knox, 1880.

[4] Handbook of Philosophical Logic, 2nd edition, Editors D Gabbay and F Guenthner, Springer.

[5] Barwise J., editor, Handbook of Mathematical Logic, Elsevier, 1977.

[6] Anderson, A.R., and Belnap, N.D., 1975, Entailment: The Logic of Relevance and Necessity, Volume I. Princeton: Princeton University Press

[7] D Gabbay and J Woods- Handbook of the History of Logic, 12 Volumes , Elsevier 2002-2015

[8] Alec Fisher, The Logic of Real Arguments, Cambridge: Cambridge University Press, 2004.

[9] Hamblin, C. L., 1970, Fallacies, London: Methuen.

[10] H. Hansen. Fallacies. In *The Stanford Encyclopedia of Philosophy*, Edward N. Zalta, ed. Metaphysics Research Lab, Stanford University, 2015. https://plato.stanford.edu/archives/sum2015/entries/fallacies

[11] Wikipedia , list of fallacies https://en.wikipedia.org/wiki/List_of_fallacies

[12] Gabbay, Dov M. and Woods, John. (2001a). The new logic. Logic Journal of the IGPL, 9, 157-190

[13] van Eemeren, F.H., Garssen, B., Krabbe, E.C.W., Snoeck Henkemans, F.A., Verheij, B., Wagemans, J.H.M. Handbook of Argumentation Theory.

[14] John Woods. *Death of an Argument: Fallacies and other Distractions*. Newport Beach: Vale Press, 2000.

[15] Avi Gabbay article. http://www.haaretz.com/israel-news/1.761985

[16] Douglas Walton: Publications related to fallacies.

*Argument Evaluation and Evidence, Cham, Switzerland, Springer, 2016.

*Handbook of Legal Reasoning and Argumentation, ed. G. Bongiovanni, G. Postema, A. Rotolo, G. Sartor and D. Walton, Springer, 2016.

*Goal-based Reasoning for Argumentation, Cambridge, Cambridge University Press, 2015. new Burden of Proof, Presumption and Argumentation, Cambridge, Cambridge University Press, 2014.

* Emotive Language in Argumentation, F. Macagno and D. Walton, Cambridge, Cambridge University Press, 2014.

* 'Why fallacies appear to be better arguments than they are,' Informal Logic, 2010, 30: 159–184.

* Methods of Argumentation, Cambridge, Cambridge University Press, 2013. Argumentation Schemes, D. Walton, C. Reed and F. Macagno, Cambridge, Cambridge University Press, 2008.

* Informal Logic: A Pragmatic Approach, second edition, Cambridge, Cambridge University Press, 2008.

* Witness Testimony Evidence: Argumentation, Artificial Intelligence and Law, Cambridge, Cambridge University Press, 2008.

* Dialog Theory for Critical Argumentation, Amsterdam, John Benjamins Publishers, 2007. Media Argumentation: Dialectic, Persuasion and Rhetoric, Cambridge, Cambridge University Press, 2007.

* Character Evidence: An Abductive Theory, Berlin, Springer, 2007. Fallacies: Selected Papers: 1972-1982, J. Woods and D. Walton, Studies in Logic, vol 7, London, King's College, 2007.

* Fundamentals of Critical Argumentation, Cambridge, Cambridge University Press, 2006. Argumentation Methods for Artificial Intelligence in Law, Berlin, Springer, 2005.

* Abductive Reasoning, Tuscaloosa, University of Alabama Press, 2004.

* Relevance in Argumentation, Mahwah, N.J., Lawrence Erlbaum Associates, 2004.

* Ethical Argumentation, Lanham, Md., Lexington Books, 2002 (sample available).

* Legal Argumentation and Evidence, University Park, Pa., Penn State Press, 2002.

* Scare Tactics: Arguments that Appeal to Fear and Threats, Dordrecht, Kluwer Academic Publishers, 2000.

* Appeal to Popular Opinion, University Park, Pa., Penn State Press, 1999.

* One-Sided Arguments : A Dialectical Analysis of Bias, Albany, State University of New York Press, 1999.

* Ad Hominem Arguments, Tuscaloosa, University of Alabama Press, 1998.

* The New Dialectic, Toronto, University of Toronto Press, 1998.

* Appeal to Expert Opinion : Arguments from Authority, University Park, Pa., Penn State Press, 1997.

* Appeal to Pity: Argumentum ad Misericordiam (SUNY Series in

Logic and Language), Albany, SUNY Press, 1997.

* Historical Foundations of Informal Logic, (co-edited with A. Brinton), Aldershot, England, Ashgate Publishing, 1997.

* Argument Structure: A Pragmatic Theory, Toronto, University of Toronto Press, 1996.

* Argumentation Schemes for Presumptive Reasoning , Mahwah, N.J., Lawrence Erlbaum Associates, 1996 (sample).

* Arguments from Ignorance, University Park, Pa., Penn State Press, 1996.

* Fallacies Arising from Ambiguity, Dordrecht, Kluwer Academic Publishers, 1996.

* Commitment in Dialogue: Basic Concepts of Interpersonal Reasoning, D. Walton and E. C. W. Krabbe, Albany, SUNY Press, 1995.

* A Pragmatic Theory of Fallacy, Tuscaloosa, University of Alabama Press, 1995.

* The Place of Emotion in Argument, University Park, Pa., Penn State Press, 1992.

* Plausible Argument in Everyday Conversation, Albany, State University of New York Press, 1992.

* Slippery Slope Arguments. Oxford, Oxford University Press, 1992.

* Begging the Question: Circular Reasoning as a Tactic of Argumentation New York, Greenwood Press, 1991.

* Practical Reasoning: Goal-Driven, Knowledge-Based, Action-Guiding Argumentation, Savage, Maryland, Rowman and Littlefield, 1990.

* Informal Logic: A Handbook for Critical Argumentation, Cambridge, Cambridge University Press, 1989.

* Question-Reply Argumentation Westport, Connecticut, Greenwood Press, 1989. Informal Fallacies (Pragmatics and Beyond Companion Series, IV), Amsterdam, John Benjamins, 1987.

* Courage: A Philosophical Investigation, Berkeley, U. of California Press, 1986.

* Argument: Critical Thinking, Logic, and the Fallacies, J. Woods, A. Irvine and D. Walton, Berkeley, U. of California Press, 1986.

* Arguer's Position: A Pragmatic Study of Ad Hominem Attack, Criticism, Refutation, and Fallacy Westport, Connecticut, Greenwood Press, 1985.

* Physician-Patient Decision-Making Westport, Connecticut, Green-

wood Press, 1985. Logical Dialogue-Games and Fallacies, Lanham, Maryland, University Press of America, 1984.

* Ethics of Withdrawal of Life Support Systems Westport, Connecticut, Greenwood Press, 1983.

* Topical Relevance in Argumentation, Amsterdam, John Benjamins, 1982.

* Brain Death: Ethical Considerations Lafayette, Indianan, Purdue University Press, 1980.

[17] John Woods: Publications related to fallacies.

* (1974) Proof and Truth. Toronto: Peter Martin Associates

* (1974) The Logic of Fiction: A Philosophical Sounding of Deviant Logic. The Hague and Paris: Mouton and Co. A second edition was published in 2009 by College Publications, ISBN 1-904987-99-0

* (1978) Engineered Death: Abortion, Suicide, Euthanasia, Senecide. Ottawa: The University of Ottawa Press/Editions de l'Université d'Ottawa. ISBN 0-7766-1020-1

* (1982) Argument: The Logic of the Fallacies. Toronto and New York: McGraw-Hill (with Douglas Walton) ISBN 0-07-548026-3

* (1989) Fallacies: Selected Papers, 1972-82. Dordrecht and Providence: Foris (with Douglas Walton). A selection was translated in French and published with a new introduction in 1992 as Critique de l'Argumentation: Logiques des sophismes ordinaires, xii, 233, Paris: Éditions Kimé

* (1992) Woods, J., 'Who cares about the fallacies?' in Argumentation Illuminated, F. H. van Eemeren, et al. (eds.), Amsterdam: 1992 SicSat, pp. 23–48.

* (2000) Argument: Critical Thinking Logic and The Fallacies. Toronto: Prentice-Hall (with Andrew Irvine and Douglas Walton). A 2nd edition was published in 2004: ISBN 0-13-039938-8

* (2001) Aristotle's Earlier Logic. Oxford: Hermes Science Publications. ISBN 1-903398-20-7 (second revised edition London: College Publications, 2014)

* (2003) Paradox and Paraconsistency: Conflict Resolution in the Abstract Sciences. Cambridge: Cambridge University Press. ISBN 0-521-00934-0

* (2003) Agenda Relevance: An Essay in Formal Pragmatics. Volume 1 of A Practical Logic of Cognitive Systems, Amsterdam: North Holland (with Dov M. Gabbay) ISBN 0-444-51385-X

* (2004) The Death of Argument: Fallacies in Agent-Based Reasoning. Dordrecht and Boston: Kluwer. ISBN 1-4020-2663-3

* (2005) The Reach of Abduction: Insight and Trial. Volume 2 of A Practical Logic of Cognitive Systems, Amsterdam: North Holland (with Dov M. Gabbay) ISBN 0-444-51791-X

* (2007) Fallacies: Selected Papers 1972-1982, 2nd edition, with a Foreword by Dale Jacquette, xvi, 322. London: College Publications, (with Douglas Walton).

* (2010) Fictions and Models: New Essays, edited, iii, 442, Munich: Philosophia Verlag.

* (2015) Inconsistency Robustness, edited, volume 52 of Studies in Logic, lxxi, 535, London: College Publications, (with Carl Hewitt).

Moreover, Woods has been a co-editor (with Dov Gabbay) of the eleven-volume Handbook of the History of Logic, published by North-Holland (now Elsevier), as well as editor, with Gabbay and Paul Thagard, of the sixteen-volume Handbook of the Philosophy of Science, by the same publisher.

[18] Copi's book has undergone many Editions.
* Copi, I. M., 1961, Introduction to Logic, (2nd ed.), New York: Macmillan. * by Irving M. Copi (Author), Carl Cohen (Contributor), Kenneth McMahon (Contributor) Introduction To Logic 14th Edition Paperback, Routledge 2016

[19] L. Powers. Equivocation, in Hansen and Pinto 1995, *Fallacies: Classical and Contemporary Readings*, University Park: Penn State Press. pp. 287–301, 1995.

[20] W. Salmon. *Logic*, Englewood-Cliffs: Prentice-Hall, 1963.

[21] T. Skura. Refutation systems in propositional logic. In D Gabbay and F Guenthner, *Handbook of philosophical logic Vol 16*, Springer 2011, pp 115–157.

[22] R. Johnson and J. A. Blair. *Logical Self-Defence*, 3rd ed., Toronto: McGraw-Hill Ryerson, 1993.

[23] D. Gabbay. What is a logical system; An evolutionary view, 1964–2014, in *Handbook History of Logic volume 9, computational logic*, Elsevier 2014, pp 41–135.

[24] G. J. Massey. *The fallacy behind fallacies*, Midwest Studies in Philosophy, 6: 489–500, 1981; page references are to reprint in Hansen and Pinto 1995, *Fallacies: Classical and Contemporary Readings*, University Park: Penn State Press. pp. 159–171.

[25] M. Wreen. A bolt of fear, *Philosophy and Rhetoric*, 22: 131–40, 1989.

[26] T. Govier. What's wrong with slippery slope fallacies? *Canadian Journal of Philosophy*, 12: 303–16, 1982.

[27] D. Walton. *Begging the Question*, New York: Greenwood, 1991.

[28] A. Brinton. The ad hominem. In Hansen and Pinto 1995, *Fallacies: Classical and Contemporary Readings*, University Park: Penn State Press. pp. 213–222.

[29] J. B. Freeman. The appeal to popularity and presumption by common knowledge. In Hansen and Pinto 1995, *Fallacies: Classical and Contemporary Readings*, University Park: Penn State Press pp. 265–273.

[30] R. C. Pinto. Post hoc, ergo propter hoc. In Hansen and Pinto, *Fallacies: Classical and Contemporary Readings*, University Park: Penn State Press pp. 302–311.

[31] J. Woods and D. Walton. *Fallacies: Selected Papers, 1972–1982*, Dordrecht: Foris, 1989.

[32] J. Woods. Who cares about the fallacies? In *Argumentation Illuminated*, F. H. van Eemeren, et al. (eds.), Amsterdam: SicSat, pp. 23–48, 1992.

[33] F. H. van Eemeren. *Strategic Maneuvering in Argumentative Discourse*, Amsterdam: John Benjamins, 2010.

[34] F. H. van Eemeren and R. Grootendorst. *Speech Acts in Argumentative Discussions*, Dordrecht: Foris, 1984.

[35] F. H. van Eemeren. *Argumentation, Communication and Fallacies*, Hillsdale: Erlbaum, 1992.

[36] F. H. van Eemeren. *A Systematic Theory of Argumentation*, Cambridge: Cambridge University Press, 2004.

[37] J. Biro. Rescuing "begging the question", *Metaphilosophy*, 8: 257–71, 1997.

[38] D. Walton *A Pragmatic Theory of Fallacies*, Tuscaloosa: University of Alabama Press, 1995.

[39] D. Walton. Why fallacies appear to be better arguments than they are, *Informal Logic*, 30: 159–184, 2010.

[40] D. Hitchcock. Do fallacies have a place in the teaching of reasoning skills or critical thinking? In Hansen and Pinto 1995, *Fallacies: Classical and Contemporary Readings*, University Park: Penn State Press pp. 319–327, 1995.

[41] J. A. Blair. The place of teaching informal fallacies in teaching reasoning

skills or critical thinking. In Hansen and Pinto 1995, *Fallacies: Classical and Contemporary Readings*, University Park: Penn State Press pp. 328–338, 1995.

[42] V. Correia. Biases and fallacies: The role of motivated irrationality in fallacious reasoning, *Cogency*, 3: 107–126, 2011.

[43] P. Thagard. Critical thinking and informal logic: neuropsychological perspectives, *Informal Logic*, 31: 152–170, 2011.

[44] H. V. Hansen and R. C. Pinto, eds. *Fallacies: Classical and Contemporary Readings*, University Park: Penn State Press, 1995.

[45] E. David D. Gabbay, G. Leshem, and Students of C.S. Ashkelon. *Logical analysis of cyber vulnerability and protection*, Submitted to Journal of cyber security, Oxford University Press.

[46] D. Gabbay, G. Rozenberg and Students of CS Ashkelon. Introducing Abstract Argumentation with Many Lives, Submitted to *Argument and Computation*, IOS press, April 2017.

[47] D. Gabbay, G. Rozenberg and Students of CS Ashkelon. Temporal Aspects of Many Lives, Draft May, 2017

[48] D. Gabbay. *Labelled Deductive Systems*, OUP, 1996.

[49] D. Gabbay. *Fibring Logics*, OUP, 1998.

[50] D. Gabbay. *Meta-logical Investigations in Argumentation Networks*. College Publications, 2013, 770pp.

[51] M. A. Finocchiaro. Six types of fallaciousness: towards a realistic theory of logical criticism, *Argumentation*, 1: 263–282, 1987.

[52] Big Data on Wikipedia (accessed June 11, 2017, 0230 hours UK) `https://en.wikipedia.org/wiki/Big_data`

[53] D. Walton. Profiles of Dialogue: A Method of Argument Fault Diagnosis and Repair, *Argumentation & Advocacy*, 52 (2), 2015, 89-106. `http://www.dougwalton.ca/papers%20in%20pdf/15profiles2.pdf`

[54] A.Tversky and D. Kahneman. Judgment Under Uncertainty: Heuristics and Biases, Science, New Series, Vol. 185, No. 4157. (Sep. 27, 1974), pp. 1124-1131

[55] Peter B.M. Vranas. Gigerenzer's normative critique of Kahneman and Tversky, *Cognition* 76 (2000) 179–193.

[56] John Woods. *Errors of Reasoning. Naturalizing the Logic of Inference*, College Publications, (Studies in Logic) Paperback, 24 Jul 2013.

[57] H. V. Hansen. The straw thing of fallacy theory: the standard definition of 'fallacy'. *Argumentation* 16: (2002), 133–155.

[58] Sharon Bailin and Mark Battersby. *Reason in the Balance: An Inquiry Approach to Critical Thinking*. Paperback 488 pages. Hackett Publishing Co, Inc; 2 edition (1 Mar. 2016).

[59] Master list of fallacies, accessed on July 15, 2017. `http://utminers.utep.edu/omwilliamson/ENGL1311/fallacies.htm`.

[60] Hans V. Hansen and Cameron Fioret. A Searchable Bibliography of Fallacies. *Informal Logic* Vol 36, No 4 , 2016.

[61] John Stuart Mill. *A System of Logic, Ratiocinative and Inductive*, Vol 1, 1843.

[62] Gilbert H. Harman. Induction, A chapter in the book *Acceptance and Rational Belief*, Marshall Swain, ed. Pp 83–99 , Synthese Library book series (SYLI, volume 26), 1970.

Appendix

A   More background on fallacies

The Johnson and Blair approach started the formal attempts to provide better analyses of fallacies, a programme pursued by a large number of researchers, including Govier [26] on the slippery slope, Wreen [25] on the ad baculum, Walton [27] on begging the question, Brinton [28] on the ad hominem, Freeman [29] on the appeal to popularity, and Pinto [30] on post hoc ergo propter hoc.

The next step came from John Woods and Douglas Walton [31], their claim is that, for many of the fallacies standard formal logic is inadequate to uncover the unique kind of logical mistakes in question — it is too coarse conceptually to reveal the unique character of many of the fallacies. To get a satisfactory analysis of each of the fallacies they must be matched with a fitting logical system, one that has the facility to uncover the particular logical weakness in question. Inductive logic can be employed for analysis of hasty generalisation and post hoc ergo propter hoc; relatedness logic is appropriate for ignoratio elenchi; plausible reasoning theory for the ad vercundiam, and dialectical game theory for begging the question and many questions. Woods [32, p. 43] refers to this approach to studying the fallacies as methodological pluralism.

This view is perfectly compatible with the former deductive views, provided we understand "deductive" as "*New logic with mechanisms*".

Modern times, second wave

Frans van Eemeren and Rob Grootendorst [34] put forward the Pragma-dialectic approach. They start with argumentation as a procedure involving two parties trying to overcome interpersonal disagreements. The procedure is a discussion having four analytical stages: a confrontation stage in which the participants become aware of the content of their disagreement; an opening stage in which the parties agree (most likely implicitly) to shared starting points and a set of rules to govern the ensuing discussion; an argumentation stage wherein arguments and doubts about arguments are expressed and recognised; and a final stage in which a decision about the initial disagreement is made, if possible, based on what happened in the argumentation stage.

In this context the fallacies are defined as "violation of any of the rules of the discussion procedure for conducting a critical discussion' [36, p. 175].

The Pragma-dialectical theory proposes that each of the core fallacies can be assigned a place as a violation of one of the rules of a critical discussion. For example, the ad baculum fallacy is a form of intimidation that violates the rule that one may not attempt to prevent one's discussion partner from expressing their views; equivocation is a violation of the rule that formulations in arguments must be clear and unambiguous; post hoc ergo propter hoc violates the rule that arguments must be instances of schemes correctly applied. Moreover, on this theory, since any rule violation is to count as a fallacy this allows the possibility that there may be hitherto unrecognised "new fallacies". Among those proposed are declaring a standpoint sacrosanct because that breaks the rule against the freedom to criticise points of view, and evading the burden of proof which breaks the rule that you must defend your standpoint if asked to do so (see van Eemeren [33, p. 194].

We note that the Pragma-dialectical rules of a critical discussion are not just rules of logic, but rules of conduct for rational discussants, making the theory more like a procedural code than a set of logical principles.[13] Accordingly, this approach to fallacies rejects all three of the necessary conditions of **SDF**: a fallacy need not be an argument, and thus the invalidity condition will not apply either, and the appearance condition is excluded because of its subjective character (Van Eemeren and Grootendorst, [36, p. 175]. See also Woods' critique in chapters 9, 10 and 11 of *The Death of Argument*, 2004, listed in [17].

A key point of this approach from our HEAL2100 point of view is the fact that the Pragma-dialectical analysis of fallacies as rule-breakings in a procedure for overcoming disagreements also takes account of the rhetorical dimension of argumentation. Pragma-dialectics takes the rhetorical dimension to stem from an arguer's wish to have their view accepted which leads dialoguers to engage in strategic manoeuvering vis-à-vis their dialogue partners. However, this desire must be put in balance with the dialectical requirement of being reasonable; that is, staying within the bounds of the normative demands of critical discussions. The ways of strategic manoeuvring identified are basically three: topic selection, audience orientation, and the selection of presentational devices, and these can be effectively deployed at each stage of argumentation (Van Eemeren, [33, p. 94]). "All derailments of strategic manoeuvering are fallacies", writes van Eemeren [33, p. 198], "in the sense that they violate one or more of the rules for critical discussion and all fallacies can be viewed as derailments of strategic manoeuvering". This means that all fallacies are ultimately

---

[13]Note however that Dov Gabbay's algorithmic point of view included in his *New logic with mechanisms and networks*, see [23], can accept certain procedures as part of logic. So according to Gabbay, Classical Logic with Resolution formulation is not the same logic as Classical Logic with Tableaux formulation. To the extent that the Pragma-dialectical approach with its procedures can be embedded/represented within *New logic with mechanisms and networks*, then we can still maintain the view that Fallacies are "*New logic with mechanisms and networks* movements/arguments" that are actually not *New logic with mechanisms and networks* correct but nevertheless do look correct".

attributable to the rhetorical dimension of argumentation since, in this model, strategic manoeuvering is the entry of rhetoric into argumentation discussions. "Because each fallacy has, in principle, sound counterparts that are manifestations of the same mode of strategic manoeuvering" it may not appear to be a fallacy and it "may pass unnoticed" ([33, p. 199]. Nevertheless, Pragma-dialectics prefers to keep the appearance condition outside the definition of 'fallacy', treating the seeming goodness of fallacies as a sometime co-incidental property, rather than an essential one.

Our point of view is to accept/ integrate (in HEAL2100) some uses of these fallacies as correct integrated moves, to be countered by other fallacies.

We note that in our New Logic 2, [23] we include argumentation and network logics as well as Algorithmic Proof theory and so the Pragma-Dialectic approach can be simulated/included in our system. However New Logic 2 supports a plurality of Logics and so it will not agree with Pragma-dialectical approach looking towards a single ideal model of argumentation. We view each argumentation procedure is another New Logic 2 system, hopefully usable in some application area.

Another important second wave approach to fallacies is the work of Biro [37, pp. 265–66]. The way we understand his examples is that in order for an argument not to be a fallacy , the assumptions are required to have factual verification or general acceptance as facts. Biro calls this epistemic seriousness. He gives the following example:
All members of the committee are old Etonians;
Fortesque is a member of the committee;
Fortesque is an old Etonian.

In this example, given the minor premise, the major cannot be known to be true unless the conclusion is known to be true. Consequently, on the approach to fallacies taken by Biro, the second argument, despite the fact that it is valid, is non-serious, it begs the question, and it is a fallacy. If there was some independent way of knowing that the major premise was true, such that it was a bylaw that only old Etonians could be committee members, the argument would be a serious one, and not beg the question. This approach does not insist that all justification must be deductive, but facts must be verifiable. Thus it allows for arguments the possibility of their being fallacies (as well as good arguments) by non-deductive standards, something precluded by **SDF**.

We consider this idea important because watching many debates on YouTube we find a lot of false unverifiable alternative facts being introduced. See Example 5.5 below.

We now address the pragmatic approach of Doug Walton. Doug Walton has written or edited over forty-five books about fallacies, analysing them one by one, following the Woods-Walton first wave view on fallacies. As we see it, Walton responded to the Pragma-dialectic approach by offering considering argumentation dialogues. On the Walton approach, a fallacy is associated with a small local sequence of dialogue called a profile of dialogue. See [53]. This paper builds the profiles of dialogue tool into a fault diagnosis method that can be applied to problematic examples of argumentation such as those involving informal fallacies. The profiles method works by comparing a descriptive graph with a normative graph. The descriptive graph represents how a dialogue sequence actually went in the example chosen for analysis. The normative graph represents an analysis of how the sequence should ideally proceed, according to the protocols (rules) for this type of dialogue. The descriptive graph is mapped into the normative graph, so that a comparison can be made to diagnose the fault in the sequence displayed in the descriptive graph. and repair it.

These are distinct normative dialectical frameworks (persuasion dialogue, inquiry dialogue, negotiation dialogue, etc.) rather than the single model of a critical discussion proposed by Pragma-dialectics. Postulating different kinds of dialogues with different starting points and different goals, Walton claims, will bring argumentation into closer contact with argumentation reality. So fallacies happen when there is an illegal shift from one kind of a dialogue to another [38, pp. 118–23], for example, using arguments appropriate for a negotiation dialogue in a persuasion dialogue.[14]

So if I am a medical expert witness and I am asked to describe what procedures I used on the patient, I might take offence and say
Are you calling me a liar?
However, if I claim at a conference that I solved an open problem in maths (say $P = NP$?), and I am asked for the idea of the proof, I cannot say
Are you calling me a liar?

The definition of fallacy Walton proposes [38, p. 255] has five parts. A fallacy:

1. an argument (or at least something that purports to be an argument) that

2. falls short of some standard of correctness;

3. is used in a context of dialogue;

4. has a semblance of correctness about it; and

5. poses a serious problem to the realisation of the goal of the dialogue.

Let us stress that Walton's approach depends on context, not on structure alone. Our tolerance of the above claim, "you are calling me a liar", depends also on context and not only on its irrelevant meta-level (personal) aspect. The Pragma-dialectic approach can string together several of Walton schemes to form a logic and then claim a fallacy if they are not put together correctly. Both approaches can be embedded in the *New logic with mechanisms and networks* concept.

---

[14]Note however that the view that fallacies are due to illicit dialogue shifts is pretty well abandoned in [38].

On Walton's definition, no inference can be fallacious, unless an inference is a solo argument in which the roles of each contending party is played by the same person.

Modern times: Issues in fallacy theory

Quoted from the scholarly and most valuable article "Fallacies" in the Stanford Encyclopaedia of philosophy (SEP by H. V. Hansen) There are four major questions to be addressed by the Fallacies research community according to SEP:

- The nature of fallacies

- The appearance condition

- Teaching of Fallacies

- The role of Biases

Since this is the view (according to SEP/H. V. Hansen) of how the current fallacies research community would like to go forward, we think it is best to simply integrate and almost quote what SEP says about these plans. In the next subsection we will present our own plans for integrating the fallacies and compare with the fallacies communities plans. We hope and look forward for co-operation. Our own comments in the quote are in boldface

The nature of fallacies

A question that continues to dog fallacy theory is how we are to conceive of fallacies. There would be advantages to having a unified theory of fallacies. It would give us a systematic way of demarcating fallacies and other kinds of mistakes; it would give us a framework for justifying fallacy judgments, and it would give us a sense of the place of fallacies in our larger conceptual schemes. Some general definition of Ôfallacy' is wanted but the desire is frustrated because there is disagreement about the identity of fallacies. Are they inferential, logical, epistemic or dialectical mistakes? Some authors insist they are all of one kind: Biro and Siegel, for example, that they are epistemic, and Pragma-dialectics that they are dialectical. There are reasons to think that all the fallacies do not easily fit into one category.

. . .

**In the community fallacies have been identified in
relation to some ideal or model of good arguments,
or good argumentation, or rationality.**

Aristotle's fallacies are shortcomings of his ideal of deduction and proof, extended to contexts of refutation. The fallacies listed by Mill are errors of reasoning in a comprehensive model that includes both deduction and induction. Those who have defended **SDF** as the correct definition of 'fallacy' take logic *simpliciter* or deductive validity as the ideal of rationality. Informal logicians view fallacies as failures to satisfy the criteria of what they consider a cogent argument. Defenders of the epistemic approach to fallacies see them as shortfalls of the standards of knowledge-generating arguments. Finally, those who are concerned with how we are to overcome our disagreements in a reasonable way will see fallacies as failures in relation to ideals of debate or critical discussions.

**We note that the authors (Gabbay–Rivlin)
approach to fallacies (which we may call *New logic
with mechanisms, networks and fallacies*
approach), is that we consider a fallacy any effective instrument of argumentation currently used
in the social media and politics which is not a *New
logic with mechanisms and networks* instrument!**

The standard treatment of the core fallacies did not emerge from a single conception of good argument or reasonableness but has rather, like much of our unsystematic knowledge, grown as a hodgepodge collection of items, proposed at various time and from different perspectives, that continues to draw our attention, even as the standards that originally brought a given fallacy to light are abandoned or absorbed into newer models of rationality. Hence, there is no single conception of good argument or argumentation to be discovered behind the core fallacies, and any attempt to force them all into a single framework, must take efforts to avoid distorting the character originally attributed to each of them.

The appearance condition

From Aristotle to Mill the appearance condition was an essential part of the conception of fallacies. However, some of the new, post-Hamblin, scholars have either ignored it (Finocchiaro, Biro and Siegel) or rejected it because appearances can vary from person to person, thus making the same argument a fallacy for the one who is taken in by the appearance, and not a fallacy for the one who sees past the appearances. This is unsatisfactory for those who think that arguments are either fallacies or not. Appearances, it is also argued, have no place in logical or scientific theories because they belong to psychology (van Eemeren and Grootendorst, [36]. But Walton (e.g., [39]) continues to consider appearances an essential part of fallacies as does Powers [19, p. 300] who insists that fallacies must "have an appearance, however quickly seen through, of being valid.' If the mistake in an argument is not masked by an ambiguity that makes it appear to be a better argument than it really is, Powers denies it is a fallacy.

The appearance condition of fallacies serves at least two purposes. It can be part of explanations of why reasonable people make mistakes in arguments or argumentation: it may be due in part to an argument's appearing to be better than it really is. The appearance condition also serves to divide mistakes into those that are trivial or the result of carelessness for which there is no cure other than paying better attention, and those which we need to learn to detect through increased knowledge of their seductive nature. Without the appearance condition, it can be argued, no division can be made between these two kinds of errors: either there are no fallacies or all mistakes in argument and/or argumentation are fallacies, a conclusion that some are willing to accept, but which runs contrary to tradition. One can also respond that there is an alternative to using the appearance condition as the demarcation property between fallacies and casual mistakes, namely, frequency: fallacies are those mistakes we must learn to guard against because they occur with noticeable frequency. To this it may be answered that Ônoticeable frequency' is vague, and is perhaps best explained by the appearance condition.

Teaching

On the more practical level, there continues to be discussion about the value of teaching the fallacies to students. Is it an effective way for them to learn to reason well and avoid bad arguments? One reason to think that it is not effective is that the list of fallacies is not complete, and that even if the group of core fallacies was extended to incorporate other fallacies we thought worth including, we could still not be sure that we had a complete prophylactic against bad arguments. Hence, we are better off teaching the positive criteria for good arguments/ argumentation, which will give us a fuller set of guidelines for good reasoning. But some (Pragma-dialectics and Johnson and Blair) do think that their stock of fallacies is a complete guard against errors because they have specified a full set of necessary conditions for good arguments/argumentation and they hold that fallacies are just failures to meet one of these conditions. Another consideration about the value of the fallacies approach to teaching good reasoning is that it will tend to make students overly critical and lead them to see fallacies where there aren't any; hence, it is maintained we could better advance the instilling of critical thinking skills by teaching the positive criteria of good reasoning and arguments (Hitchcock, [40]). In response to this view, it is argued that, if the fallacies are taught in a non-perfunctory way which includes the explanations of why they are fallacies — which normative standards they transgress — then a course taught around the core fallacies can be effective in instilling good reasoning skills (Blair [41]).

We have a new method of teaching called DADI (Data Driven Instruction) which can be used for teaching about the Fallacies. See Appendix C

Biases

Recently there has been renewed interest in how biases are related to fallacies. Correia ([42]) has taken Mill's insight that biases are predisposing causes of fallacies a step further by connecting identifiable biases with particular fallacies. Biases can influence the unintentional committing of fallacies even where there is no intent to be deceptive, he observes. Taking biases to be Òsystematic errors that invariably distort the subject's reasoning and judgment,' the picture drawn is that particular biases are activated by desires and emotions (motivated reasoning) and once they are in play, they negatively affect the fair evaluation of evidence. Thus, for example, the Òfocussing illusion' bias inclines a person to focus on just a part of the evidence available, ignoring or denying evidence that might lead in another direction. Correia ([42, p. 118]) links this bias to the fallacies of hasty generalization and straw man, suggesting that it is our desire to be right that activates the bias to focus more on positive or negative evidence, as the case may be. Other biases he links to other fallacies.

Thagard [43] is more concerned to stress the differences between fallacies and biases than to find connections between them. He claims that the model of reasoning articulated by informal logic is not a good fit with the way that people actually reason and that only a few of the fallacies are relevant to the kinds of mistakes people actually make. Thagard's argument depends on his distinction between argument and inference. Arguments, and fallacies, he takes to be serial and linguistic, but inferences are brain activities and are characterized as parallel and multi-modal. By "parallel" is meant that the brain carries out different processes simultaneously, and by "multimodal" that the brain uses non-linguistic and emotional, as well as linguistic representations in inferring. Biases (inferential error tendencies) can unconsciously affect inferring. "Motivated inference", for example, "involves selective recruitment and assessment of evidence based on unconscious processes that are driven by emotional considerations of goals rather than purely cognitive reasoning" [43, p. 156]. Thagard volunteers a list of more than 50 of these inferential error tendencies. Because motivated inferences result from unconscious mental processes rather than explicit reasoning, the errors in inferences cannot be exposed simply by identifying a fallacy in a reconstructed argument. Dealing with biases requires identification of both conscious and unconscious goals of arguers, goals that can figure in explanations of why they incline to particular biases. "Overcoming people's motivated inferences", Thagard concludes, "is therefore more akin to psychotherapy than informal logic" [43, p. 157], and the importance of fallacies is accordingly marginalized.

In response to these findings, one can admit their relevance to the pedagogy of critical thinking but still recall the distinction between what causes mistakes and what the mistakes are. The analysis of fallacies belongs to the normative study of arguments and argumentation, and to give an account of what the fallacy in a given argument is will involve making reference to some norm of argumentation. It will be an explanation of what the mistake in the argument is. Biases are relevant to understanding why people commit fallacies, and how we are to help them get past them, but they do not help us understand what the fallacy-mistakes are in the first place — this is not a question of psychology. Continued research at this intersection of interests will hopefully shed more light on both biases and fallacies.

B   Applications: Internet of things

This is a possible application. It is not essential or influential to our new concept of 2100-logic, but it is related and who knows what its future impact could turn out to be.

From Wikipedia: `https://en.wikipedia.org/wiki/Internet_of_things`

> "The Internet of things (IoT) is the inter-networking of physical devices, vehicles (also referred to as "connected devices" and "smart devices"), buildings, and other itemsÑembedded with electronics, software, sensors, actuators, and network connectivity that enable these objects to collect and exchange data. In 2013 the Global Standards Initiative on Internet of Things (IoT-GSI) defined the IoT as "the infrastructure of the information society." The IoT allows objects to be sensed or controlled remotely across existing network infrastructure, creating opportunities for more direct integration of the physical world into computer-based systems, and resulting in improved efficiency, accuracy and economic benefit in addition to reduced human intervention. When IoT is augmented with sensors and actuators, the technology becomes an instance of the more general class of cyber-physical systems, which also encompasses technologies such as smart grids, smart homes, intelligent transportation and smart cities. Each thing is uniquely identifiable through its embedded computing system but is able to interoperate within the existing Internet infrastructure. Experts estimate that the IoT will consist of almost 50 billion objects by 2020.

Typically, IoT is expected to offer advanced connectivity of devices, systems, and services that goes beyond machine-to-machine (M2M) communications and covers a variety of protocols, domains, and applications. The interconnection of these embedded devices (including smart objects), is expected to usher in automation in nearly all fields, while also enabling advanced applications like a smart grid, and expanding to areas such as smart cities.

Equipped with HEAL2100 we can offer better logic at the service of the IOT. The IOT systems are complex inter-related components each of which are intelligent to some degree and based on logic. The need of HEAL2100 for IOT is a necessity not just another application!

We give an Example:
Imagine we want to improve protection against Phishing. If we use traditional logic in building protection we use rules, as in the following case: Email filters: a message you receive is analysed by the mail program which then adds to the subject line a warning that this may be phishing or spam. Such warnings already exist.

If we open the email and we see a very convincing service message from Paypal that our account has they paid $30 to an unfamiliar company, we then have to consider whether the message is or is not malicious. However, our reaction to this unexplained apparent disappearance of money from our account is emotional, faster and more immediate than reasoning. Worried that even more money will disappear and seeing a button saying "cancel transaction" it is quite likely that we would be panicked into clicking on it almost before we realise what is happening.

What we need is an equally emotional warning, like a flashing button in red and yellow blinking the message "SPAM—STAY AWAY!" It may not be difficult for a mail program to do this if it realises the underlying principles of our HEAL2100 logic — that is that the object is TO WIN, not to reach a consensus.

C DADI: Data Driven Instruction, A new method of teaching logic and fallacies

We developed a new teaching method capable of writing joint research papers with first year students as joint authors. The philosophy of the teaching/research approach is outlined below.

It is especially suited for teaching logic and fallacies.

We observed that PHd students conducting research for the purpose of writing a thesis, need to go through four stages:

1. read and familiarise themselves with a relevant area of research;

2. have a good new idea for pushing the frontier of the area forward;

3. develop the details of the idea;

4. write it down as a paper/thesis, and this includes knowing the scientific language and structure for writing their ideas.

The received wisdom about Phd studies is that you need 3 (BSc)–4 (MSc) years of university study to be able to approach a thesis.

We conjectured that the 3–4 years are required for item 1 above.

We asked ourselves, what if the area where the research to be done is so familiar that first year student already have the background knowledge to move to item 2 above?

Can first year students have a good new idea leading to a research paper?

Of course first year students do not know how to write a paper nor do they know any research methodology, but neither do PHd students — for this we have a supervisor. So all we need to experiment with this idea is to choose a topic which

• First year students are familiar with

• Good ideas are forthcoming

• It connects with known international research area.

Then all we need to do is for say, Dov, to present the question to the students and let them develop a model. This is no different from offering a topic of research to a new Phd student. Dov Gabbay accepted teaching at Ashkelon Community College and conducted such courses.

Description of the experiment:

Class of 2015/2016. In 2015 Gabbay was teaching general logic to a first year class of 15 students. At the time there was much debate in the media and politics in UK and Israel, about young couples unable to join the housing ladder. In simple words:

Flats are too expensive and young couples cannot get the initial minimum funds to enable them to get an affordable mortgage to buy a home. The political solution was to offer such young couples cheap mortgages and help.

Gabbay asked the student to formulate principles (known from the media) for qualifying for this help Using one arrow connective

(If x is true and y is true) $\Rightarrow$ do z.

Gabbay formulated known rules from the media based on government data. Then he posed a problem to the class: How to stop young couples from using the benefits and buying two flats in parallel? The class participated in modifying the rules to stop such abuse. There were creative in cheating the system as well as creative in fixing it.

The reader will observe that Gabbay developed action logic and cyber security principles for guarding against hacking. Gabbay applied the systems we got for the cyber security intelligent home and wrote the paper [45]. We were invited to submit to an OUP international Journal, on Cyber Security. The key to this is that the students knew how to write rules and knew how to cheat the system — they understood the need to get your own flat and were creative in dealing with it.

Class of 2016/2017. This year the first class had 49 students. We again chose a familiar topic to the students. This year the media and the law was concerned in Israel and UK with sex offenders. Many famous figures were accused by victims of sex offences and every week there was a new scandal discovered. The students had detailed knowledge of such cases. Dov Gabbay presented the question of **How many complaints does it take for us to decide that there is need to investigate?**

The view was of a survival game of the sex offenders where each offender had a number of lives before it is dead. The students were also familiar with T.V. survival games. So we started developing a model based on their knowledge of the numerous sex offender cases going on in the media. We developed a basic model in the area of argumentation. We wrote papers [46, 47] and are invited to submit to the IOS journal Argument and Computation.

The students are able to develop points *1–*3 above and the teacher needs to write point *4.

NOTE THAT THE DADI METHOD IS ESPECIALLY SUITABLE FOR TEACHING FALLACIES BECAUSE NOWADAYS THE SOCIAL MEDIA IS FULL OF DEBATES AND POLITICS AND ISIS, ETC USING FALLACIES AS WEAPONS. THE STUDENTS ARE VERY FAMILIAR WITH THEM.

Limitations of the method.

1. The students cannot deal with abstraction. So if they construct a model for a certain area (with which they are familiar), they cannot recognise the same abstract model in another area, even when the similarities are clearly pointed out to them.

   The students recognised and defined the many lives abstract argumentation model in the sex offenders area. The same model applies in the nutrition area, where various foods (e.g. alcohol) attack parts of the body (e.g. the liver). This was pointed out to them and they were given a lecture by a nutritionist and yet they did not see the connection.

2. The students found difficulty in understanding abstract set theoretical definitions but could easily understand definitions by algorithms. So to define a set we must give an algorithm for constructing it.

3. The best approach to teaching/developing a theory or a model is to present it as an algorithmic game or a puzzle.

4. We plan to address the use of some of the fallacies (ad hominem) in the class of 2017/2018. The students are familiar with political debates, personal attacks and counterattacks especially in the Trump era. It is a strategic survival game and we shall see if next year's students can model it.

CHAPTER 4
INTRODUCING ABSTRACT
ARGUMENTATION WITH MANY LIVES

1 Orientation: The many lives idea

Our starting point is to view argumentation networks (of the form (S, R)) as representing a survival game. The players are the elements of S and the relation R is the attack relation. The various traditional Dung semantics for subset of S can be viewed as defining extensions in the form of possible survival groups $E \subset S$. The survival sets E (which are the traditional extensions) are groups of players which are conflict free and able to protect themselves. So far we have a different point of view on extensions which is compatible with the traditional Dung formal mathematical machinery. However, given the survival point of view we can generalise and add additional features to the traditional argumentation networks:

1. The new features are:

   (a) We can add to each x in S a many lives value $M(x)$, meaning how many live attackers are needed to force x to be out (i.e. x to become dead).

   (b) We associate with each attack pair (y, x) in R a value $K(y, x)$, meaning how many lives are taken out of $M(x)$ should the attack of y on x be successful (i.e. y is alive). The value $K(y, x)$ may be, or may not be, correlated or even related to the number of lives $M(y)$ which y has.

   (c) The traditional concept of conflict free set is that of a set whose members do not attack one another. With many lives available we look at "living together" sets, using a concept of being able to stay alive together. Members can attack but not able to kill one another. In fact we could introduce different strengths of attack, one when attacking inside a "living together" set and possibly another when a "living together" set protects itself.

(d) We can now investigate semantics for such systems (S, R, M, K).

2. The ideas of adding M and K arise from our research into the argumentation/logic behaviour of mulitiple complaints. Thus the semantics and additional features of argumentation that we study are inspired by real life applications.

   In fact, to protect an alleged offender x against attacks from a group of complainers/victims E, x needs to present much stronger counter attacks, and furthermore the public will tolerate a little bit of inconsistencies among E (i.e. E need not be completely conflict free). This observation led us to the idea that to present a formal argumentation system we need to define three types of attacks, $\alpha_{\mathbf{a}}, \alpha_{\mathbf{d}}$, and $\alpha_{\mathbf{p}}$, in increasing strength. For E to attack x we use the $\alpha_{\mathbf{a}}$ attack. For Z to protect x, Z must use the $\alpha_{\mathbf{p}}$ attack and for E (resp. for Z) to be considered conflict free its members must not $\alpha_{\mathbf{d}}$ attack one another (though we may tolerate them $\alpha_{\mathbf{a}}$ attacking one another). Furthermore, the attacks can be defined using the basic attack relation R in a more complex manner. For example $z\ \alpha_{\mathbf{a}}$ attacking x can be defined as $(zRx \wedge ((\forall u)(uRz \rightarrow zRu)))$.

3. We discuss our results and compare with other papers on the numerical and ranking aspects of argumentation.

According to Dung an argumentation network (S, R), where S is a non empty set (of arguments) and R is a binary relation on S. When $(x, y) \in R$ holds we say that x (geometrically) attacks y. Dung [6] (see Section 3) introduced several concepts related to (S, R), among them the concept of:

D1. A subset E of S attacks a node $y \in S$ iff (for some $e \in E$ we have eRy).[1]

D2. A subset E of S is conflict free iff (for no e_1, e_2 in E do we have $e_1 R e_2$).

D3. A subset E of S protects a node $x \in S$ iff (for all y, if yRx then E attacks y).

D4. A subset E of S is admissible iff E is conflict free and it protects all its members.

D5. A subset E is a complete extension iff E is admissible and contains all nodes it protects.

The above concepts were defined by Dung using the geometrical single attack, between x (the attacker) and y (the target), namely $(x, y) \in R$.

Our generalisation to to the above is to change D1. We introduce a function $M(x)$, for $x \in S$, giving a natural number value ≥ 0, for each x, and using M to introduce the new notion of many lives argumentation network, as the system (S, R, M) and modifying the definition D1 into the new DM1 below:

DM1. A subset E of S attacks a node $y \in S$ iff (for some $e_i \in E$ we have $e_i R y$, where $i = 1, ..., M(y)$ and where $i \neq j$ implies $e_i \neq e_j$, for all $0 \leq i, j \leq M(y)$).

The function $M(x)$ gives the many lives of x, meaning how many live attackers of x we need in order to kill x.

The change of DM to DM1 necessitates changes in the other DM clauses. In other words, we need to define new corresponding clauses DM2–DM5.

To give our readers an idea of the nature of this Chapter and the relation of its contribution to formal argumentation, we answer some questions:

---

[1]The perceptive mathematical reader will see that D1 is not used in the following D2-D5. It is included here for reasons of Socratic exposition. See for example the next item DM1 and Remark 2.4.

Question 1: Is traditional Dung network a special case of our new networks?

Answer to question 1: Yes, because we can let $M(x) = 1$, for all x and define the semantics options for M in such a way that they agrees with the Dung semantics options. However, we must be careful how to define DM2–DM5, so that they also conform to the special case. It may be, however, that we will judge that it is more natural not to force restrictions on M and try to get the Dung semantics as a special case but rather to allow us to depart from the Dung semantics options even in the case that $M(x)$ is always 1.This decision may depend on the needs of the multiple complaints offender application area and on general mathematical smoothness properties which it can offer.

Questions 2: What happens with the concept conflict freeness? When arguments have many lives they may be attacked but still be alive , so in what sense can a set of arguments be conflict free? Consider a single point e which attacks itself and has 2 lives, i.e. we have $S = \{e\}$. $R = \{(e, e)\}$ and $M(e) = 2$. e is not dead because it suffers only one attack and it takes 2 attacks to kill it.

- is $\{e\}$ conflict free?

- How many lives does e have (after the attack)?

Answer to question 2: Let us move carefully here. We have that e geometrically attacks itself (that is $(e, e) \in R$) but cannot kill itself. Of course we can set up our system to allow e to repeatedly attack itself again and again, in which case e will kill itself after two rounds, but we may choose to allow attackers only one only one attempt at attacking. In this case no matter how we look at it, e cannot be dead or undecided. It is alive with one life left. To overcome this lack of clarity, let us talk about "geometrical attack" and "successful attack" of a set E on a node x. The set E geometrically attacks x if for some y in E we have $(y, x) \in R$, (R being the graph "geometry" on S). The set E successfully attacks x if it manages (according to our agreed definition of this notion) to reduce the many lives of x to 0. Given a subset E, we can also talk about the old Dung concept E as being "geometrical conflict free" and introduce a new concept of E as being "at peace" or as "able to survive together".

So according to these new concepts $E = \{e\}$ with $M(e) = 2$, does geometrically attack e, but e is able to survive together with itself because it cannot kill itself. We can also reasonably say that e has one life left now after having attacked itself.

Question 3: What is a complete extension? The example in Question 2 creates a problem because we get a new network with e geometrically attacking itself, where e has one life (one life left). Why don't we allow e to carry on attacking?

Answer to question 3: We are therefore forced to say that the new concept of a complete extension of any one network is another network. Section 4 discusses how it is identified. So to be clear, given a network (S, R, M, K) and a notion of "semantics" for such networks, the output of this notion is a family of networks of the same type. In comparison for the case of Dung networks of the form (S, R) the output of a semantics is a family of subsets E of S. Note that since any such E is conflict free we can regard it as a network with the empty attack relation, $(E, \varnothing)$. So the network $N = (\{e\}, \{(e, e)\}, M(e) = 2)$, has the single complete extension which is the network $N' = (\{e\}, \{(e, e)\}, M(e) = 1)$ which has the traditional complete extension $N'' = \varnothing = \{e = \text{undecided}\}$. We immediately ask: Is this concept compatible with the old Dung concept of extension?. The answer is yes, it is.

Question 4: How do we view our Chapter?

    i Is it a contribution to the area of Numerical Argumentation (by introducing the functions M and K)?

    ii Is it a contribution to Ranking of arguments? something comparable to Grossi and Modgil [8, 24]?

iii Is it part of the Equational Approach [9]?

iv Is it arising from some application area? If indeed it is connected with an application area and is not a purely technical Chapter, then we further ask: Does it model some part of the application area or does it just draw ideas from the application area and offers another formal argumentation system to (i) or to (ii) which can approximate some features of the application area ?

Answer to question 4:   The Chapter draws ideas from several application areas, as described below, which have the many lives feature in common, and is inspired to formulate a sample formal argumentation theory which connects with the formal areas of numerical argumentation and of ranking. The formal systems suggested are good and flexible enough to be adapted to modelling more accurately any of the application areas which inspired them.

The old Dung concept of a complete extension E is a set but E can be can be viewed as another network because it is conflict free so it is a network with the empty attack relation. So the new concept contains the old concept. This is OK.

In familiar everyday life we have many examples of the many lives/tolerance/ resilience function $M(x)$ of x. These include:[2]

1. How many complaints against x can be tolerated/covered-up/ ignored before action needs to be taken

2. How many applications/demonstrations/hints/pressure/ repeated nuisance, can be tolerated before compliance/giving-in.

3. How many witnesses are needed legally to establish a fact in law

4. How many violations are sufficient to cross a legal threshold to the next legal level.

---

[2]The idea of many lives actually arose from our argumentation modelling of sex offender's Therapy [1, 2].

There have been many cases in the UK where public figures and celebrities were accused by several complainers of alleged misconduct. All these cases and accusations had a similar pattern.

Let x be the accused. First a y_1 would come forward with allegations against x. Naturally x would deny any wrong doing and dismiss y_1's accusations. Then more and more accusers come forward, say $y_2, y_3, \ldots, y_n$. At some point, say at accuser n, the public perception will change and action/response is taken. It usually starts with increased activity in social networks and may end up in social pressure on the accused to resign or pressure on the police to investigate the complaints and press charges. The next scenarios can vary from case to case; they include:

1. The public figure x resigns and disappears from the news and that is the end of the story.

2. The police investigates and the accused might end up with a prison sentence

3. Any outcome between the outcomes (1) and (2) above.

   We now address items (i)–(iv).

   It is true that the function $M(x)$ associates a numerical value with each node x and it is also true that this value is seen in relation to the number of geometrical attackers of x. So on the face of it, there seems to be a connection with Numerical argumentation and the Ranking of arguments. However, the way we use this number $M(x)$ in producing extensions is different. It is metal-level. We want at least $M(x)$ live ("in") attackers of x in order for x to be "out".

   If we look at Figure 2, the node y has two lives but its attack on x is counted as one attack. Furthermore geometrically x has two attackers (so its ranking is 2) but in order to be considered "out" in an extension these two attackers must be live ("in").

   So the use of these values is different.

   As for item (iii), the equational approach is itself meta-level. It derives from the annotated graph a system of equations, solves the equations and derives the extensions from the solutions. This procedures can be done for our many lives graphs as well. We need to find and motivate the right equations.

As for item (iv), we confirm that we looked at various ways of dealing with multiple complaints and devised the system of this Chapter as generic, showing what kind of features and technical moves to expect, and allow the system to be adapted/refined/-expanded for modelling the more specific complaints application areas.

We conclude our answer to question 4 by directing the reader to Remark 2.3 in Section 2.

2 Semi-formal discussion of the many lives idea in the complaints context

This section presents the many lives idea in a slightly more precise (semi-formal) way, in order to prepare the readers from the informal argumentation community for the later formal sections. We assume such readers have some minimal background in Mathematics.

Our readers from formal argumentation theory can skip this section, after reading the next formal Definition 2.1, of what is a many lives network.

Definition 2.1 (Many lives network). *A general many lives network has the form (S, R, M, K), where S is a non-empty set of arguments, R is the binary attack relation on S, M is a function giving each x in S a natural number of how many lives it has, (including possibly 0) and K is a strength function defined on ER, giving a positive natural number for each (y, x) in R, such that it is at most $M(y) + 1$.*

We now discuss and motivate Definition 628-DJ1.

Remark 2.2 (Motivating M). *Let us first focus on the number $M(x) = n$ of the many number of lives of the node x and consider it as the resilience of x to attacks.*

162

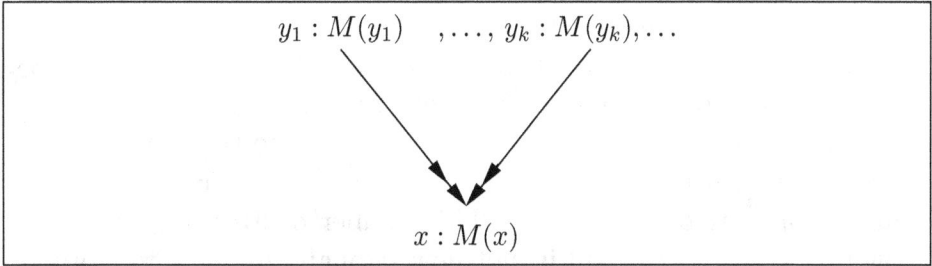

$$y_1 : M(y_1) \quad , \ldots, y_k : M(y_k), \ldots$$

$$x : M(x)$$

Figure 1: General attack formation, where "$\twoheadrightarrow$" denotes attack.

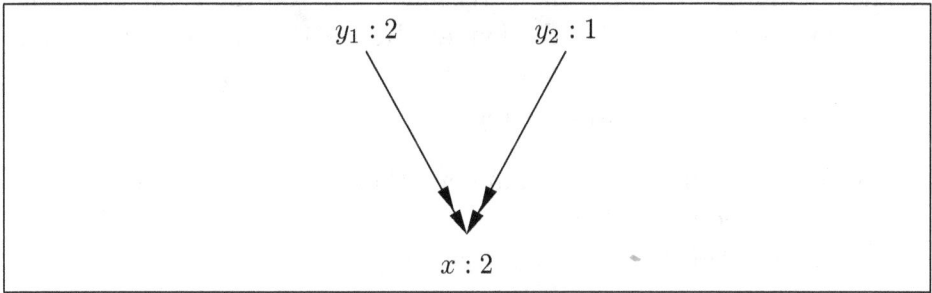

$$y_1 : 2 \qquad y_2 : 1$$

$$x : 2$$

Figure 2

$M(x) = $ how many live attackers does it take to kill x.[3]

Figure 1 indicates this basic situation. Figure 1 is a general schematic description and Figure 2 is a particular case of it. Note that is this Chapter double headed arrows "$\twoheadrightarrow$" denote attacks.

In this figure we assume that each node z has a value $M(z)$ of number of lives and that $y_1, \ldots, y_k$ attack x. If we have that all the $y_1, \ldots, y_k$ are alive then x would be dead if $M(x) \leqslant k$. In fact we can write a formula for the new value $M^*(x)$, which is obtained after the attack of $y_1, \ldots, y_k$ is carried out. The value is

$M^*(x) = M(x) - k$, if $k < M(x)$ and 0 otherwise.

In particular for mathematical reasons we are going to allow the M function to give values 0. This would force us to say that $M(z) = 0$ means that z is "dead" for any z.

---

[3]We use here the informal words "live", "dead" and "kill". We ask the reader to understand them intuitively in this motivating section. Formal definitions will be given later in the formal sections.

So in Figure 2 the node x has 2 lives. If for example the node x had 4 lives then it could survive the attack of the nodes y_1 and y_2, but its number of lives would have been reduced from 4 to 2, because it withstood the attacks of 2 live attackers. Note that although the attacker y_1 has two lives in Figure 2, its attack on x reduces x's number of lives by 1 life only. The number of lives of y_1 indicates how many attacks can kill it, not how strongly y_1 can attack others. Note also that we allow y_1 and y_2 to attack only once and not to attack again and again. This is reasonable if you think of the attack as a complaint on an alleged offender. Repeating the same complaint again and again is still the same attack.

We note the first two principles we are adopting here:

PP1: Every element x has a number $M(x)$ of lives (including possibly the value 0). To really kill x you need to kill it $M(x)$ times.[4] In particular non-attacked elements retain all their many lives intact and have the capability of attacking other elements (reducing the target's number of lives) if their value is not 0.

PP2: Although an element y may have $M(y)$ lives, when attacking any x it can kill only one of x's lives.

Remark 2.3 (Motivating K). *The reader may wonder at the strong over-simplification of principle* **PP2**. *Surely even if an alleged offender like say a minister or a president would normally require maybe 6 or seven complaints to be "killed" (i.e. to create enough of a public pressure to force resignation or prosecution), a particularly nasty complaint may reduce the number from 6 to much less! Our answer is that we are simplifying for the sake of simpler mathematics. We are not completely modelling reality in this Chapter but we are just approximating it. We admit that in real examples of complaints y against alleged offender x (namely $y \twoheadrightarrow x$), the strength of attack is not necessarily only 1 (i.e. killing only one of the lives of x).*

---

[4]For example the case of Israeli minister Sylvan Shalom 2015 (see [26]). Apparently he had 6 lives. After 6 complaints of alleged offences he resigned. We have never heard his name in public since.

*The perceptive reader might feel that we are simplifying too much.
Two strong complaints can kill maybe 3 lives. We can perhaps agree
to a more realistic model and allow the annotation for y in the model
to be of two numbers,*

1. *$M(y)$ the number of lives which y has.*

2. *$K(y)$, the strength of attack of y or in other words, how many
lives does y take when attacking. $K(y)$ can be related to $M(y)$.
The rationale being that if $M(y)$ is higher then y is stronger,
because y is harder to kill, therefore its attack is stronger. The
notation $K(y)$ assumes that the strength of attack of y is the
same, no matter whom y attacks. This is still a simplification.
We realise that K should also be dependent on the x attacked.
If the attack on x, for example is a complaint of y against x,
then y might feel more strongly about x than about another x',
therefore its attack on x will be stronger than its attack on x'.
If we want to make the strength of attack also depend on the
target of y, we need to make K a function of the pairs (y, x)
where y attacks x. We can write it as $K(y \twoheadrightarrow x)$ or $K(y, x)$, for
$(y, x) \in R$.*

*So according to this model, Figure 1 will become Figure 3 and the new
value $M^*(x)$ of x after the attack from all y_i would be*

$$M^*(x) = M(x) \mathbin{\dot{-}} \sum_{i=1}^{k} K(y_i, x).$$

Where the symbol "$\dot{-}$" is truncated substraction, namely.

$$\alpha \mathbin{\dot{-}} \beta = \begin{cases} \alpha - \beta, & \text{if } \alpha \geqslant \beta \\ 0, & \text{if } \alpha \leqslant \beta \end{cases}$$

*So for example in Figure 4, we have that y has attack strength 1, z
has 2 and u has 3.*

*The number of lives of x is 7. So after the attack the new number
of lives of x in Figure 4 is $M^*(x)$.*

$$M^*(x) = 7 - (1 + 2 + 3) = 7 - 6 = 1.$$

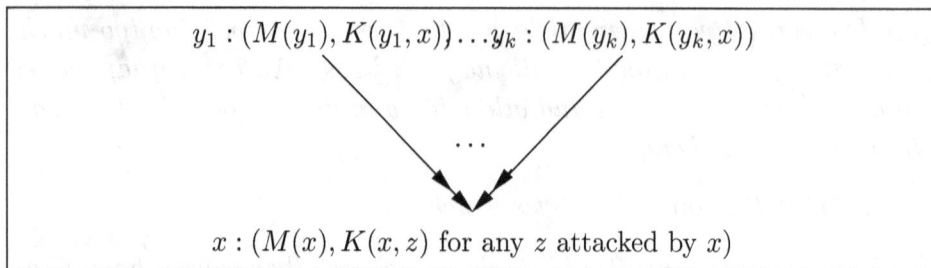

Figure 3

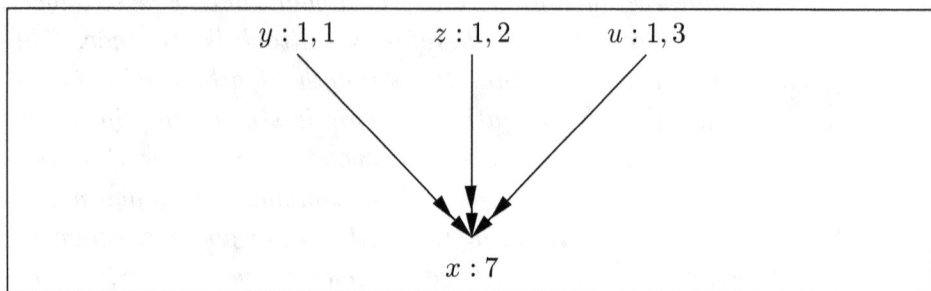

Figure 4

Note that we get a new network which is with the same geometrical graph as that of Figure 4, and the same strengths and lives for the top nodes y, z and u but for node x the number of lives is 1.

We hope the reader with experience in dealing with multiple complaints (e.g. student's complaints about a lecturer), can see that there is no end to our improving the model, getting the mathematics more and more complicated. Surely, we can further refine our model, saying, for example, that having one attack of strength 3 should be weaker than 3 attacks of strength 1. We must give extra bonus in recognition that there are more complaints (attacks) on x. We thus can continue and further agree to deduct one more life if the number of attacks is more than 2. Let us show the reader what it would look like to write this formula.

Let k be a natural number. Define $\beta(k)$ (β for bonus) to be

$$\beta(k) = \begin{cases} 1, & \text{if } k > 2 \\ 0, & \text{otherwise} \end{cases}$$

The calculation for $M_\beta^*(x)$ of Figure 4 with bonus β is

$$M_\beta^*(x) = 7 - (1 + 2 + 3 + 1) = 0.$$

Thus with the bonus we get that x is dead. Let us go on and further improve the model and get the mathematics even more complicated.

We further remark that we have not addressed in detail the question of a node x attacking more than one other node. For example x may attack node y and also node z. We associated the strength of attack to node x, so the attack of x on y will have the same strength as the attack on the node z. This is not true for all possible applications. In many other complaints contexts, the strength of the complaints of x against y may not be as solid and strong as the complaints on z. This means that the strength of attack needs to be associated not with the node x itself but with the attack arrows emanating from x, giving possibly different strengths to different arrows.

There are examples where the strength of attack is done by associating a number with x itself. In a survival game where the attack is done by shooting a gun, then x has a gun and x shoots always the same strength.

We can now, for the time being, formulate our new principle for the case of strength attached to nodes:

PP2 new: Given a network of the form (S, R, M, K) and a node x in S with k live attackers $y_1, \ldots, y_k$ of x with attack strength $K(y_1), \ldots, K(y_k)$[5] respectively, i.e. we have $M(y_i) > 0$ for $i = 1, .., k$ and m dead attackers $z_1, ..., z_m$, with $M(z_j) = 0$, for $j = 1, ..., m$, and given $M(x)$ as the number of lives of x, then the new number of lives $M^*(x)$ after the attack is given by the formula[6]

$$M_\beta^*(x) = M(x) \dotminus [\beta(k) + \sum_{i=1}^{k} K(y_i)] \qquad (*)$$

---

[5]If the strength of attack is associated with arrows we replace "$K(y)$" by "$K(y \twoheadrightarrow x)$".

[6]The formal definition we give in Section 4 is slightly more general. See Definition 4.1.

Remark 2.4.

1. *We now discuss the idea of a different strength of attack required for protection. Assume for example that student y attacks professor x by accusing x of being verbally abusive to y. Assume z comes to the protection of x by accusing y of being a liar and a cheat and that y is attacking x for leftist political reasons. In this context we might expect the attack of z on y to be extra strong. Had y attacked x for something else, say for lack of clear course notes, then, without the verbal abuse context, the attack of z on y might not have been expected to be as strong.*

   Let us sum up and say that it is possible to take the view that to protect a node x against the attack from node y you need much stronger killing attacks then just an ordinary attack to kill.

2. *Similarly suppose we have a set E of elements trying to live together. The concept of conflict free is that the different elements of E do not attack to kill one another. To be on the safe side we might take the view that to be sure that the elements of E can indeed live together then even if the attack available is stronger than ordinary attack, (but still weaker or equal the protective attack), then still, the elements of E cannot kill one another.*

3. *We thus have 3 types of attacks, which we call $\alpha_{\mathbf{a}}, \alpha_{\mathbf{p}}$ and $\alpha_{\mathbf{d}}$, meaning respectively, $\alpha_{(\text{ordinary attack})}, \alpha_{(\text{attack to protect})}, \alpha_{(\text{attacks used in the context of living together})}$, with the restriction that $\alpha_{\mathbf{p}}$ is stronger or equal to $\alpha_{\mathbf{d}}$ which is stronger or equal to $\alpha_{\mathbf{a}}$. So for example of three such attacks we can have:*

   (a) *A set of nodes $Y\alpha_{\mathbf{a}}$ attacks a node z if for at least one element $y \in Y$ attacks z*

   (b) *A set of nodes $Y\alpha_{\mathbf{d}}$ attacks a node z if for at least two different elements $y \in Y$ attack z*

   (c) *A set of nodes $Y\alpha_{\mathbf{p}}$ attack a node z if for at least three pairwise different element $y \in Y$ attack z.*

3 Background and concepts from abstract argumentation with additional methodological remarks

This section presents, for the convenience of the reader, some basic concepts of what we called traditional argumentation theory. Such systems contain attacks only. We refer to such system as Dung Argumentation with Attack only (see [6]). We shall then add methodological remarks and explain in what way the systems developed for this Chapter depart from the traditional ones.

There are two traditional ways to present the semantics for the traditional Dung argumentation with attack, the traditional set theoretical approach and the Caminada labelling approach.[7] For the mapping connections between the two approaches, see [7]. Let us briefly quote the traditional set theoretic approach:

Definition 3.1.

1. *We begin with a pair* (S, R), *where* S *is a nonempty set of points (arguments) and* R *is a binary relation on* S *(the "attack" relation).*

2. *Given* (S, R), *a subset* E *of* S *is said to be conflict free if for no* x, y *in* E *do we have* xRy.

3. E *protects an element* $a \in S$, *if for every* x *such that* xRa, *there exists a* $y \in E$ *such that* yRx *holds.*

4. E *is admissible if* E *is conflict free and protects all of its elements.*

5. E *is a complete extension if* E *is admissible and contains every element which it protects.*

---

[7]Actually there are more ways of calculating the extensions

3. The equational approach of Gabbay [9]

4. The algorithmic approach, see [1]

Various different semantics (types of extensions) can be defined by identifying different properties of E. For example we might define that E is a stable extension if E is a complete extension and for each $y \notin E$ there exists $x \in E$ such that xRy or the grounded extension as the unique minimal extension or a preferred extension, being a maximal (with respect to set inclusion) complete extension. The above properties give rise to corresponding semantics (stable semantics, grounded semantics and preferred semantics).

It can be proved that extensions satisfying items (1)- -(5) of Definition 3.1 do exist. The proof is set- theoretical using fixed points. It is easy to see how the above conditions on extensions E can be interpreted as defining a survival group. The members of the group do not attack one another and attack anyone who attacks one of them. The group also adds to itself all candidates it can protect. This is a group of nodes taking a maximal defensive position.

Remark 3.2. *Definition 3.1 uses geometrical properties (the "attack" arrow $\twoheadrightarrow$, to define survival concepts. Since later we are going to generalise the concept of one life to many lives, it is helpful already at this point to rewrite Definition 3.1 in survival terms.*

The clause numbers here correspond to the clause numbers in Definition 3.1

1. *Given (S, R), where S is a nonempty set of points and R is a binary relation on S, a subset E of S is said to attack a point x in S if for some y in E we have that yRx holds.*

2. *A subset E of S is said to be able to survive together, if for no subset Y of E and no point x in E do we have that Y attacks x.*

3. *E protects an element a in S if whenever a set X attacks a, then the set $X - Y$ does not attack a, where Y is the set*

$$Y = \{y | y \text{ in } X \text{ and } E \text{ attacks } y\}.$$

4. *E is admissible if E is able to survive together and E protects all of its elements.*

Note for example that if we allow for many lives then if a → b and b → a and each of {a, b} have two lives, then the set {a, b} is able to survive together, because neither of its elements can kill the other.

5. *E is a complete extension if E is admissible and contains every element which it protects.*

We can also present the complete extensions of $A = (S, R)$, using the Caminada labelling approach, see [7].

Definition 3.3. *A Caminada labelling of S is a function $\lambda : S \mapsto$ {in, out, und} such that the following holds.*

(C1) $\lambda(x) = in$, if for all y attacking $x, \lambda(y) = out$.

(C2) $\lambda(x) = out$, if for some y attacking $x, \lambda(y) = in$.

(C3) $\lambda(x) = und$, if for all y attacking $x, \lambda(y) \neq in$, and for some z attacking $x, \lambda(z) = und$.

A consequence of (C1) in Definition 3.3 is that if x is not attacked at all, then $\lambda(x) = $ in. Any Caminada labelling yields a complete extension and vice versa. Any {in, out} Caminada labelling (i.e. with no "und" value) yields a stable extension and vice versa. Set theoretic minimality or maximality conditions on extensions E correspond to the respective conditions on the "in" parts of the corresponding Caminada labellings, see [7].

Remark 3.4. *Let us summarise the comparison of the Caminada λ function (and hence the notion of the traditional Dung extension which is equivalent to it) with the many lives function $M(x)$:*
We can understand the Caminada labelling function $\lambda(x)$ a partially defined function M, giving values in {0, 1} satisfying certain restrictions. If we write $M(x) = $ undefined when M is not defined on x, and write $M(x) = $ in, to mean $M(x) = 1$ and $M(x) = $ out to mean $M(x) = 0$, then the conditions (C1), (C2) and (C3) of Definition 3.3 become the restrictions on M.

This observation is of methodological importance. We are offering a new many lives system and we need to show how the traditional Dung system fits in as a special case. We have just shown that if we allow M to be partial function and put conditions on M in terms of R we can get the traditional Caminada Dung semantics as a special case.

4 Formal set theoretic semantics

This section formally defines the notion of many lives networks for our Chapter and, following Dung [6], develops set theoretic semantics for it.

Definition 4.1 ($MK\beta$ annotation for a network). *Let (S, R, M, K, β_m) be an annotated network as follows:*

1. *(S, R) is a network with $S \neq \varnothing$ and $R \subseteq S \times S$.*

2. *M is a function on S giving for each $x \in S$ a natural number in $\{0, 1, 2, 3, \ldots\}$ called the number of lives of x.*

3. *$K(y, x)$ is a function giving each attack $(y, x) \in R$ a natural number value in $\{1, 2, 3, \ldots\}$ called the strength of the attack.*

4. *β_m for m a natural number or ∞ is a function k we have $\beta_m(k) = 0$ if $k \leqslant m$ and $\beta_m(k) = 1$ if $k > m$.*

5. *Let $\delta(x)$ be Kronecker δ function, namely*

$$\delta(x) = \begin{cases} 0, & \text{if } x = 0 \\ 1, & \text{if } x \neq 0 \end{cases}$$

6. *Let $Attack(x)$, for $x \in S$ and subsets E of S be the set $\{y | yRx\}$. Let E be any subset of S and let $Attack\,(E, x)$ be $\{y | y \in E \wedge yRx\}$.*

7. *Let E be a non-empty subset of S. Let $M^*(E,x)$ be defined for $x \in S$ as the function derived from M, satisfying the implicit equation (*) for any subset E of S and any $x \in S$:*[8]

$$M^*(E,x) = M(x) \doteq [\beta_m(\textstyle\sum_{y \in E \wedge yRx} \delta(M^*(E,y))) +$$
$$\textstyle\sum_{y \in E \wedge yRx}(\delta(M^*(E,y))K(y,x))] \tag{*}$$

8. *Let (S,R,M,K) be a system with $K(x,y) = 1$ for all (x,y) in R and $M(x) \leqslant 1$ for all x in S and no β present. We say that this system has a numerically balanced M labelling iff $M^* = M$.*[9]

Definition 4.2. *Given an $MK\beta$ network as in Definition 4.1, with a set S of nodes and a relation R on S, let us define the notion of a non-empty subset E attacking a node x, Notation $\alpha_a(E,x)$, as follows:*

---

[8]The perceptive reader might ask why we have M^* in the right had side of the equation in item 7. We explain this by example. Take the network of Figure 15 with nodes $\{a,b,c\}$ and attacks $a \twoheadrightarrow b$ and $b \twoheadrightarrow c$. Let K and β play no role. Let $E = \{a,b,c\}$. So we have only M, and let $M(a) = 2, M(b) = 1$ and $M(c) = 2$. Test the equation of item 7 on this network. We get

$$M^*(E,a) = 2$$
$$M^*(E,b) = M(b) - \delta(M^*(E,a)) = 1 - 1 = 0)$$
$$M^*(E,c) = M(c) - \delta(M^*(E,b)) = 2 - 0 = 2).$$

If we do not put M^* on the right hand side we get for $M^*(b)$ the value 1. The definition of M^* is to yield the many lives values of the nodes following the the propagation of the attacks.

[9]The conditions on (M,K) of item 8 makes the network practically a traditional network with "in" and "out" annotation. If the network is acyclic a numerically balanced labelling exists. Note that we allow $M(x) = 0$ even for x which is not attacked (i.e. even when all attackers are non-existent or have M value 0). If we insist that $M(x) = 1$ in such cases (note this is one of the Caminada conditions), then M will still be numerically balanced but M will yield the grounded stable extension in the acyclic case. Consider a three point acyclic network of Figure 15, with $S = \{a,b,c\}$ and $R = \{(a,b),(b,c)\}$, (that is the network $a \twoheadrightarrow b \twoheadrightarrow c$). Consider the numerically balanced $M(a) = 1, M(b) = M(c) = 0$. This M does not give rise to a Dung grounded extension but M' with $M'(a) = M'(c) = 1$ and $M'(b) = 0$, which is also numerically balanced does give a Dung grounded extension.

(♮) $\alpha_{\mathbf{a}}(E, x)$ holds iff by definition $M^*(E, x) = 0$, where $M^*(E, x)$ is as defined in item 7 of Definition 4.1.

It is very important to note that for any E, E' and x we have:

- E attacks x and E is a subset of E' then E' attacks x.

- E does not attack x and E' is a subset of E then E' does not attack x.

- Note that the attack $\alpha_{\mathbf{a}}$, is defined using item 7 of Definition 4.1, and is therefore dependent on M and on K. If we use another many lives function N and another strength of attack function L, we will get a different attack relation., which we can call for example by the name $\alpha_{\mathbf{p}}$. Note further that if for all x, y we have that $M(x, y) \leqslant N(x, y)$, and/or $L(x, y) \leqslant K(x, y)$ then $\alpha_{\mathbf{p}}$ is a stronger attack than $\alpha_{\mathbf{a}}$, namely if E can $\alpha_{\mathbf{p}}$ kill x then E can $\alpha_{\mathbf{a}}$ kill x.

Definition 4.3. *Let (S, R) be a given geometrical network. Imagine we have several possible functions of the form $M(x), K(x, y)$ and β defined on (S, R). We can use different functions M, K, β to define different kinds of attacks as done in Definition 4.2.*

Let $\alpha_{\mathbf{a}}, \alpha_{\mathbf{d}}$, and $\alpha_{\mathbf{p}}$, be three such attacks as defined in Definition 4.2. Assume the relative strength of these attacks is as follows:

(s1) *If $Y \alpha_{\mathbf{p}}$ attacks z then $Y \alpha_{\mathbf{d}}$ attacks z*

(s2) *If $Y \alpha_{\mathbf{d}}$ attacks z then $Y \alpha_{\mathbf{a}}$ attacks z.*

1. *We say that E is at peace iff for no Y, a in E do we have $\alpha_{\mathbf{d}}(Y, a)$ holds ("at peace" means "able to live/survive together" where the attack does not kill, compare with Definition 4.7 and Remark 2.4).*

2. *E protects x if for every Y such that $\alpha_{\mathbf{a}}(Y, x)$ holds we have that for some subset Y' of Y the protecting set E successfully $\alpha_{\mathbf{p}}$ attacks all elements of Y' and that the remaining elements of Y, namely the set $Y - Y'$, does not successfully $\alpha_{\mathbf{a}}$ attack x.*

3. E is $(\mathbf{a}, \mathbf{p}, \mathbf{d})$ admissible if E is at peace and protects its elements

Lemma 4.4. If E admissible and protects x then $E \cup \{x\}$ protects itself.

Proof. This is true because E protects all elements of $E \cup \{x\}$ so $E \cup \{x\}$ does it (i.e. protects) as well because of the monotonicity condition. $\square$

Lemma 4.5. If E is at peace and protects its elements and E protects x then $E \cup \{x\}$ is at peace.

Proof. Assume that $E \cup \{x\}$ is not at peace, get a contradiction. We immediately see that x is not in E.

Let $Y \subseteq E \cup \{x\}, z \in E \cup \{x\}$ be such Y successfully $\alpha_{\mathbf{d}}$-attacks z. Then by our assumptions Y also successfully $\alpha_{\mathbf{a}}$ attacks z. We distinguish several cases:

Case 1. $x \notin Y, x \neq z$. This case contradicts E at peace.

Case 2. $x \notin Y, z = x$. We have Y successfully $\alpha_{\mathbf{d}}$ attacks x and therefore also successfully $\alpha_{\mathbf{a}}$-attacks x. Since E $\alpha_{\mathbf{p}}$-protects x, E must successfully $\alpha_{\mathbf{p}}$-attack some elements $y_1, \ldots, y_k$ such that $Y - \{y_1, \ldots, y_k\}$ does not successfully $\alpha_{\mathbf{a}}$-attack x. Since Y does successfully $\alpha_{\mathbf{a}}$-attack x, there must be at least one y_1 in Y (and therefore y_1 is not x) such that E successfully $\alpha_{\mathbf{p}}$-attacks y_1. Since by our assumptions say that $\alpha_{\mathbf{d}}$ attacks are stronger than $\alpha_{\mathbf{p}}$ attacks (this is assumption (s1)), we get that E $\alpha_{\mathbf{d}}$ attack y_1. Thus we have found a y_1 in E which is successfully $\alpha_{\mathbf{d}}$ attacked by E, a contradiction.

Case 3. $x \in Y$ and x is different from z. Let Y_o be a subset of E and assume that $Y = Y_o \cup \{x\}$. So we have that $Y_o \cup \{x\}$ successfully $\alpha_{\mathbf{d}}$-attacks z and $z \neq x$. Since $z \in E$, E $\alpha_{\mathbf{d}}$-attacks elements of $Y_o \cup \{x\}$. E cannot attack any elements from Y_0 so E attacks x but this is now case 2, which is impossible.

Case 4. $x \in Y, z = x$. so we have $Y_o \cup \{x\}$ $\alpha_{\mathbf{d}}$ attacks x. Therefore it $\alpha_{\mathbf{a}}$ attacks x. Since E protects x, E attacks $Y_0 \cup \{x\}$ but E cannot attack any of its elements. $\qquad\square$

Lemma 4.6. *There exists an admissible set $E \subseteq S$ s.t. $E = $ all elements it protects.*

Proof. Start with $\varnothing$. It protects its elements and is at peace. Suppose $\varnothing$ protects x then $\{x\}$ protects x and is at peace.

Continue to increase the set using Lemma 4.4, until we reach a maximal st. This is the set E we need. $\qquad\square$

Definition 4.7. *Let (S, R, M, K, β) be the network defined in Definition 4.1, and assume that we have the notion of $\alpha_{\mathbf{a}}, \alpha_{\mathbf{p}}$, and $\alpha_{\mathbf{d}}$ -attack to go with it. Using the notion of such attacks we can identify the family of sets E which are admissible and are equal to the set of all the elements E protects. Let E be such a set. E may $\alpha_{\mathbf{d}}$ attack some of its elements but such attacks are not successful. This is why E is at peace, precisely because the attacks of E on its elements, $x \in E$, are not successful, i.e. these attacks cannot reduce to 0 the many lives $M(x)$ of x. We can now use the notion of $\alpha_{\mathbf{d}}$-attack to update the number of lives of each element x in E. Let x be any element x in E such that E $\alpha_{\mathbf{d}}$-attacks x. Let the new annotation of x be $M^*(E, x)$ of item 7 of Definition 4.1. If x is not $\alpha_{\mathbf{d}}$ attacked by E, leave its annotation unchanged.*

*Let M_E be the new annotation on E. We refer to the system $(E, R$ restricted to E, M_E restricted to $E)$ together with the **a** respective attacks restricted to E, as an E complete extension of the original system.*

We thus can define the set of all E-complete extensions of the original system.

Example 4.8. *Let us illustrate the concepts of Definition 4.7 using the network of Figure 4. In the network of this figure, (with nodes $\{y, z, u, x\}$ and where y has attack strength $1, z$ has 2 and u has 3 and the number of lives of x is 7). In the network of this figure, the*

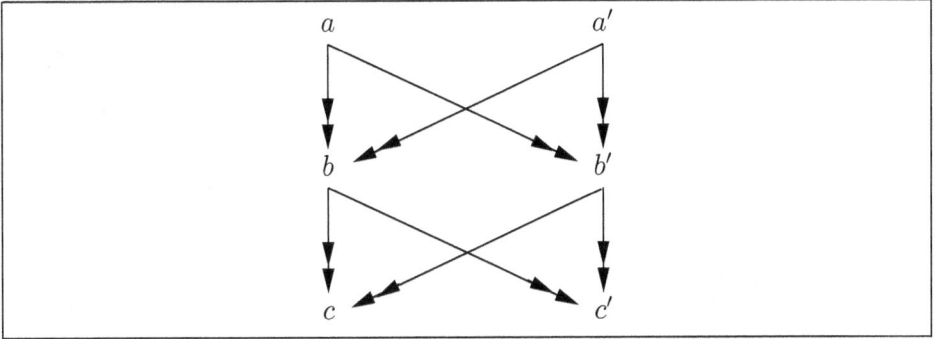

Figure 5

entire set E of nodes is at peace with itself, because y, z and u are not attacked and although they all attack x, x can survive the attack with 1 life left. Thus here M_E is actually the calculated M^*.

Remark 4.9. *We make a few key points related to the definitions of this section.*

1. *Note what the network with M, K, β looks like when M is a fixed number $m > 1$ for all nodes (say $m = 2$) and K is always 1 and $\beta = 0$. This means we have a network where all nodes require 2 attackers to be dead. To appreciate this case, consider the simple Figure 5. Assume $m = 2$. So we are giving each node in Figure 5 two lives. In this case we get that $\{a, a', c, c'\}$ are "in" with two lives each and $\{b, b'\}$ are "out" with zero lives each. If we let $m = 3$ we get a new network with the same graph figure but with different lives distribution. a, a' have 3 lives, and b, b', c', c' have 1 life each.*

2. *Suppose in item (1) above we adopt a geometrical point of view and say that to be killed we need 3 geometrical attackers. We do not assign life to nodes, just say to be "out" you need 3 "in" attackers. This is in the spirit of [1]. In this case we simply get that all are "in". We do not subtract the number of live attacks from the many lives of the target. When the target is attacked, it is either killed or if not enough live attackers are present, then it stays as is.*

3. *The reader might ask why in Definition 4.2 we were talking about $\alpha_\mathbf{a}$-attacking, what is the role of the index "$\mathbf{a}$"? We have this index because we might have more than one type of attack, say we might have also another kind of attack which we might call $\alpha_\mathbf{p}$. The two notions of attacks, $\alpha_\mathbf{a}$ and $\alpha_\mathbf{p}$ might play different roles in calculating extensions. For example the nodes $\{a, a'\}$ are geometrically protecting the nodes $\{c, c'\}$ because they are geometrically attacking the nodes $\{b, b'\}$ which are the geometrical attackers of $\{c, c'\}$. We might make a distinction and say that attacks are usually $\alpha_\mathbf{a}$ attacks, but when protecting elements we must use $\alpha_\mathbf{p}$ attacks.*

*For this to work smoothly we need to require that the relationships of Definition 4.3 to hold between $\alpha_\mathbf{p}$ and $\alpha_\mathbf{a}$. In fact we can also add a another notion of $\alpha_\mathbf{d}$ attacks and say that for a set E to be considered conflict free (at peace) we want that no subset $Y \subseteq E$ can $\alpha_\mathbf{d}$ attack any $e \in E$. To work smoothly we need condition (**s**1) of Definition 4.3 to hold for $\alpha_\mathbf{d}$ and $\alpha_\mathbf{p}$.*

4. *Note that Definition 4.3 and the Lemmas and proofs following it, do not use the exact definitions of the attacks but only their relative strengths, being conditions (**s**1) and (**s**2). This allows us to give possibly completely different definitions of attacks in our paper [38].*

5. *The ideas of different types of attacks was introduced in Section 8.3 of [1, p. 1855]. The reader can see more discussion in item (2) of the comparison with the literature Section 5 below and in the methodological and concluding Section 6.*

5 Comparison with the literature

We compare several related papers.

(1). Comparison with the universal distortion paper [1].
This paper deals with thinking distortions of sex offenders in particular and of general thinking distortions in general. Part of this

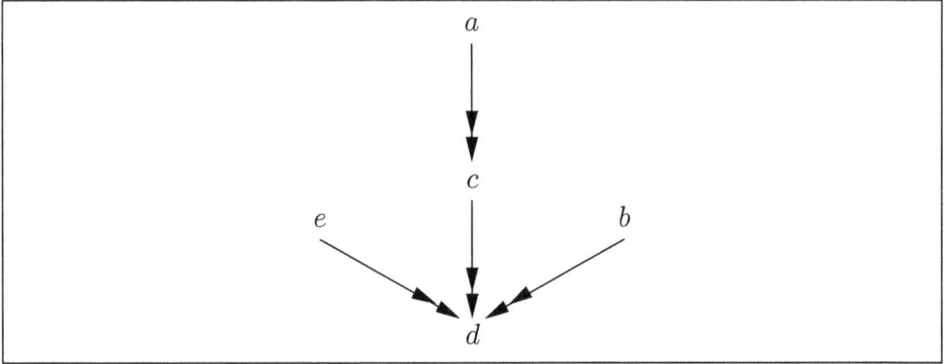

Figure 6

paper is the observation that the idea of many lives can be used in argumentation. A simple model is given in the paper and some semantics is described. The full analysis and study of many lives was postponed to the present paper and other papers [27].

(2). Comparison with graded acceptability of arguments paper [8] and [24]. These papers (among other results) propose a framework with a view of distinguishing between nodes that are out because of, say, two successful attacks, as opposed to nodes that are out because of, say, one successful attack. So for example, in Figure 6 which describes a traditional network, d is more "dead" than c because d is attacked by two living attackers while c is attacked by only one.

The authors are trying to bring this difference out by defining a predicate $d_n^m(X)$, where X is a set of nodes (intended to be an admissible set) and $d_n^m(X)$ is the set which X protects. For the purpose of comparison with our own Chapter, we use the definition for the case $d_2^1(X)$, because this is sufficient to bring out the differences with our Chapter and our notion of many lives. So X is a set of nodes and $d_2^1(X)$ defines the set of points which X protects.

We now quote and rewrite Definition 5 of [8] for the case $m = 1, n = 2$

$$d_2^1(X) = \{x | \neg \exists_{\geqslant 1} y ([y \twoheadrightarrow x \wedge \neg \exists_{\geqslant 2} z (z \to y \wedge z \in X)]\}$$

We can rewrite the above as the following:

$$d_2^1(X) = \{x | \forall y[(y \twoheadrightarrow x) \rightarrow \exists_{\geqslant 2} z(z \twoheadrightarrow y \wedge z \in X)]\}.$$

We can again rewrite as the final version $\sharp$:

$$d_2^1(X) = \{x | \forall y[(y \twoheadrightarrow x) \rightarrow$$
$$\exists z_1, z_2(z_1 \neq z_2 \wedge (z_1 \twoheadrightarrow y) \wedge (z_2 \twoheadrightarrow y) \wedge z_1 \in X \wedge z_2 \in X)]\}. \tag{$\sharp$}$$

This formula says that x is protected by X iff every attacker y of x, that is, $(y \twoheadrightarrow x)$ is itself attacked by two different members of X.

The above formula $d_2^1(X)$, which describes how X can protect a node x, looks very related to our two lives concept. However, it is not the same as a 2 lives. To see this, consider Figure 15. Let us apply $d_2^1(\varnothing)$ to a. This will determine whether a is alive or not. Substituting a for x in the rewritten formula, we find that $d_2^1(\varnothing)$ holds for a because a has no attackers. Thus a is alive.

Let us now consider the node b. b attacks c. In order for c to be defended, b, being the attacker of c, must be attacked by two live attackers. Such attackers are not available in the figure.

However, b is being attacked by a and to get b dead it is enough to have one live attacker of b which cannot be defended.

The important point here is that we cannot assign a simple number of lives to b. For b the attacker of c, b has number of lives 2. For b the victim being attacked, the number of lives is 1. This is why d_2^1 has two indices "1" and "2". We can, however, assign two numbers to b, one for it being attacked as a victim and one for being attacked as an attacker. This is what we discussed in item 3 of Remark 4.9.

Further note that in our paper [38] we study the notion of forward looking attacks and semantics. We prove in [38] that the many lives semantics is forward looking, while the Grossi Modgil semantics/attack is not.

We now summarise the comparison of our many lives approach with the Grossi and Modgil approach of [8, 24]:

1. From the technical mathematical point of view, given a network (S, R) where S is the set of arguments and R is the geometrical

attack relation, we can simulate the system d_n^m of [8, 24] (with n greater or equal m) using two many lives functions M and N, and two types of attacks $\alpha_{\mathbf{a}}$ (to kill you need $M(x) = m$ live attackers) and $\alpha_{\mathbf{p}}$ (to protect you need to attack the attacker with n live attackers). We can even add as a bonus another many lives function $G(x) = k$, with k greater or equal to n, and define $\alpha_{\mathbf{d}}$ attacks. We can thus do a triple index Grossi-Modgil geometrical function $d[m, n, k]$.

2. Note that the attack $\alpha_{\mathbf{p}}$ can be required to have different strengths for different nodes. Our machinery naturally allows for this. So in the terminology of [8, 24], the function d_n^m can be different for each of the nodes involved.

3. From the conceptual point of view the two approaches, the many lives approach and the Grossi Modgil ranking approach, S are independent and have different origins and goals. The many lives idea comes from, and is inspired by, the offender/complaints/ survival point of view and is to be tested by its ability to adapt and serve its intended application areas. The d_n^m approach of [8, 24] comes from the geometrical ranking approach of pure formal argumentation, catering for the intuition of

(*) x being more "in" or more "out" than y.

The extent to which [8, 24] succeed in addressing this intuition is not relevant to our comparison in this Chapter (it is discussed, however, in our paper [10]). However, it is relevant to the question of to what extent many lives can also be applied to the same ranking question (*).

4. We appreciate the fact that implicit in the Grossi and Modgil attempt in [8] to address (*) is the idea that to protect we can ask for a stronger attack,compatible with our developing networks with a progression of connected attacks. Note that the idea of different types of attacks also appears in Section 8.3 of [1, p. 1855].

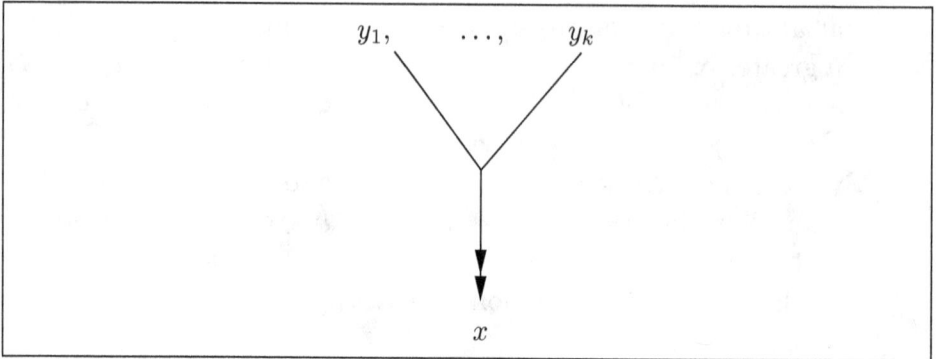

Figure 7

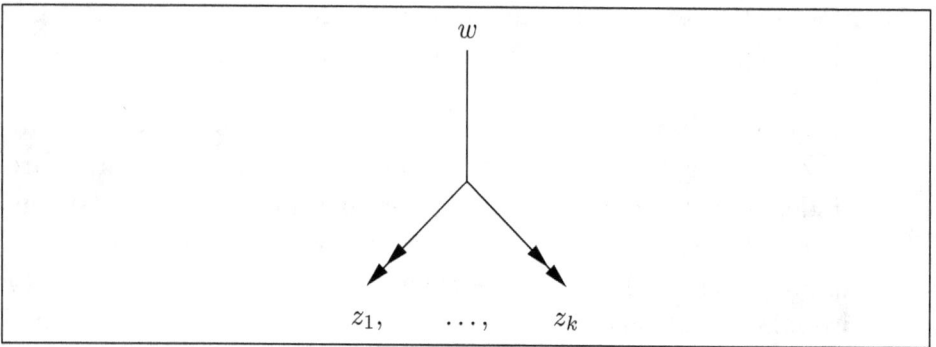

Figure 8

We shall study [8, 24] critically elsewhere, see [10] and Section 8.3 of [1].

(3). Comparison with joint (also called Collective) attacks [17] and [4, Chapter 7]. The idea of joint attacks introduced in [17] and also studied in [4, Chapter 7] is explained in Figure 7. In this figure the set $Y = \{y_1, \ldots, y_k\}$ jointly attacks the node x. The meaning is that only when all $\{y_i\}$ are live (in) do we have that x is dead (out). Nielsen and Parsons in [17] use a set to point relation $\mathbb{R}$ for such an attack. So they consider networks of the form $(S, \mathbb{R}$ where $\mathbb{R} \subseteq 2^S \times S$. In Figure 7 we have $(Y, x) \in \mathbb{R}$. The notation of Figure 7 is used by [4, Chapter 7], who also allows for disjunctive attacks of the form of Figure 8.

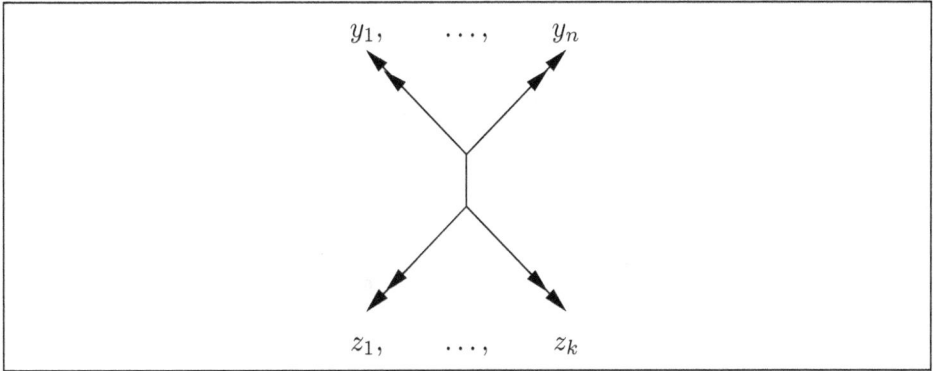

Figure 9

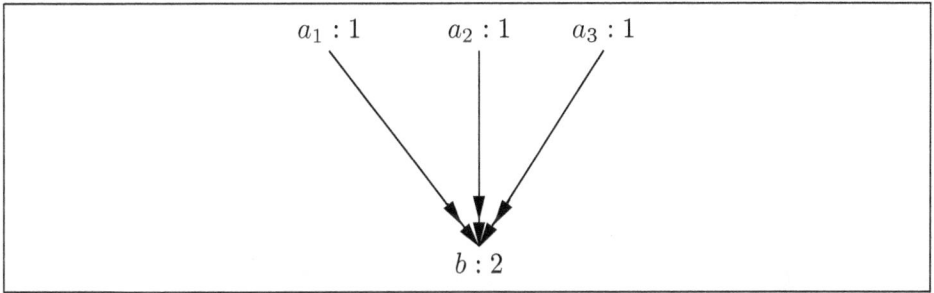

Figure 10

This means that if w is alive (in), then one of $Z = \{z_1, \ldots, z_k\}$ must be out. We can also have conjunctive–disjunctive attacks of the form of Figure 9

This means that if all of $\{y_i\}$ are in then one of $\{z_j\}$ must be out. See [18]. This can be written as a relation between sets Y and Z.

It is important to realise that the attacks of sets E on nodes e in this Chapter are not joint attacks but an aggregation of single attacks. Not all members of the set E need to mount a successful attacks. This is why we have the monotonicity rule, that if E mounts successful attack on e so does any superset of E. The connection with m lives is explained in Figures 10 and 11.

In Figure 10, a_1, a_2 and a_3 attack b. b has 2 lives and so for b to be out, at least two of its attackers must be in. Now suppose that only a_1 is in and a_2 and a_3 are out. What do we say now about b?

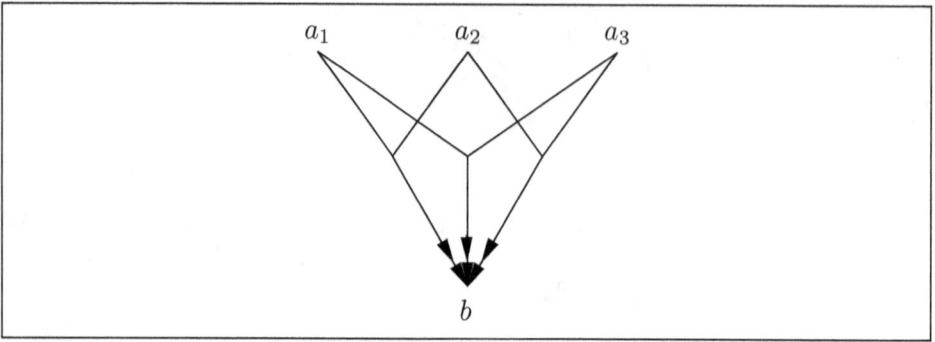

Figure 11

We say two statements

1. b is not out, b is still in.

2. The number of lives of b is 1 (reduced by 1).

It is statement (2) about b in Figure 10 which cannot be properly captured/translated/reduced, by using joint attacks. Statement 1 can be translated into conjunctive attacks as shown in Figure 11 with $\{a_1, a_2\}\mathbb{R}b$, $\{a_1, a_3\}\mathbb{R}b$ and $\{a_3, a_2\}\mathbb{R}b$, but statement (2) is not represented in Figure 11. There is a more severe way of bringing out the difference. Joint attacks still operate within the framework that each node x is either in or out (or undecided). It may take a joint attack of $m > 1$ live/in nodes to kill x, but still m cannot be reduced to $m - 1$ if we have only single live attack.

Furthermore, the translation from Figure 10 to Figure 11 does succeed in translating statement (1) about b when b has only 2 lives, but what do we do with the case of b having 4 lives? What do we write? There are not enough attacking nodes to make any distinctions.

We are fairly confident that in general we cannot (prove a theorem that we can always) translate a many lives network into a single life network with joint attacks . In other words we believe the many lives concept cannot be reduced to the concept of joint attacks. Note that the network of Figure 10 can be reduced/translated to a network with joint attacks only, namely Figure 11 with b with one life only. We ask

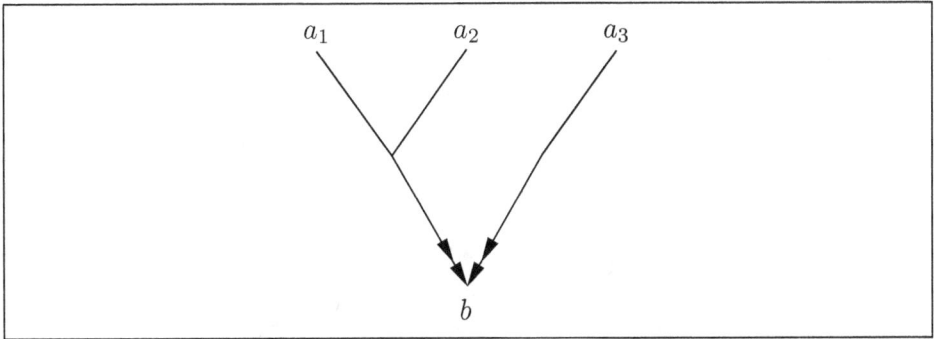

Figure 12

however, how would a translation go if b were to have 10 lives in Figure 10?

Let us summarise as follows:

The semantics for many lives networks is to yield other many lives networks whose nodes have less lives. The semantics of networks with joint attacks is to yield subsets of nodes which are in or out (the rest being undecided).

Let us now ask about the other direction. Can the many lives model simulate joint attacks?

Consider Figure 12. This is a figure with two types of joint attacks.

For the attack of a_3 to succeed, we need b to have one life. For a_1 or a_2 alone not to succeed we need b to have two lives. The problem is that the joint attacks can be mixed, with different joint attacks having a different number of attackers. We can perhaps compensate by adding strength of attack to $a_3 \twoheadrightarrow b$ and get Figure 13

This may work in this case but the reader can see that the two ideas, joint attacks and many lives are different intuitions.[10]

---

[10]One of the referees made the following remark, we quote:

"The idea of many lives in not new since in the literature it has been somehow captured by collective attacks. Of course, the approach followed was quite different but the purpose is the same. Personally, I prefer the encoding via collective attacks. The reason is that the number $M(x)$ is independent from the attackers, namely from their strength, their relations to each others, their relevance to the target, etc. This may lead to counter-intuitive results. Assume for instance

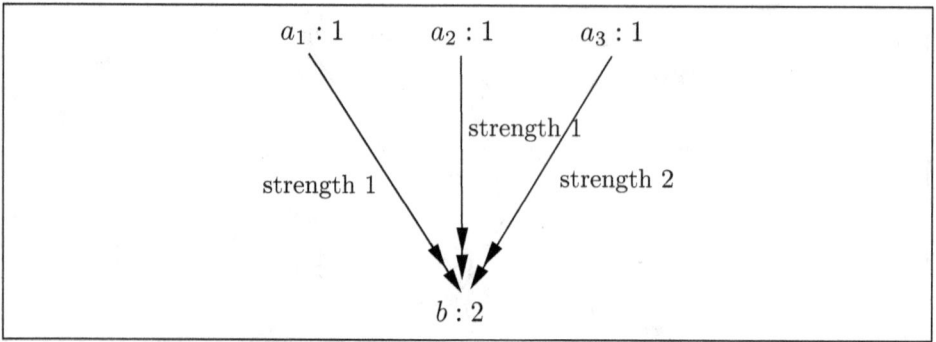

Figure 13

Example 5.1 (Can many lives be reduced to joint attacks?). *Consider Figure 14. We have a_2, a_2, a_3 attacking b which has 2 lives. Assume this figure is part of a much larger network and so it is not known what values $\{a_1, a_2, a_3\}$ get, ("in" or "out"). We further assume that $\{a_i\}$ are the only attackers of b and that the larger network is finite acyclic and so has only the ground extension and it is stable (no undecided). Assume also that b is the only argument in the entire network which has two lives.*

Our objective is to represent this figure within the traditional

---

that $N(x) = 3$ and surprisingly x is attacked by 3 non-attacked arguments y_1, y_2, y_3. According to the formalism proposed in the paper x will be rejected independently of y_i. Suppose that y_i are all similar (or logically equivalent). x would be rejected while it should not."

Our answer to this remark is as follows:

1. The default assumption in argumentation is that different letters for arguments denote completely independent arguments. Otherwise we have the same problem in ordinary argumentation. For example if we have $x \twoheadrightarrow a, y \twoheadrightarrow a$ and $b \twoheadrightarrow x$, then we never ask if x is equivalent to y and so perhaps then b protects a? The default is that x is independent of y.

2. We already remarked that apart from the fact that the idea of many lives in different from that of joint/collective attacks, we do not believe that technically we can reduce the machinery of many lives to that of joint attacks.

Such a reduction is possible if we base argumentation on linear logic and use the attack as information input, see [40].

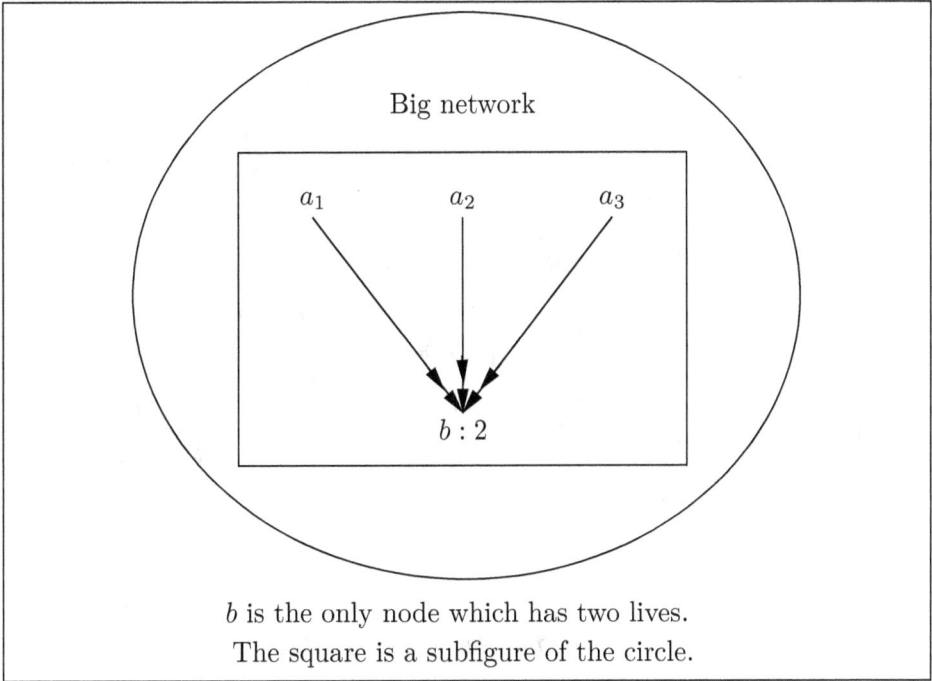

b is the only node which has two lives.
The square is a subfigure of the circle.

Figure 14

framework of Dung. We try to do that using common sense and see what happens. The traditional framework cannot represent numbers, so let us duplicate b and introduce b_1, b_2 and write in the meta-level that $b_1 = b_2$. This gives b two lives.

Then we replace $\{b\}$ by $\{b_1, b_2\}$ in the larger network.

The problem is that we have a_1, a_2, a_3 each attacking b and we need to say how they attack $\{b_1, b_2\}$.

We know that

1. *If only one of $\{a_i\}$ is in then b is not out, but its life is reduced from 2 to 1.*

2. *If two of $\{a_i\}$ are in then b is out.*

So if we split/replace b to b_1, b_2 we must then satisfy:

(1*) If only one of $\{a_i\}$ is in then only one of $\{b_1, b_2\}$ is out

187

(2*) If two of $\{a_i\}$ are in then both $\{b_1, b_2\}$ are out.

To implement the above we face a technical problem:

- *What attack arrows do we draw from the a_i to the b_j?*

Let us perform a detailed analysis of our options.

(p1) We do not draw any attack from a_1.

This is not possible because if $a_1 = $ in and $a_2 = a_3 = $ out then we need exactly one of $\{b_i\}$ to be out.

(p2) OK. Let $a_1 \twoheadrightarrow b_1$. Clearly we will not have that a_1 also attacks b_2 because then if $a_1 = $ in and $a_2 = a_3 = $ out both b_1, b_2 will be out.

(p3) How about a_2?

a_2 must attack one of the b_i, otherwise if $a_2 = $ in and $a_1 = a_2 = $ out then none of b_i would be out. OK, then a_2 must attack b_2, i.e. $a_2 \twoheadrightarrow b_2$ (it cannot attack b_1).

(p4) Now we have an impossibility. What does a_3 attack? If it does not attack at all, then if $a_3 = $ in and $a_2 = a_3 = $ out, the none of $\{b_i\}$ is out. This is wrong.

If a_3 does attack say $a_3 \twoheadrightarrow b_1$ and not attack b_2, then if $a_1 = a_3 = $ in then only b_1 will be out and not both of $\{b_i\}$. Again not good.

If a_3 also attacks b_2 then if $a_3 = $ in and $a_1 = a_2 = $ out then we get that both $\{b_i\}$ are out, again not correct.

OK, so what do we do now? It is natural to follow a continuation idea.

Let us form a set $\{b_1, b_2\}$ and let all a_1, a_2, a_3 each attack the set.

Namely

$$a_1 \twoheadrightarrow \{b_1, b_2\}$$
$$a_2 \twoheadrightarrow \{b_1, b_2\}$$
$$a_3 \twoheadrightarrow \{b_1, b_2\}$$

Let us say that to attack a set is to attack one of the members. We again have a problem.

- *If each a_i says explicitly which member it attacks we are back to the previous dilemma.*

- *If a_i does not say which member it attacks then we cannot prevent all a_i attacking b_1 and we gain nothing.*

- *If we say all attackers must attack separate members and otherwise (if there are no more un-attacked members) not attack, then this is a fancy language basically repeating the much simpler original numerical $b : 2$ representation.*

What we want is a better representation for the traditional Dung network, which with a small change *will represent the many lives generalisation.*

To summarise:

- **We need an inspiration**

This is a problem for another paper. It is possible to do using linear logic and using ideas from paper [40].

(4). Comparison with abstract dialectical framework (ADF) [14]–[16], [23] and [39]. ADF is a powerful system which can express practically anything you can throw at it. It can express the joint attacks easily. For example the condition on b of Figure 12 can be written as

$$b \leftrightarrow (\neg a_1 \lor \neg a_2) \land \neg a_3.$$

So our adding many lives to ordinary Dung style networks can be easily added to ADF. Take for example Figure 13. ADF simply uses its functions to give the kind of semantics required. The simplest way of doing it is to allow ADF to talk about nodes with numerical annotations, (in other words the basic units are pairs (node, number)), and allow it to associate conditions on the combinations of attacking nodes again with their numerical annotations. So we can say (see Figure 13):

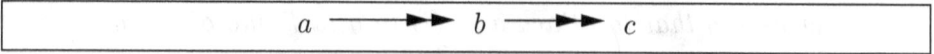

$$a \longrightarrow\!\!\!\!\rightarrow b \longrightarrow\!\!\!\!\rightarrow c$$

Figure 15

- The new annotation of node $(b : 2)$ is $(b : 99)$, if its attackers all have annotations which are not prime numbers.

Why prime numbers? Well ADF is mathematical. In mathematics we only increase what we can do and not try to insist on things we should not be allowed to do!. If we insist that ADF uses only (in, out, undecided) annotation, then it cannot simulate the many lives annotation. The discussion of Figure 10 in comparison (3) above holds in this case as well.

More interesting is the other direction, can the many lives approach simulate ADF or fragments of it? The answer is that the many lives model is monotonic, namely

- E attacks x and $E \subseteq E'$ then E' attacks x.

ADF does not have this restriction. We can write in ADF an acceptance condition which is not monotonic, for example:

- E attacks x if the number of elements in E is even.

(5). Comparison with papers with the idea of graduality, e.g. [11]–[13]. These papers and many others like them want to pay attention to the number of attackers on x and the number of attackers on attackers, etc. Paying attention to such distinctions allows us to say that some nodes are "more in" than other nodes. For example in Figure 15, a is "more in" than c. This is a different theme but we can use many lives as another instrument to measure this feature. This is best explained by an example. Consider Figure 15. If we give all nodes one life we get:

One life: $a = 1, b = 0, c = 1$
Two lives: $a = 2, b = 1, c = 1$
Three lives: $a = 3, b = 2, c = 2$.

So if the network has n nodes, go in sequence up to n lives and see what you get. The differences will show in the sequence. Like the difference between a and c in the sequence for Figure 15.

(6) Comparison with numerical argumentation. We really need not compare our Chapter with any of the purely technical numerical argumentation publications. Our Chapter is not intended as such. The relevant papers for comparison are really the ones described in items 1 and 2 above. There is a comparison from the numerical point of view in our paper [10]. Nevertheless, we are including some comparative discussion (of papers [19, 20, 21, 23, 25, 33], and [34] in Appendix 2).

6 Methodological discussion and conclusion

This section summarises what is going on and indicates what is more to be done. Given a geometric network of the form (S, R) with $S \neq \varnothing$ and $R \subseteq S \times S$, we propose that we view it as a base/a carrier to be used to define an argumentation network for a target application. We should not be locked into the view that R is the network attack relation. R could be some important relation in the target application from which we derive the relevant attacks and supports. Once we accept this view about a network (S, R), we can turn it into an argumentation network in many different ways, by adding extra structure to it and defining the basic argumentation notions on top of the structure. Our paper [38] discusses some such specific examples obtained by using the relation R, meanwhile, let us give an example generalising the many lives approach.

Let us add a numerical function

$$\mathbf{f} : S \mapsto [0, 1]$$

i.e. for each $x \in S, \mathbf{f}(x)$ is a real number $1 \geqslant \mathbf{f}(x) \geqq 0$.

The system $(S, R, \mathbf{f})$ is very general. $\mathbf{f}$ can be interpreted i many ways. It can be:

- Fuzzy value, introducing argumentation networks.

- Probabiity function, introducing probabilistic argumentation

- Measure of strength, introducing numerical argumentation

- A $[0,1]$ solution to some equations in the equation approach generating a many lives function

$$M(x) = \frac{x}{1-x} = \frac{1}{1-1/x}$$

$\mathbf{f}(x) = 0$ means 0 lives $(M(x) = 0)$.

$\mathbf{f}(x) = 1$ means immortal lives $M(x) = \infty$.

Further note that we can add other types of functions, not necessarily numerical. For example we can add a structured logic function $\Delta(x)$, giving for each $x \in S$, a logical theory or a formula $\Delta(x)$ from some logic **L**. **L** could be classical logic or intuitionistic logic of logic programming or some nonmonotonic logic. If we do that we need to define what it means for one logical theory to attach another logical theory. this can lead to systems like Aspic$^+$ or Assumption based argumentation or argumentation as information input, etc.

Let us continue with the numerical function $\mathbf{f}$ and let us turn to the system $(S, R, \mathbf{f})$ into an argumentation network. We need to define some additional basic notions. Let us elaborate using a question and answer dialogue:

Q1. What can be attacked?

Answer 1:

- Individuals $x \in S$ or

- subsets $Y \subseteq S, Y \neq \varnothing$

- Geometrical arrows (i.e. elements of R, giving rise to a higher level attacks)

Q2: Who are the attackers?

Answer 2:

- Subsets $E \subseteq S \ E \neq \varnothing$ or

- individuals $x \in S$.

- Geometrical arrows (i.e. elements of R, giving rise to a higher level attacks attacking other attacks)

Q3: What is the nature of the attacks?

Answer 3: Let us take three types of attacks $\alpha_{\mathbf{a}}, \alpha_{\mathbf{p}}, \alpha_{\mathbf{d}}$, such that the following holds:

- If $Y \subseteq S, \alpha_{\mathbf{p}}$ attacks x then $Y \alpha_{\mathbf{d}}$ attacks x.

- If $Y \subseteq S, \alpha_{\mathbf{d}}$ attacks x then $Y \alpha_{\mathbf{a}}$ attacks x.

Note to Answer 3: Note hat we can define the attacks in many different ways. Some we already did using the many lives functions M, N, G. We can define many other types of attacks using the function **f**. We shall give different examples in our paper [38], namely Geometrical Attacks, (these are attacks defined in first or higher order logic of the language of (S, R)).

The attacks on x make $\mathbf{f}(x)$ smaller.

Q4: What corresponds to the notion of protecting?

Answer 4: We use the attack $\alpha_{\mathbf{p}}$. E protects x if $E \ \alpha_{\mathbf{p}}$ attacks any set Y which $\alpha_{\mathbf{a}}$ attacks x. There can be more variations on this, see Appendix 2.

Q5: What corresponds to the notion of conflict freeness of a set E?

Answer 5: We use $\alpha_{\mathbf{d}}$. E is at peace (i.e. $\alpha_{\mathbf{d}}$ conflict free) if it does not $\alpha_{\mathbf{d}}$ attack its elements.

Q6: How do we define semantics for $(S, R, \mathbf{f})$?

Answer 6: The general notion of semantics is a function $\mathbb{F}$, giving for a given system $(S, R, \mathbf{f})$ a family of new systems $\{(E_i.R_i, \mathbf{f}_i)\}$. There are four main methods of defining semantics:

1. The Dung like set theoretical method.

2. Translating into classical, intuitionistic, modal or some other logic and taking suitable models (in the semantics of that logic) and translating back.

3. Using the equational approach. Generating equations from $(S, R, \mathbf{f})$ and solving them and the solutions generating semantics.

4. Giving direct algorithms on $(S, R, \mathbf{f})$, using the attacks to run around (S, R) and redefining new $(E_i, R_i, \mathbf{f}_i)$. This is the Algorithmic Approach.

Given a family of $\{(E_i, R_i, \mathbf{f}_i)\}$ we an seek to prove completeness theorems, answering the question of which methods from (1)–(4) can produce this family.

For example we can ask for the case of traditional Dung extensions, which equations and which algorithms can yield exactly all the preferred extensions?

Q7: What does this Chapter do?

Answer 7: We are inspired by the many lives phenomena in the complaints area to define a function $\mathbf{f}$ and look at suitable and compatible networks $(S, R, \mathbf{f})$ with attack functions and give a set theoretical semantics. As a result of looking at the multiple complaints application area we reached the conclusion that an argumentation network must have 3 different attack relations (all geometrically defined using R) in increasing strength, all participating in defining extensions.

We compared with other related papers in the literature.

Q8: What is your opinion of the significance of this Chapter?

Answer 8: I think there are two aspects to our contribution:

1. Introducing many lives with a view of applications to the "complaints" areas of applications as well as many other notions motivated by the many lives idea.

2. It seems that it is time to introduce some methodological order to the chaotic jungle of formal argumentation publications. The recent publication of volume 1 of the *Handbook of Argumentation* and the planned material for volume 2 together (I personally believe) the methodological view of this section and the stimulus generated by the Grossi and Modgil papers is the starting trigger point.

Q9: What is your next paper in this area?

Answer 9: We completed paper [27] (also the next Chapter in this book), dealing with the temporal aspects of multiple victims complaining one after another as the case develops in the media. We are writing [31]. We observed that when an alleged offender is attacked by one or two complaints, all of a sudden many more complaints come forward. See the story of [32] for a very famous example. Our paper [27] cannot deal with that. Ordinary traditional temporal logic or any variations based on it cannot deal with sudden avalanche of simultaneous triggered changes. The proper way of modelling this is via Reactive Attacks. We envisage a set S of offenders and victims and a binary relation R on S of inactive attacks from victims to offenders. When a victim x comes forward to complain about offender y, the attack xRy becomes active. The other attacks are dormant and are activated when several attacks coming to life. Exactly how many complaints trigger the activation of all the others depends on circumstances. Temporal logic is not the right way to handle this. What is happening is that because of the first one or two brave attackers coming forward all the other gain courage to join and attack.

For the idea of Reactivity and its applications see [29]. For the details of reactive grammars we want to use to handle this type of formal argumentation see [30]. The paper itself will be [31].

Acknowledgements

We thank Pietro Baroni, Matthias Thimm, Leon van der Torre and Bart Verheij, as well as the three anonymous referees of the *Journal of Argument and Computation* and the *IfCoLog Journal of Applied Logics*, and the referees for this Springer Volume for most valuable comments.

References

[1] D. Gabbay, G. Rozenberg and L. Rivlin. Reasoning under the influence of universal distortion. *Ifcolog Journal of Logic and Their Applications*, 4(6), July 2017, pp. 1769–1900.

[2] D. Gabbay and Gadi Rozenberg. Reasoning schemes, expert opinion and critical questions: Sex offenders case study. *Ifcolog Journal of Logic and their Applications*, 4(6), July 2017, pp. 1687–1769.

[3] P. Baroni, M. Giacomin, G. Guida. SCC-recursiveness: a general schema for argumentation semantics. *J. Artificial Intelligence*, 168(1):162–210, 2005.

[4] D. Gabbay. Meta-logical Investigations in Argumentation Networks. College Publications, 2013, 770pp.

[5] H. Prakken and G. Vreeswijk. Logics for Defeasible Argumentation, in Handbook of philosophical logic vol 4, D Gabbay and F Guenthner, editors, Springer 2002, Pages 219-318

[6] P M Dung, On the acceptability of arguments and its fundamental role in nonmonotonic reasoning, logic programming and n-person games Artificial Intelligence Volume 77, Issue 2, September 1995, Pages 321-357

[7] M. Caminada and D. Gabbay. A logical account of formal argumentation. Studia Logica, 93(2-3): 109-145, 2009.

[8] D. Grossi and S. Modgil. On the graded acceptability of arguments. In *Proceedings of IJCAI'15*, tAAAI Press, 2015, pp. 868–874.

[9] D. M. Gabbay Equational approach to argumentation networks *Argument and Computation* 3 (2-3):87 - 142 (2012)

[10] D. Gabbay and O. Rodrigues, An Equational Approach for Ranking-Based Semantics, original version 2015, short version in *COMMA 2016*, current version March 2018

[11] C. Cayrol and M.-Ch. Lagasquie-Schiex. Graduality in argumentation. *Journal of Artificial Intelligence Research*, 23:245–297, 2005.

[12] S. Egilmez, J. Martins, and J. Leite. Extending social abstract argumentation with votes on attacks. In *Proc. 2nd Int.Workshop on Theory and Applications of Formal Argumentation*, pages 16–31, 2013.

[13] C. Cayrol and M.-C. Lagasquie-Schiex. Gradual acceptability in argumentation systems. In *Proc 3rd CMN (International workshop on computational models of natural argument)*, Acapulco, Mexique, pages 55–58, 2003.

[14] Gerhard Brewka and Stefan Woltran. Abstract dialectical frameworks. In *Proc. KR'10*, pages 102–111. AAAI Press, 2010.

[15] G. Brewka, S. Ellmauthaler, H. Strass, J. P. Wallner, and S. Woltran. Abstract dialectical frameworks revisited. In *Proceedings of the Twenty-Third international joint conference on Artificial Intelligence* (pp. 803–809). AAAI Press. (2013, August).

[16] S. Polberg. Extension based semantics of abstract dialectical frameworks. In STAIRS 2014: *Proceedings of the 7th European Starting AI Researcher Symposium* (Vol. 264, p. 240). IOS Press. (2014, August).

[17] Nielsen, S. H. and Parsons, S. 2006, A generalization of Dung's Abstract Framework for Argumentation: Arguing with Sets of Attacking Arguments. in N Maudet, S Parsons and I Rahwan (eds), *Argumentation in Multi-Agent Systems (ArgMAS)*, 2006, pp 54–73

[18] D. Gabbay and M. Gabbay. Disjunctive attacks in argumentation networks, *Logic journal of IGPL* Volume 24, Issue 2, Pp. 186–218. http://jigpal.oxfordjournals.org/content/early/2015/09/10/jigpal.jzv032.full.pdf

[19] Sylvie Coste-Marquis, Caroline Devred, Pierre Marquis:Symmetric Argumentation Frameworks. *ECSQARU 2005*: 317-328

[20] Claudette Cayrol, Caroline Devred, Marie-Christine Lagasquie-

Schiex:Acceptability semantics accounting for strength of attacks in argumentation. *ECAI 2010*: 995-996

[21] Paul E. Dunne, Anthony Hunter, Peter McBurney, Simon Parsons, Michael Wooldridge:Weighted argument systems: Basic definitions, algorithms, and complexity results. *Artif. Intell.*, 175(2): 457-486 (2011)

[22] Gerhard Brewka, Stefan Woltran: GRAPPA: A Semantical Framework for Graph-Based Argument Processing. *ECAI 2014*: 153-158

[23] Joerg Puehrer:Realizability of Three-Valued Semantics for Abstract Dialectical Frameworks. *IJCAI 2015*: 3171-3177

[24] Davide Grossi and Sanjay Modgil. On the Graded Acceptability of Arguments in Abstract and Instantiated Argumentation, full paper , preprint February 23, 2018

[25] Diego C. Martinez, Maria Laura Cobo, and Guillermo Ricardo Simari, A Petri Net Model of Argumentation Dynamics, in U. Straccia and A. Cal'o (Eds.): *SUM 2014*, Springer LNAI 8720, pp. 237–250, 2014

[26] Sylvan Shalom case 2015: `https://www.timesofisrael.com/pm-shalom-made-the-right-decision-to-quit-over-sex-scandal/` (accessed on April 18, 2018, 1000 hours, Luxembourg time)

[27] D. Gabbay, G. Rozenberg and Students of CS Ashkelon. Temporal aspects of many lives. First Draft January 2017. Current final draft is the next Chapter in this book.

[28] Pietro Baroni, Dov Gabbay, Massimiliano Giacomin and Leon van der Torre, eds. *Handbook of Formal Argumentation*. 1028 pages. College Publications, 2018.

[29] Dov Gabbay. *Reactive Kripke Semantics,Theory and Applications*, 450 pp, Research Monograph Springer 2013, ISBN 978-3-642-41388-9 ISBN 978-3-642-41389-6 (eBook)

[30] H. Barringer, D. Rydeheard, D. Gabbay. (2014) Reactivity and Grammars: An Exploration. In: Dershowitz N., Nissan E. (eds) *Language, Culture, Computation. Computing - Theory and Technology*. Lecture Notes in Computer Science, vol 8001. Springer, Berlin, Heidelberg DOI `https://doi.org/10.1007/978-3-642-45321-2_6`

[31] D. Gabbay, Lendert van der Torre, Howard Barringer and David Rydeheard. Reactive virtual attacks in formal argumentation, 2018. Submitted to the *Journal of Applied Logics*.

[32] `https://en.wikipedia.org/wiki/Harvey_Weinstein_sexual_abuse_allegations`, accessed 1200 hours UK time , on August 22,

2018

[33] H. Barringer, Dov Gabbay and John Woods. Temporal Dynamics of Argumentation Networks In *Volume Dedicated to Joerg Siekmann*, D. Hutter and W. Stephan, editors. Mechanising Mathematical Reasoning, Springer Lecture Notes in Computer Science 2605, pp. 59-98, 2005.

[34] H. Barringer, Dov Gabbay and John Woods. Temporal, Numerical and Metalevel Dynamics in Argumentation Networks, *Argument and Computation*, 2012, vol 3 issues (2-3), pp 143-202

[35] https://en.wikipedia.org/wiki/T-norm (accessed September 17, 2018, 0515 UK time)

[36] D. Martinez, A. Garcia, G. Simari. An abstract argumentation framework with varied-strength attacks. In: *Proceedings of the 11th International Conference on Principles of Knowledge Representation and Reasoning (KRŎ08)*, 2008, pp 135-143.

[37] T. Bench-Capon. Value Based Argumentation Frameworks https://arxiv.org/ftp/cs/papers/0207/0207059.pdf (Submitted on 15 July 2002)

[38] D Gabbay. Geometric Attacks in Argumentation Networks. (Paper 627, September 2018.)

[39] Gerhard Brewka, Stefan Ellmauthaler, Hannes Strass, Johannes P. Wallner, Stefan Woltran; Abstract Dialectical Frameworks. An Overview, *IFCoLog Journal of Logics and Their Applications*, Vol. 4 No. 8 2017 , pp 2263–2317

[40] D. Gabbay and M. Gabbay. Argumentation as information input. 2015. Short version published in *Proceedings COMMA 2016, Computational Models of Argument*, Pages 311–318 DOI10.3233/978-1-61499-686-6-311 Volume in Series Frontiers in Artificial Intelligence and Applications IOS press Volume 287: Full version published in the College Publications G. Simari Tribute: *Argumentation-based Proofs of Endearment: Essays in Honor of Guillermo R. Simari on the Occasion of his 70th Birthday* Paperback. 7 Nov 2018. Editors Carlos I Chesnevar, Marcelo A Falappa, Eduardo Ferme, Alejandro J. Garcia, Ana G. Maguitman, Diego C. Martinez, Maria Vanina Martinez, Ricardo O. Rodriguez, Gerardo I. Simari pp 145-197.

A   Appendices

Appendix 1: Discussion of $d[m, n, k]$

It is useful to introduce a familiar story as an example for $d[m, n, k]$, the story of the party. To help us appreciate the story let us distinguish three types of attacks for a traditional network (S, R) illustrating $d[1, 2, 2]$. The examples deals with traditional attack and protect but changes the notion of conflict free:

Definition A.1. *Let (S, R) be an argumentation network, We define two notions of attack of a subset Y of S on a node x using R as follows:*

- *$Y\alpha_{\mathbf{a}}$-attacks x of a subset Y of S on a node x yRx. This is the traditional Dung attack notion.*

- *$Y\alpha_{\mathbf{d}}$ (respectively $Y\alpha_{\mathbf{p}}$)-attacks x iff for two different y_1 and y_2 in Y we have that y_1Rx and y_2Rx (respectively same condition for $Y\alpha_{\mathbf{p}}$).*

Example A.2. *We are planning a party and we have a set S which is the maximal set of all relatives friends, colleagues, etc. who can be invited to the party. The problem is that some of them do not get along/hate some others. So we have a relation R, where xRy (which we might denote by $x \twoheadrightarrow y$) means that if x is invited, y must not be invited. Let us also assume that for the sake of fairness, if any candidate has no people objecting to him, the candidate should be invited. For example if the party is a diplomatic event, then certainly all diplomats should be invited unless there is a problem. With this view the problem becomes an ecological kind of network (if you are not attacked you are alive). With this understanding we get here a traditional argumentation network with attack relation R. The complete extensions are possible groups of people we can invite.*

*These are the traditional Dung extensions obtained by using the **a**-attack notion. If the party is a wedding, we can invite whom we please. So even if someone is not objected to, we can choose not to*

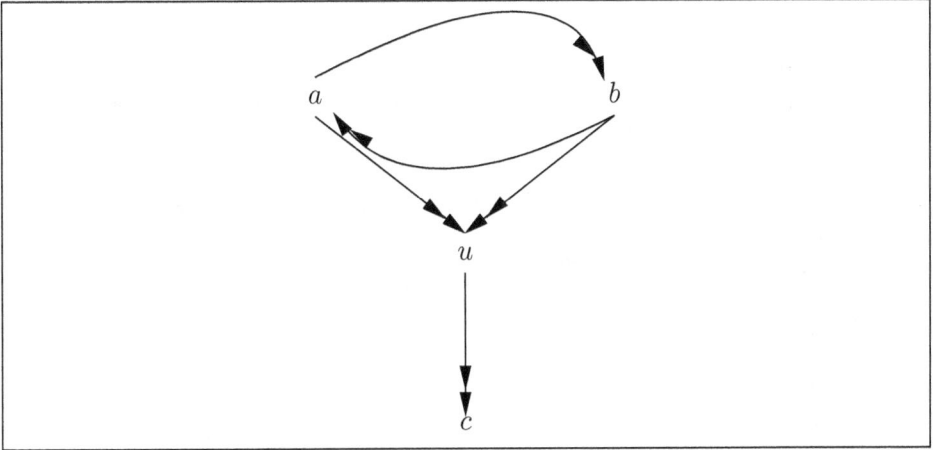

Figure 16

*invite him. So if we have $S = \{a, b, c\}$ with $a \twoheadrightarrow b \twoheadrightarrow c$, we can invite b and not invite a and c, using the symmetrical closure of the given notion of attack. We cannot get this $\{b\}$ extension if we use the **a** notion of attack the R attack only. We do need its symmetric closure. However, other problems may arise at a wedding scenario. Let $S = \{a, b, c, u\}$ with $R = \{(a, b), (b, a), (a, u), (b, u), (u, c)\}$. Think of xRy to mean x Hates y. So we have a married couple/parents $\{a, b\}$ who hate each other and hate the uncle u and a child c who is hated by the uncle u. The wedding is of the child. See Figure 16: Common sense wants to invite the child c and the parents a, b.*

Traditional Dung extensions semantics can invite only one parent from $\{a, b\}$ and invite c. Even if we use the Grossi Modgil $d[m, n]$ approach, the set $\{a, b\}$ is not conflict free. If we use our proposed approach with the attacks of Definition A.1, the parents set will be at peace, the uncle u will be out and c will be invited.

Appendix 2: Comparison with argumentation networks with numerical annotations

Our starting point is a network (S, R), with additional annotations to the nodes $L(x), x \in S$, where $L(x)$ is any labelling function on S

giving values from some labelling domain.

The labels $L(x)$ can be used in the computation of extensions. The labels can restrict or supplement R in some way, or can be transferred to extensions. The following are some examples:

1. $L(x)$ can be either "high" or "low" with the understanding that any xRy is disregarded if $L(x) =$ "low" and $L(y) =$ "high". (See [37].)

2. $L(x)$ can be a numerical value (say in $[0, 1]$) and we can disregard any set $E \subseteq S$ provided that $\Sigma_{x \in E} L(x) \leqslant \varepsilon$ where $0 \leqslant \varepsilon < 1$. (See [21].)

3. $L(x)$ can be the many lives function $M(x)$ as it is used as in this Chapter.

4. $L(x)$ can be a probability distribution on S and it can be transferred as a probability distribution to extensions.

5. Note that in examples 2 and 3 the numerical value is used to modify the attack relation R and is not used as strength of attack. We could however use $L(x)$ as part of the attack itself. Assume we have the nodes $y_1, \ldots, y_k$ attacking the node x. The nodes z in S are annotated with a numerical number $L(z)$. In the general case $L(z)$ can be any real number. It is convenient to assume that the number $L(z)$ is in the interval $[0, 1]$. We can thus adopt the view that $L(z)$ represents the strength of attack which z can generate. 1 is full strength and 0 is no strength at all.

6. The most general point of view of numerical attacks was taken in papers [33] and [34]. In these papers we considered the most general uninterpreted case of numerical attacks. Assume that y attacks x, (that is $y \twoheadrightarrow x$) with $L(y) = a$ and $L(x) = b$. Mathematically what we have here is two numbers, with one number a attacking another number b, namely $a \twoheadrightarrow b$ and we asked what number b' we get as a result of this attack. The papers discuss various possibilities, among them the obvious $b' =$

$b(1-a)$. If we interpret $a = 1$ as true (or full strength) an $a = 0$ as false, we get compatibility with Dung like attacks. There is a connection with fuzzy logic and t-norms [35].

7. Note that the number annotation $L(z)$ is used in the attack. Compared with the many lives approach, the number $M(z)$ is not used in the attack. So if y has many lives $M(y) = 100$ and y is attacking x with $M(x) = 2$ then the attack strength of y on x is just 1. The number $M(y)$ reflects resilience to attacks on y and not strength of y as an attacker on x.

8. We can imagine a prosecutor trying to decide, given the complaints of $y_1, \ldots, y_k$ on x, whether to press charges on x by putting forward as evidence all the complaints of $y_1, \ldots, y_k$ or perhaps (expecting a counter attack from x) by putting forward only the more resilient y's (with $M(y)$ a large number). In this case we are using $M(y)$ as a weight and we exclude the attack of y if $M(y)$ does not pass a threshold. This is what paper [21] does, as described in item (2) above. Paper [21] also contains a comparison with papers [33] and [36].

9. The papers [33, 34] and [20, 21, 36] all use the numerical annotation in the attack. The qualitative difference between these papers and our current many lives paper is manifested technically in the handling of loops. Loops are welcome and are easier to handle in numerical attacks context, as they naturally lead to fixed point equations. Consider the situation of item (6) above, namely where $y \twoheadrightarrow x$ and assume also that $x \twoheadrightarrow y$. Thus the attack relation is symmetrical. We use symmetrical R in Ecologies, where different species attack one another and we seek to identify states of equilibrium. (For analysis of networks with symmetrical R, see [19]). Also note that when an offender x is attacked by a victim y, the offender immediately attacks back, so the relation is always symmetrical). When numerical strength annotations are present, we solve equations. For example for x

and y above we solve the two equations

$$b' = b(1 - a') \text{ and } a' = a(1 - b')$$

we get that the new equilibrium labels are

$$b' = b(1 - a)/(1 - ab) \text{ and } a' = a(1 - b)/(1 - ba)$$

10. The annotations can be quite complex, as in the paper [25]. We must be careful, however, to keep our systems closely related to application areas and not embark on pure mathematical extensions. I think there is a connection with [31].

Chapter 5

Evolutionary Temporal Logic for Modelling Many-Lives Argumentation Networks

1 Introduction

This Chapter deals with the temporal aspects of the many-lives argumentation networks. The many-lives idea comes from modelling the reasoning behaviour of sex offenders, which required argumentation systems where each argument x has a natural number $M(x)$, indicating how many live attackers are needed to ensure that x is out. The temporal aspect associated with such applications (in general: how many complaints are required to take x out) is that the attackers come at different times. It seems that traditional temporal logic is unable to properly deal with such behaviour and a new type of temporal logic is required.

We call it "evolutionary temporal logic". Thus argumentation many-lives systems inspire new developments in temporal logic.

This introductory section explains the ideas and results of this Chapter. It describes two related but independent components in formal argumentation. The idea of many-lives and the idea of evolutionary temporal argumentation. From the formal point of these two ideas are independent, from the applicational/pragrmatic point of view they are strongly related in the sense that they appear strongly intertwined in a major application area of reasoning and modelling the argumentation logic of sex offenders.

The two components are the following:

Many-lives component. In formal argumentation developed and studied Dung style [10, 11], there is the notion of attack of argument x on argument y (notation $x \twoheadrightarrow y$) and the property that if one attacker (say x) is live ("in") then the target y is dead ("out"). The notion of many-lives is an index given to any target argument y (a notation $M(y)$) which requires that at least $M(y)$ live ("in") attackers $x_1 \twoheadrightarrow y, \ldots, x_{M(y)} \twoheadrightarrow y$ on y to be able to render y dead ("out").

Evolutionary temporal logic component. Intuitively for the purpose of this Introduction, think of a time sequence of finite logical databases, $\Delta_1, \Delta_2, \ldots, \Delta_n, \ldots$ in which the constraints on the logical properties of Δ_{n+1} depends on the nature of $\{\Delta_1, \ldots, \Delta_n\}$. So there must be some algorithmic function $\mathbb{F}(\Delta_1, \ldots, \Delta_n)$ which constrains Δ_{n+1}. Examples will be given in the next subsection.

These two components strongly appear in argumentation, interacting in many ways. The simplest example for such an interaction is that when argument y with $M(y) > 1$ lives is attacked by an "in" argument x (i.e., $x \twoheadrightarrow y$) at time n, then at time $n + 1$ we have $M'(y) = M(y) - 1$.

Such interaction is very common in the area of complaints and sex offender abuse allegations. We know that one complaint is not sufficient to open a formal investigation but in many cases more and more independent complaints show up in time and there is a number M of complaints which will force action to be taken.

1.1 Motivating examples

Example 1.1 (Mr Malkinson Case). *This case is real and actually happened in 2023 (see next Example 1.1).*

Seventeen years ago, in the year 2006, Mr Malkinson was convicted of committing rape and was sentenced to prison, on the basis of circumstantial evidence and one witness. There was no DNA evidence at the time.

This year (2023) new DNA evidence emerged and on the basis of this new evidence, Mr Malkinson was declared not guilty (backwards from the year 2006) and released from prison (in the year 2023).

The law required prisoners to pay rent to the government but if they are guilty and are imprisoned, they do not have to pay rent.

Since in 2023 Mr Malkinson was declared not guilty from 2006 the prison system asked for rent for providing free lodging for him. This outraged public opinion and the law was cancelled retroactively backwards in time.

Example 1.2 (Story of the article in the Daily Mail).
https://www.dailymail.co.uk/news/article-12377133/ Innocent-man-wrongly-jailed-17-years-rape-didnt-commit-WONT-pay-staying-prison-Justice-Secretarys-intervention-following-outrage-shocking-miscarriage-justice.html

Wrongly convicted people WON'T have to pay for staying in prison after shock miscarriage of justice involving man jailed for a rape he didn't commit sparked ministerial intervention. JACK WRIGHT, PUBLISHED: 00:00 BST, 6 August 2023 | UPDATED: 09:27 BST, 6 August 2023.

The innocent man who was wrongly jailed 17 years for a rape he did not commit will not have to pay living costs covering his time in prison following a dramatic intervention by Rishi Sunak's Government. Justice Secretary Alex Chalk KC made the change covering wrongly convicted people with immediate effect on Sunday after the miscarriage of justice case of Andrew Malkinson sparked outrage.

Mr Malkinson spent 17 years in prison for a rape he did not commit, and appeal judges quashed his conviction last week after DNA linking another man to the crime was produced. The 57-year-old expressed concern that the rules meant expenses could be deducted from any compensation payment he may be awarded to cover the costs of his jail term. Downing Street indicated that the Prime Minister believed the deductions were unfair amid demands to drop the charges.

Mr Chalk has now updated the guidance dating back to 2006 to remove them from future payments made under the miscarriage of justice compensation scheme. The reform to eligible cases was broadly welcomed, but there were calls to pay back the money already deducted from wrongly convicted individuals

Let us now describe the flow of events of this story. Since the present Chapter is on Temporal Evolutionary aspects of argumentation networks, we will use argumentation notation and use this example to illustrate the technical details of this Chapter.

Remark 1.3.

1. *An argumenation network with attack and (deductive) support has the form* $(\mathbf{S}, \mathbf{R}, \mapsto)$ *where* $\mathbf{S}$ *is a set of arguments atoms* $\mathbf{R} \subseteq \mathbf{S} \times \mathbf{S}$ *is the attack relation and* $\mapsto$ *is a deductive support relation.* $\mapsto$ *is a subset of* $\mathbf{S} \times \mathbf{S}$. $\mapsto$ *is really a logic, say classical propositional logic provability* $\vdash$.

   We hasten to comment that we are giving here a very special case definition of bipolar argumentation network (i.e., with attack and support) where the support is deductive support. This is sufficient for analysing our example.

2. *To be able to explain/formalise better the Mr Malkinson example, we also add to our argumentation language the classical negation* $\neg$ *and the classical conjunction* $\wedge$.

   So we can write $\neg x$ *and* $x \wedge y$, *when* x *and* y *are arguments. Of course we must impose the restriction that only one of* $\{x, \neg x\}$ *can appear in any network.*

3. *Thus if* x, y, z *are arguments and we write*

   $$(x \wedge y) \mapsto z.$$

   We mean that $(x \wedge y) \vdash z$ *in classical logic.*

   When we write

   $$(x \wedge y) \twoheadrightarrow z$$

   we mean that $\{x, y\}$ *jointly attack* z

   Note that logic is not involved in this attack, the attack is generated from the meaning/content of x, y, z, *as they appear in the application to be modelled.*

Thus for $\{x, y\}$ to force z to be "out", then both x and y must be "in".

Also note that if we have a network with arguments

$$\{\neg x, x \twoheadrightarrow y\}$$

we have by the meaning of $\neg x$ that x must be "out" and therefore y is "in". This is not the traditional Dung language but equivalent to the network

$$x \twoheadleftrightarrow \neg x$$
$$x \twoheadrightarrow y$$

with the choice $\neg x$ to resolve the loop.

4. *Note that our purpose in this subsection 1.1 is only to provide a motivating example for the machinery of evolutionary argumentation networks. We chose a real example from the Daily Mail to formalise. For this we need to present the example as a network in time with attack and deductive support.*

Since the example is a simple real life example we do not need to develop formally the argumentation theory of attack + (deductive) support.

Definition 1.4.

1. *Let the following denote arguments related to the Daily Mail story of Example 1.2.*

    1.1. *GM = Mr Malkinson is guilty of rape*

    1.2. *LPR = The law says that prison residents have to pay rent to the prison authorities*

    1.3. *PM = Mr Malkinson is a resident in prison*

    1.4. *RM = Mr Malkinson has by law to pay rent for his residence in prison.*

1.5a. *By law, resident prisoner must pay rent. This is represented by* $(PM \wedge LPR) \mapsto RM$. *We need jointly PM and LPR to get RM.*

So if the law is changed and we have $\neg LPR$ *then we cannot get RM.*

1.5b. *Guilty prisoners need not pay rent (This is represented in argumenation by the attack arrow* $\twoheadrightarrow$ *in*

$$GM \twoheadrightarrow RM$$
$$(PM \wedge LPR) \mapsto RM$$

1.6. *DNA = DNA was discovered proving that Mr Malkinson did not commit the rape (and is therefore not guilty of rape).*

2. *The following is the time flow of events in terms of the argumentation networks* $(\mathbf{S}, \mathbf{R})$ *available at each time interval.*

2.1. *[2006–2022]*

$\mathbf{S}_{2006-2022}$ = *existing arguments at the time =*
$= \{LPR, PM, RM, GM\}.$

$\mathbf{R}_{2006-2022}$ = *attacks at the time:*

$$\left\{ \begin{array}{c} GM \twoheadrightarrow RM \\ (P \wedge LPR) \mapsto RM \end{array} \right\}$$

- *The extension at the time is* $\{LPR = in, PM = in,$ $GM = in, RM = out\}.$

2.2. *[2003]*

$\mathbf{S}_{2003} = \{LPR, \neg PM, RM, GM, DNA\}.$

$$\mathbf{R}_{2023} = \left\{ \begin{array}{c} DNA \twoheadrightarrow GM \twoheadrightarrow RM \\ (LPR \wedge PM) \mapsto RM \end{array} \right\}$$

- *The extension in 2023 is therefore*

$\{DNA = in, GM = out, LPR = in, PM = out, RM = in\}.$

Remark 1.5. *There is a problem with the description of what happens the minute the DNA proves the innocence of Mr Malkinson in 2023.*

1. *Clearly Mr Malkinson will immediately leave his residence in prison. Thus in argumentation terms PM is no longer an element of $\mathbf{S}_{2023}$.*

   Technically this means that elements of $\mathbf{S}_{t1}$ can leave the network in $\mathbf{S}_{t2}, t_1 < t_2$.

2. *To help us express this fact more clearly we can take out PM from the network and put in instead $\neg PM$.*

   This is a better way of doing it, as we shall see later.

3. *The second problem with the situation in 2023 is that "DNA" attacks "GM", not only in the year 2023, but the attack goes backwards in time since the conviction in 2006.*

   So in 2023, the view of history changes.

   This requires modelling in two dimensional temporal logic. The first dimension is the time of the point of view and the second dimension is the time of the event (according to the point of view). To clarify this point, we need to write the history explicitly.

   We caution the reader that the general theory of two dimensional temporal logic does not require any evolution in the view of history from time t to time $t+1$. In comparison, the view of history at time t in our example depends on the views and data of what happened in earlier times. So evolutionary temporal logic is a special case of two dimensional temporal logic and in fact it is so special that it needs to be formalised directly without any input/use of the theory two dimensional temporal logic.

   In fact the theory of evolutionary temporal argumentation is even more of a special case of the theory of evolutionary temporal logic.

4. *We remark here that for the clarity of the example we assume that in 2024 the law LPR was cancelled (see Example 1.2. The law was actually changed later in 2023 but we do not want to split the year into two parts. So for clarity, we use 2024)*

Definition 1.6.

1. *Let $\mathbb{P}$ be a set of labels denoting points of view. It can be names of people, or moments of time, etc.*

2. *For each $\pi \in \mathbb{P}$, and each moment of time t, let $(\mathbf{S}_t^\pi, \mathbf{R}_t^\pi, \mapsto)$ be a deductive argumentation network (at time t from the point of view of π).*

3. *When $\mathbb{P}$ is also a set of moments of time, we can talk about evolution of the network at a fixed time t_0, through times $s_1, s_2, s_3, \ldots$. We look at $(\mathbf{S}_{t_0}^{s_j}, \mathbf{R}_{t_0}^{s_j}, \mapsto)$ for $j = 1, 2, 3, \ldots$*

Example 1.7. *Let us use Definition 1.6 to trace the evolution in the case of Mr Malkinson.*

$$\mathbf{S}_{2022}^{2022} = \{LPR, PM, RM, GM\}$$
$$\mathbf{R}_{2022}^{2022} = \left\{ \begin{array}{l} GM \twoheadrightarrow RM \\ (LPR \wedge PM) \mapsto RM \end{array} \right\}$$
$$\mathbf{S}_{2023}^{2023} = \{LPR, \neg PM, RM, GM, DNA\}$$
$$\mathbf{R}_{2023}^{2023} = \left\{ \begin{array}{l} DNA \twoheadrightarrow GM \twoheadrightarrow RM \\ (LPR \wedge PM) \mapsto RM \end{array} \right\}$$

We have a problem representing $\mathbf{R}_{2022}^{2023}$.

Intuitively the set of arguments $\mathbf{S}_{2022}^{2023}$ remains the same as $\mathbf{S}_{2022}^{2022}$.

However, we know in 2023 that the DNA argument makes Mr Malkinson not guilty (i.e. $DNA \twoheadrightarrow GM$). So GM must be out. But in the language of 2022, DNA is not present! So how can we see that GM is out?

For this reason we allowed negation $\neg$ in the language. We can write

$$\mathbf{S}_{2022}^{2023} = \{LPR, PM, RM, \neg GM\}.$$

The attacks and supports remain the same

$$\mathbf{R}_{2022}^{2023} = \left\{ \begin{array}{l} GM \twoheadrightarrow RM \\ (LPR \wedge PM) \mapsto RM \end{array} \right\}$$

Since $\neg GM$ is available, RM is not attacked and since in 2022 the w2023 point of view accepts that PM is "in" and the law "LPR" is *also in. We get that M is "in" meaning that Mr Malkinson has to pay rent. This is why public opinion forced the cancellation of the law. Thus from the point of view of 2024 we have $\neg LPR$ present in $\mathbf{S}_{2022}^{2024}$.*

We get that the following

$$(\mathbf{S}_{2022}^{2024}, \mathbf{R}_{2022}^{2024})$$

$$(\mathbf{S}_{2023}^{2024}, \mathbf{R}_{2023}^{2024})$$

and

$$(\mathbf{S}_{2024}^{2024}, \mathbf{R}_{2024}^{2024})$$

cases are the same as the $(\mathbf{S}^{2023}, \mathbf{R}^{2023})$ cases except that the arguement LPR is replaced by $\neg LPR$. So in $(\mathbf{S}_{2022}^{2024}, \mathbf{R}_{2022}^{2024})$ RM is out, because $(LPR \wedge PM) \mapsto RM$ cannot be used to get RM.

Example 1.8. *This example gives a possible imaginary variation on the Mr Malkinson story, a variation which will illustrate what role the many-lives idea can play in the example.*

According to the real story, a DNA test showed conclusively in the year 2023 that Mr Malkinson was not the rapist and he was found not guilty backwards from the year 2006.

Let us consider a different possible scenario where in 2023 a new witness shows up which casts doubt about the 2006 conviction. The prosecutor is not going to hasten and find Mr Malkinson not guilty on the basis of just one new witness. The conviction has inertia and possibly many more witnesses (many-lives) are required.

Let $M = 6$ be a reasonable number of additional new independent and solid witnesses that would force the prosecution to re-open the case and possibly declare Mr Malkinson not guilty.

Imagine then the following variation on the sequence of temporal data:

[2022]   GM, (with $M(GM) = 6$)

[2023]   GM (with $M(GM) = 5$), W1 and $W1 \twoheadrightarrow GM$.

[2024]   GM($M(GM) = 4$), W1, W2 with $W1 \twoheadrightarrow GM; W2 \twoheadrightarrow GM$

$\vdots$

[2028] GM is out ($M(GM) = 0$), and $W1, \ldots, W6$, with

$$W1 \twoheadrightarrow GM$$

$$\vdots$$

$$W6 \twoheadrightarrow GM$$

So only in 2028 will Mr Malkinson be declared not guilty, backwards to 2006.

Remark 1.9. *This is a methodological remark about what is the kind of evolutionary temporal logic we want to use for hte case of argumentation for legal cases.*

We do not need to consider the full two dimensional temporal model. In terms of Example 1.7 we need only the following:

$$\left(\mathbf{S}_{2022}^{2022}, \mathbf{R}_{2022}^{2022}\right)$$

$$\left(\mathbf{S}_{2023}^{2023}, \mathbf{R}_{2023}^{2023}\right)$$
and
$$\left(\mathbf{S}_{2024}^{2024}, \mathbf{R}_{2024}^{2024}\right)$$

In other words, we need to compute

$$\left(\mathbf{S}_{time\ k}^{time\ k}, \mathbf{R}_{time\ k}^{time\ k}\right)$$

for $k = 1, 2, \ldots$, now.

We do not need to know what time $t2$ thought about time $t1$, where $t1 < t2$.

This will simplify our notation considerably and we can use a single graph (argumentation network) where each element $x \in S, y \in S$ and each attack arrow $x \twoheadrightarrow y$ is annotated by the moments of time up to "now" in which they exist. Example 1.1 discusses how to do it for the main Mr Malkinson example, for the years 2022, 2023, 2024.

Remark 1.10.

1. *We are going to give a better notation for the temporal evolution of argumentation network. Before we do that, let us distill the essential features of what was going on in the Mr Malkinson real (with DNA) example.*

2. *In 2022, we had a situation of*

    1. *guilty* ↠ *pay rent*
    2. *guilty*
    3. *law in force and resident* ↦ *pay rent*

In 2023 the DNA evidence was introduced and we got ¬ *guilty because backwards in time we have*

$$DNA \ of \ 2023 \ \twoheadrightarrow \ guilty \ of \ 2022.$$

So we got in 2023 that Mr Malkinson has to pay rent using the rule:

$$law \ in \ force \wedge resident \mapsto pay \ rent.$$

In 2024 the law in force of 2022 was cancelled. Let us represent this as an attack on the law in force by a new argument called "cancellation".

So we have

$$cancellation \ of \ 2024 \ \twoheadrightarrow \ law \ in \ force \ of \ 2023\text{--}2023$$

The above presentation shows that the reason of evolution in time are the backwards in time attacks.

3. *Our model is to timestamp attackes and arguments for the time they are active. Figure 1 shows what we get.*

    Let us try to read this figure by looking at each argument and its time of existence/validity.

    cancellation: [2024]
    DNA: [2023–2024]
    Law in force: [2022–2024]
    Pay rent: [2022–2024]
    Resident: [2022]

Note the following: The argument "Resident" is a statement of fact. It cannot be attacked. The others are legal constructs. They can be changed.

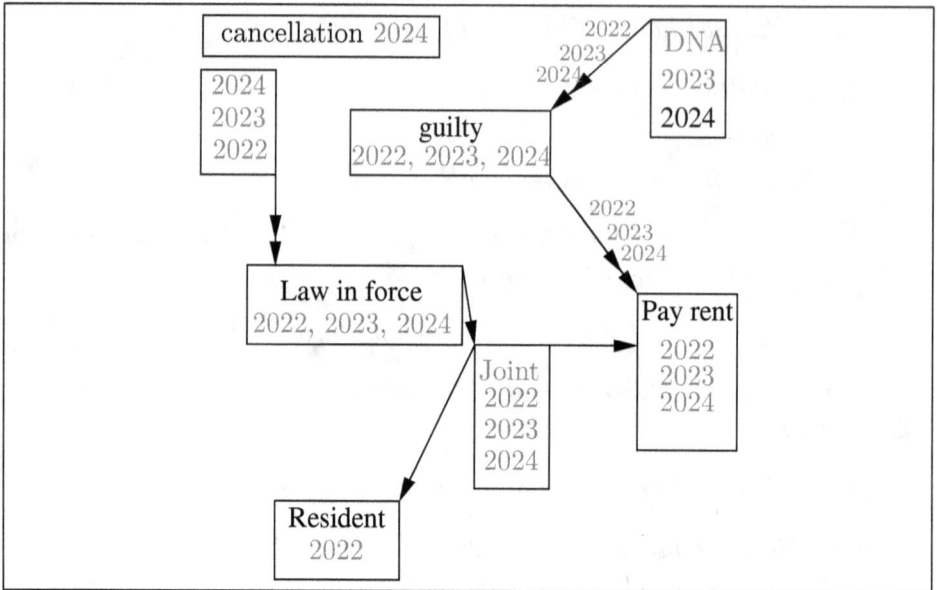

Figure 1

1.2 Technical introduction

Let $G = (S, R)$ be an argumentation network. This means that S is a non-empty set and $R \subseteq S \times S$ is the (attack) binary relation on S. For the purpose of this introduction, let us view $G = (S, R)$ as a directed graph R on the set S.

There are various possible operations we can perform on $G = (S, R)$. One such a general operation is a semantic function **SM** operating on $G = (S, R)$ and extracting from it a subset **CSM** $(G) \subseteq S$ using an algorithm α and a choice function on the properties of R. (See, for example, item 9 of Definition 2.1.) So $\textbf{SM}_\alpha G$ yields a family of subsets of S and $\textbf{CSM}_\alpha(G)$ chooses one of them.

Now we imagine that we have a temporal linear sequence of such networks. Let t be a temporal index running over $\{1, 2, 3, \ldots\}$ and for each such a t let $G_t = (S_t, R_t)$ be the network at time t of the sequence.

We can assume for the moment (but not in general) that $S_t \subseteq S_{t+1}$.

We also get a sequence $\textbf{CSM}_\alpha G_t, t = 1, 2, \ldots$ of subsets.

From the point of view of argumentation, there are three ways of looking at the sequence $\{G_t\}, t = 1, 2, 3 \ldots$.

Timed view. This view (discussed in [4]) regards time as annotating the arguments of S, to form for each $x \in S$, an annotated argument (x, H_x), where H_x is $H_x = \{t | x \in S_t\}$. Let $S_\infty = \bigcup_t S_t$. Let $S^\sharp = \{(x, H_x) | x \in S_\infty\}$.
Define an attack relation $F^\sharp$ on $S^\sharp$ by

- $(x, H_x) R^\sharp (y, H_y)$ iff there exists a $t \in H_x \cap H_y$ such that $x R_t y$.

We thus get a timed network

$$G^\sharp = (S^\sharp, R^\sharp).$$

According to the Timed View, we are interested in studying $G^\sharp$ in a traditional Dung style way.

A major application for this view is in the area of Laws and Regulations, which keep on changing and we need to be confident that they do not contradict one another at any given moment of time.

Modal view. This is the view of [5]. We regard the time flow $\{1, 2, 3, \ldots\}$ as a modal possible world model of the form $(T, <)$, where $T = \{1, 2, 3, \ldots\}$ and $<$ is the relation of smaller than among numbers. With each "possible world" $t \in T$ we associate a classical model $(S_t, R_t) = G_t$. We use modal operators and temporal operators on this system $(T, < S_t, R_t), t \in T$ and possibly define arguments and attacks using this modal language and S_t and R_t.

Evolutionary view. This view regards the future as open and has not happened yet and we can influence it by stipulating some actions and rules. For example, if an element $x \in S_1$ has not attacked any other elements in $t = 1, 2, 3, \ldots, 10$ then we stipulate that x is a "peaceful" element and expect that it not be attacked at time 11.

Let us now continue and focus on the evolutionary view of the sequence $G_t = (S_t, R_t)$.

A major application for this view is in laws and regulations which keep on changing and we need to be sure that they do not contradict one another at any given moment.

Ordinary linear temporal logic can deal with the sequence $G_t = (S_t, R_t), t = 1, 2, 3, \ldots$ viewed as a temporally changing graph or with the sequence $\mathbf{CSM}(G_t), t = 1, 2, 3, \ldots$ viewed as a temporally changing classical model.

This is the modal view. However, if we want to deal with both sequences and deal with the effect that G_{t+1} has on G_t via the algorithm α and the choice function $\mathbf{CSM}$, then we are dealing with a case of evolutionary temporal logic.

This is best explained by an example. Consider the network of Figure 15 (where $\twoheadrightarrow$ denotes attack). The problems with the temporal analysis of this figure are discussed at Example 3.3. For the purpose of this introduction, it is sufficient to say that for the purpose of continuity we must require that the algorithm gives two possible choices, $\{a\}, \{b\}$ for the figure at time one and the choice function for time 1 is allowed to choose one of these two options for example $\mathbf{CSM}$(time 1) $=\{a\}$. If this is the case then for the sake of continuity (if we wish to stipulate continuity) we expect $\mathbf{CSM}$ to yield $\{a, c\}$ for the figure at time 2. However, without the continuity principle it could choose $\{b, c\}$.

This coherence–continuity–rationality postulate becomes important when the graph is annotated. Imagine a graph $\{S, R, M\}$ where M is an annotation function giving a natural number $n = 0, 1, 2, \ldots$ for any $x \in S$. $M(x)$ means how many-lives x has (which algorithmically implies under the principle that each living attacker can take away one life) how many living attackers are needed to "neutralise it" (i.e. to reduce $M(x)$ to 0). The algorithm α for networks with many-lives then works on (S, R, M) taking into account the values of M and of R and yields several choices of a new network of the form (S, R, M_i^*) (i is the index of several choices). The choice function chooses one of them, say $\mathbf{SCM}(S, R, M) = (S, R, M^*)$.

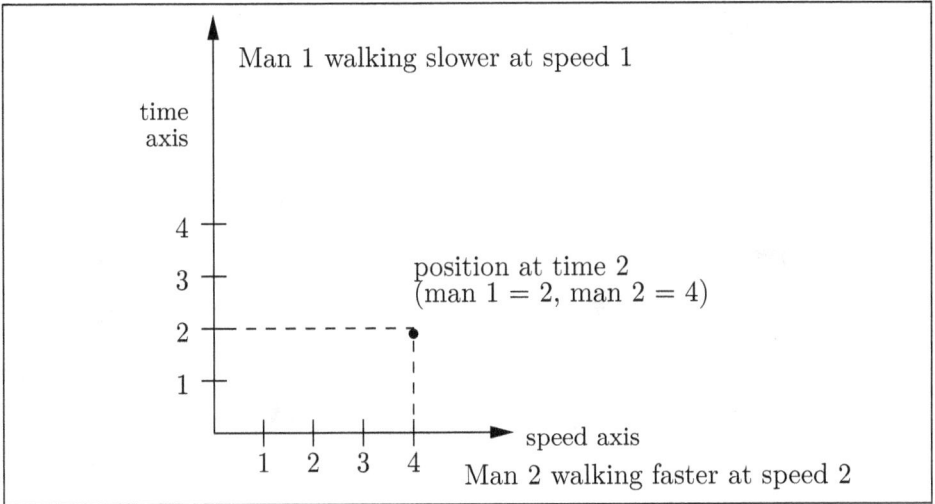

Figure 2

Again, if we have a sequence of (S_t, R_t, M_t), we get a sequence of $\mathbf{SCM}(S_t, R_t, M_t) = (S_t, R_t, M_t^*)$. We need to formulate principles of continuity and choice connecting the two sequences.

A natural view of both sequences is to consider their Tensor product sequence:

$G_1 \otimes \mathbf{CSM}(G_1)$,   Time 1

$G_2 \otimes \mathbf{CSM}(G_2)$,   Time 2

$\vdots$

Evolutionary temporal logic can talk about this sequence.

Example 1.11. *To further illustrate the need for evolutionary temporal logic is the notion of flow product [8]. Figure 2 explains it all. It is the tensor product of the two axis, the horizontal for Man-1 and the Vertical for Man-2.*

There are two men walking at different speeds. Man 1 at speed 1 meter per second and Man 2 at speed 2 meters per second.

Remark 1.12. *We now offer a methodological remark for the perceptive reader. The reader may ask, why introduce a new concept of*

Time 2	$\mathbf{d}(c)$ in(a) out(b)	$\mathbf{d}(c)$ in(a) out(b)	$\mathbf{d}(c)$ out(a) in(b)	$\mathbf{d}(c)$ out(a) in(b)
Time 1	$\neg\mathbf{d}(c)$ in(a) out(b)	$\neg\mathbf{d}(c)$ out(a) in(b)	$\neg\mathbf{d}(c)$ out(a) in(b)	$\neg\mathbf{d}(c)$ in(a) out(b)
History	H_1	H_2	H_3	H_4

Figure 3

evolutionary temporal logic, when all we need is the well-known two dimensional temporal logic?

This has already been discussed in item 3 of Remark 1.5 in the context of the Malkinson Example as well as Figure 1, but it is better to revisit this discussion again.

My answer to that is that the "two" dimensions is deceptive. If you consider the case of Figure 15, it is actually a choice of two paths out of four possible histories. You need a dimension for each possible choice.

Consider Figure 3. We use the notation:

Time 1, Time 2, ... to indicate points in the Time axis. (Compare with Figure 2.) At each of such time points the following predicates get truth values:

$\mathbf{d}(x)$: *x exists*
$in(x)$: *x is in*
$out(x)$: *x is out*
H_i: *temporal history i.*
Time *t:*
Path i: composite path i.

With the principle of continuity only H_3 and H_1 are allowed.

To consider what history is allowed, we need to list all histories! So we do not have here the traditional two dimensional temporal logic.

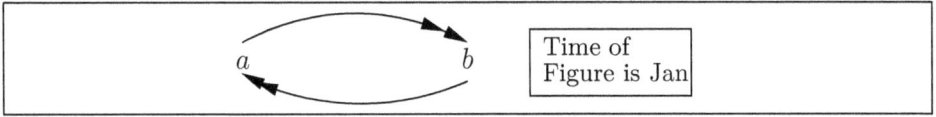

Figure 4

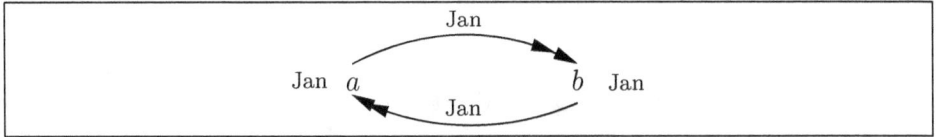

Figure 5

We now offer a better evolutionary temporal language, better than the two dimensional tensor product. We begin with a running example.

Example 1.13. *Consider another example from argumentation. The temporal story is as follows.*

1. *At January there are two arguments attacking each other. The graph for January is Figure 4. We can take the view that the figure has a graph plus a box indicating the time of its existence.*

   So, if the figure came into existence at time "January". This includes the nodes a and b and the attacks a ↠ b and b ↠ a. So we can annotate components by the time each component exists and this way we do not need the box. We get Figure 5.

2. *In February a new attacker came into existence, call it c. The figure for February is Figure 6.*

   We can write the time of existence of components in Figure 7.

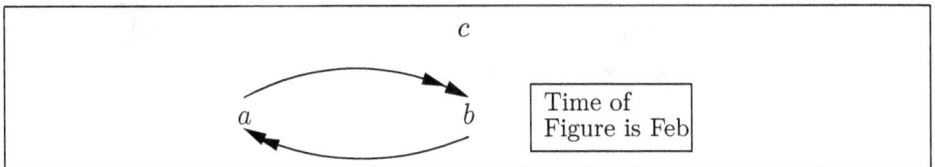

Figure 6

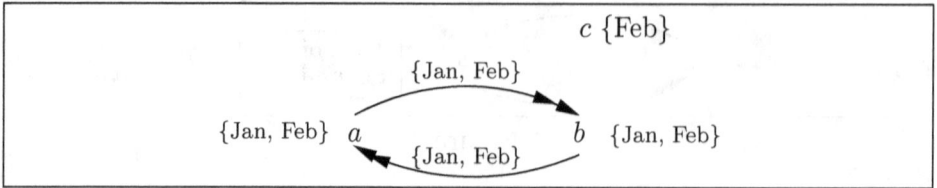

Figure 7

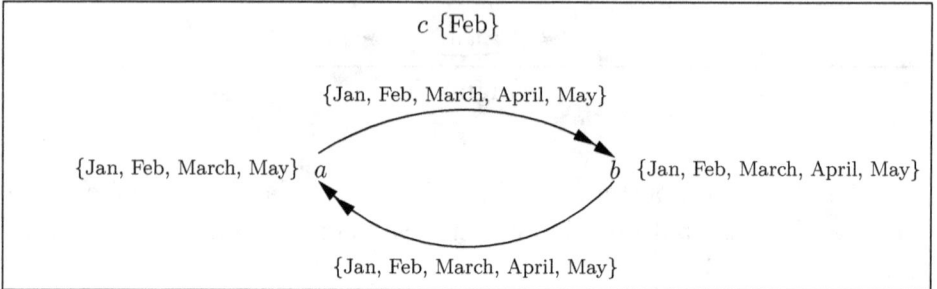

Figure 8

Figures 4 and 6 can be retrieved from Figure 7 by respectively collecting all items labelled Jan (resp. Feb) and forming the respective graphs.

3. *Let us offer you a narrative for the above graph and continue to time of March. In January a attacked b claiming that b raped a. b counter-attacked a by claiming that there was consent. Thus the January Figure 4 was created. In February a new possible witness appeared on the scene, ready to possibly testify against b (attack b). b's lawyers quickly paid c off and in March c was no longer available to testify. Furthermore by April, b settled with a as well and a simply went abroad, but did not withdraw her accusation. Thus in April a was not in the graph. However, public opinion and pressure forced a to come back in May, but did not apply any pressure on c to show up again in May because it has not attacked b. See also Example 3.13*

We get the following May graph annotation in Figure 8.

Definition 1.14.

1. *An evolutionary argumentation network has the form (S, R, T), where (S, R) is a directed graph, namely S is a non empty set of nodes and R is a binary relation on S, and T is a function giving to each element x is $S \cup R$ a set $T(x)$ of numbers from $\{1, 2, \ldots\}$.*

2. *Given an evolutionary argumentation network and a natural number n, we define the network existing at time n as the network $G_n = (S_n, R_n)$ where*

$$S_n = \{x \in S | n \in T(x)\}$$
$$R_n = \{(x, y) \in R | n \in T(x, y)\}$$

Remark 1.15.

1. *Note that according to item 2 of Definition 1.14, we may have in (S_n, R_n) attack arrows without any attacker or target or both.*

2. *Note that the traditional Dung machinery for defining extensions works also with networks of Definition 1.14 despite Item 1 in this Remark.*

3. *Note that item 1 of definition 1.14 may as well define a timed argumentation network, for the Timed View p. 13. However item 2 of this definition already moves towards the Evolutionary View p. 13.*

Example 1.16. *Let us give one more example showing how the sequences of $G_n = (S_n, R_n)$ and $\mathbf{CSM}(G_n)$ interact. We allow for many-lives.*

Consider the following two sequences in Figure 9. The sequences give general graphs where the arguments x also have many-lives $M(x)$

Think of a court case where b is accused of being a sex offender by a. b says a is lying. The court believes b and passes a verdict that b is innocent (option 2). Now at time 2, we get another victim which attacks b. Our questions are the following:

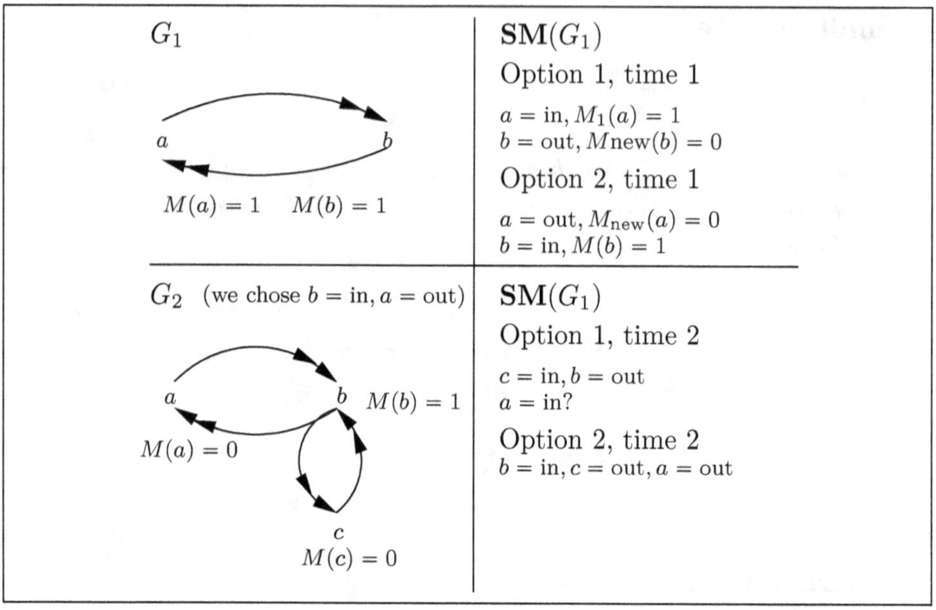

G_1

$M(a) = 1 \quad M(b) = 1$

SM(G_1)

Option 1, time 1

$a = $ in, $M_1(a) = 1$
$b = $ out, Mnew$(b) = 0$

Option 2, time 1

$a = $ out, $M_{new}(a) = 0$
$b = $ in, $M(b) = 1$

G_2 (we chose $b = $ in, $a = $ out)

$M(a) = 0$

$b \quad M(b) = 1$

c
$M(c) = 0$

SM(G_1)

Option 1, time 2

$c = $ in, $b = $ out
$a = $ in?

Option 2, time 2

$b = $ in, $c = $ out, $a = $ out

Figure 9

- *Do we believe c?*

- *Do we rely on the verdict at time 1 and say we decided b is innocent so the case is closed and we do not believe/dismiss c?*

- *What if 10 other $c_1, \ldots, c_{10}$ come at time 2 and attack b? Do we now believe that b is guilty?*

- *Do we give b more lives and believe $\{c_i\}$ only if there are at least two c's? (I.e., $(M(b) = 2$ at time 2)?*

We can formulate new policies depending on what options we choose in the past time to decide what we choose at the present time.

Remark 1.17. *We conclude this Section by explaining why linear evolutionary temporal logic is sufficient for our considerations as opposed to say open future tree temporal logic.*

The reason is very simple. Even when we have branching time, at any point on any branch the evolutionary aspect have to do with looking into the past, and the past on any branch is always linear.

Also, even when we have loops in the past , because we are dealing with argumentation, the loops will be resolved by choice at the time in the past which enable us to continue without this loop into the future.

2 Background and orientation

This section gives background from abstract argumentation and from temporal logic and identifies why traditional temporal logic cannot deal with the many-lives argumentation, evolutionary aspects and explains intuitively our proposed possible solutions.

2.1 Background and concepts from abstract argumentation

This subection presents, for the convenience of the reader, some basic concepts of what we called traditional argumentation theory. Such systems contain attacks only. We refer to such system as Argumentation with Attack only. One can also add support to the system and in this case we get systems of Argumentation with Attack and Support. We shall then explain in what way the systems required for this Chapter depart from the traditional ones.

There are two ways to present the semantics for argumentation with attack, the traditional set theoretical approach and the Caminada labelling approach. For the mapping connections between the two approaches, see [10, 11]. Let us briefly quote the traditional set theoretic approach:

Definition 2.1.

1. *We begin with a pair (S, R), where S is a nonempty set of points (arguments) and R is a binary relation on S (the "attack" relation, we read xRy as x attacks y). In the diagrams and figures we use the notation $a \twoheadrightarrow b$, to denote aRb.*

2. *Given (S, R), a subset E of S is said to be conflict free if for no x, y in E do we have xRy.*

3. *E protects an element $a \in S$, if for every x such that xRa, there exists a $y \in E$ such that yRx holds.*

4. *E is admissible if E is conflict free and protects all of its elements.*

5. *E is a complete extension if E is admissible and contains every element which it protects.*

6. *A subset E is a stable extension if E is a complete extension and for each $y \notin E$ there exists $x \in E$ such that xRy.*

7. *E is the grounded extension if it is the unique minimal complete extension (it exists, see Lemma 2.2).*

8. *E is a preferred extension, if E is a maximal (with respect to set inclusion) complete extension.*

9. *A Semantics is a (metalevel) property **S** of extensions, such as being stable, or being grounded or being preferred or being complete. Thus we can talk about $\mathbb{S}$-Semantics, (stable semantics, grounded semantics and preferred semantics or complete semantics) where we consider only **S**- extensions.*

Lemma 2.2. *For any network (S, R) there exists a grounded extension (which may be empty).*

Proof. This can be proved, using set theoretical methods, see [11]. $\square$

We can also present the complete extensions of $A = (S, R)$, using the Caminada labelling approach, see [11].

Definition 2.3. *A Caminada labelling of S is a function $\lambda : S \mapsto \{in, out, und\}$ such that the following holds.*

(C1) $\lambda(x) = in$, *if for all y attacking x, $\lambda(y) = out$.*

(C2) $\lambda(x) = out$, *if for some y attacking x, $\lambda(y) = in$.*

(C3) $\lambda(x) = und$, *if for all y attacking x, $\lambda(y) \neq in$, and for some z attacking x, $\lambda(z) = und$.*

Lemma 2.4.

1. *A consequence of (C1) is that if x is not attacked at all, then $\lambda(x) = in$.*

2. *Given an extension E let λ_E be defined by $\lambda_E(x) = \{$ in if $x \in E$, out if for some $y \in E$ we have yRx, and undecided otherwise$\}$. Conversely given a λ, define E_λ to be $\{x|\lambda(x) = in \}$.*

3. *Any Caminada labelling yields a complete extension and vice versa.*

4. *Any $\{in, out\}$ Caminada labelling (i.e. with no "und" value) yields a stable extension and vice versa.*

5. *Set theoretic minimality or maximality conditions on extensions E correspond to the respective conditions on the "in" parts of the corresponding Caminada labellings.*

Proof. See [11]. □

Remark 2.5 (Convenient Notation). *In anticipation of future examples and discussions and sometimes for the sake of language and expression or in anticipation of the concept of many-lives , we also use instead of the "in", "out", "extension " words used in Definition 2.1 and Definition 2.3 we use the the words below (think of a cat having nine "lives" and can "survive" 8 "deaths" and still be "alive"):*

(*)   *$x = $ "out", or x is "out", or x is "dead", or $x = $ "out/dead", or x has "0 lives", or simply $x = 0$.*

(**) *$x = $ "in", or x is "in", or x is "alive", or $x = $ "in/alive", or x has "more than 0 lives", or simply $x > 0$.*

(***) *"complete extension" $= $ "Survival (picture)"*

2.2 Traditional discrete linear temporal models

Our starting point is a model for the classical propositional calculus with a set of atomic propositions Q and the evolutionary connectives $\{\neg, \wedge, \vee \rightarrow\}$.

A model for this calculus is a function h giving for each $q \in Q$ a value $h(q) \in \{0, 1\}$. "0" is false ($\perp$) and "1" is true ($\top$).

The assignment function h is arbitrary, and there are no restrictions on h. In fact the set of theorems of classical propositional logic rely on this fact. If we impose restrictions on h, coming possibly from some application area, we may get a more restricted set of theorems. See Remark 2.6, where we give restriction on h coming from the area of argumentation networks.

Remark 2.6. *Note that given an argumentation network* $\mathbb{A} = (S, R)$, *which always has some extensions, we can regard each extension E of A as generating a classical propositional model h_E for the set of atoms $Q_S = S$. For x in S we define $h_E(x) = 1$ iff x is in E (i.e., iff $x = in$). So if x is out or if x is undecided then $h_E(x) = 0$.*

We can use the network A as a restriction on what assignments we can give to the atoms of $Q_S = S$.

We can turn classical propositional logic into a temporal system by adding a flow of time $(T, <)$ and making h time dependent (see [24, 25] for an extensive coverage of this area).

Let us take $T = \{1, 2, \ldots\}$ the set of natural numbers and let $<$ be the usual "smaller than" relation on the numbers. Thus the function h becomes time dependent, giving for each $t \in T$ and $q \in Q$ a truth value $h(t, q) \in \{0, 1\}$. We also write $h_t(q)$, to stress that h is dependent on the time $t \in T$.

In general, we can make any system $\mathbb{S}$ dependent on time in a methodological way. Let $\mathbb{S}$ be a system with components $\{\mathbb{C}_i\}$. We add a parameter $t \in T$ to each of the components, denoting the time dependent component by $\mathbb{C}_{i,t}$ and turning the system $\mathbb{S}$ into the time dependent system $\mathbb{S}_t = \{\mathbb{C}_{i,t}\}$.

Example 2.7. *Let us add a time parameter to an argumentation system of the form* $\mathbb{A} = (S, R)$, *where S is the set of arguments and*

$R \subseteq S \times S$ *is the attack relation. We take a flow of time to be, say* $(\{1, 2, 3, \ldots\},)$, *and let* (S_t, R_t) *be time dependent networks and let* $\mathbb{A}_t = (S_t, R_t)$.

What else does temporal logic do to the time dependent system $\mathbb{S}_t$, thus defined? Let us illustrate for the case of the classical propositional calculus.

Definition 2.8. *A traditional temporal logic starts with a given flow of time of the form* $(T, <)$, *where* T *is the set of moments of time and* $<$ *is the transitive, irreflexive, earlier-later binary relation on* T. *In addition to the classical connectives, Temporal Logic adds temporal connectives to the classical language, for example the connectives* $\{\mathbf{F}, \mathbf{G}, \mathbf{P}, \mathbf{H}, \mathbf{J}, \mathbf{Y}, \mathbf{T}\}$ *with the following truth conditions, where* $t \models \varphi$ *means that the temporal formula* φ *(written using Q) and* $\{\wedge, \vee, \neg, \rightarrow, \mathbf{F}, \mathbf{G}, \mathbf{P}, \mathbf{H}\}$ *holds at* $t \in T$ *under h which is an assignment* $h(t, q)$ *dependent on both time t, and atomic q. Note that h is arbitrary function without restrictions.*

- $t \models_h q$, *if* $h(t, q) = 1$ *for* $q \in Q$

- $t \models_h \varphi \wedge \psi$ *iff* $t \models_h \varphi$ *and* $t \models_h \psi$.

- $t \models_h \varphi \vee \psi$ *iff* $t \models_h \varphi$ *or* $t \models_h \psi$.

- $t \models_h \neg\varphi$ *iff* $t \not\models_h \varphi$.

- $t \models_h \varphi \rightarrow \psi$ *iff* $t \not\models_h \varphi$ *or* $t \models_h \psi$.

- $t \models_h \mathbf{F}\varphi$ *iff for some* $s, t < s$ *we have that* $s \models_h \varphi$.

- $t \models_h \mathbf{P}\varphi$ *iff for some* $s < t$ *we have that* $s \models_h \varphi$.

- $t \models_h \mathbf{G}\varphi$ *iff for all* $s, t < s$ *implies* $s \models_h \varphi$.

- $t \models_h \mathbf{H}\varphi$ *iff for all* $s < t$ *we have that* $s \models_h \varphi$.

- $t \models_h \mathbf{J}\varphi$ *iff we have that* $s \models_h \varphi$, *where s is the first element of the time flow if a first element exists and otherwise* $s = t$.

- $t \vDash_h \mathbf{Y}\varphi$ *iff we have that* $s \vDash_h \varphi$, *where s is the immediately preceding element of t in the time flow (i.e. the Yesterday element) if such an element exists and otherwise* $s = t$.

- $t \vDash_h \mathbf{T}\varphi$ *iff we have that* $s \vDash_h \varphi$, *where s is the immediately following element of t in the time flow (i.e. the Tomorrow element) if such an element exists and otherwise* $s = t$.

Remark 2.9. *Traditional (as opposed to evolutionary) temporal logic is concerned with mathematical and logical properties of temporal models and languages for a variety of flows of time. In other words, the temporal connectives want to talk about variations in time of various components of the system. So for example in the case of a time dependent argumentation network of the form* (S_t, R_t) *temporal logic will talk about time variations in S and R, but it is not meant to, and possibly not able to, talk about extensions and how they vary in time.*

As we shall see later, for temporally dependent such networks, this is a problem because we really do want to talk about extensions and how new and old arguments in time can affect extensions. To be able to do that we need to define what we call "Evolutionary Temporal Logic for Argumentation".

In general talking about variations in time of system components $\mathbb{C}_t$ is quite valuable.

Indeed, evolutionary temporal logics have wide applications in philosophy, general logic, theoretical computer science, artificial intelligence and the formal analysis of language.

However, as we said, traditional temporal logic is not suitable for argumentation (despite papers [26, 27] which followed traditional methodology), for the following two reasons, which are certain features of traditional temporal logic:

($\sharp$1): The models $\mathbf{h}_t$, involved in temporal logic, given for each time t, come from some application area and are fixed. We are not given any details of how they are computed. So formally, our choice of the assignment h_t is arbitrary and given by us in the meta-level.

(♯2):   The future temporal connectives, such as $\mathbf{F}\varphi$ are reduced to the temporal behaviour of φ in the model. They are not considered as atomic, with independent values.[1]

(♯3):   There are no global restrictions on the assignments to atomic q's beyond what is forced by axioms on the connectives. For example the axiom

$$[q \wedge Gq \wedge Hq]$$

forces q to be true at all moments of time, but the axiom

$$\{t|h(t,q) = 1\} \text{ is finite}$$

is not expressible using the connectives and thus cannot be enforced. Similarly see the restrictions mentioned in Remark 2.6.

2.3   The many-lives networks. A quick formal reminder

We give quick definitions of how to define extensions for many-lives argumentation. The exact details are not important for the investigation of the temporal aspects. It is given here just for the record. See [2]. We assume that the networks (S, R) we deal with are acyclic for the purpose of certain inductive definitions.

Definition 2.10 (Labelling annotation for a network).

1. *Let (S, R, M) be an annotated network as follows: (S, R) is a finite acyclic argumentation network.*

   M is a function on S giving for each $x \in S$ a natural number in $\{1, 2, 3, ...\}$ being the number of lives of x (in argumentation terms, to ensure that x will be labelled out, we need at least $M(x)$ nodes e such that eRx and $M(e) > 0$, i.e. e is labelled in, see next item).

---

[1]$\mathbf{F}\varphi$ is true now if φ will be true in the future. So if we know the values of φ we know the values of $\mathbf{F}\varphi$. Compare this with the connective $\mathbf{B}\varphi = $ "I believe φ". I can believe or not believe φ independent of whether φ is true or not.

2. *Let Attack(x), for x ∈ S be the set of all y in S such that yRx holds.*

3. *Let M^* be defined for $x \in S$ using structural induction on the finite acyclic network as the function derived from M, satisfying the implicit equation (*1) and (*2) as follows:*

   (*1) $M^*(x) = M(x)$, if there is no y in S attacking x

   (*2) $M^*(x) = \max\{0, (M(x)$ — the number of elements y in Attack(x) such that $M^*(y) > 0)\}$.

4. *Using M^* we can give Caminada like in, out labelling of the nodes of (S, R, M), following our calculation in item 3 above:*

   x *is out if* $M^*(x) = 0$
   x *is in with remaining lives* $M^*(x)$ *if* $M^*(x) > 0$.

Example 2.11. *For practical examples of many-lives consider the following:*

1. *How many complaints of students against a lecturer can we tolerate before we open a case (hearing) against the teacher? (Probably maybe 5–8, certainly not just one.)*

2. *Driving licence example, see Example 4.1.*

Example 2.12. *We illustrate the computation of M^* as in Definition 2.10 and make an important point about this definition. Consider Figure 10, in this figure $M(z)$ for each node is 1.*
   We now calculate M^*:

- $M^*(x) = M(x) = 1$

- $M^*(b) = 0$, *because it is attacked by x. We do not need to care about $M^*(a)$, (which also attacks b because no matter what $M^*(a)$ is, $M^*(b)$ must be 0). See Remark 2.15.*

- $M^*(c) = 1$. *This is so not because c is not attacked but because it is given that $M(c) = 1$. Had we given $M(c) = 0$ we would have had $M^*(c) = 0$ and not $M^*(1)$ even though $M^*(b) = 0$. See clause (*2) in Definition 2.10.*

The number of lives of each node in this figure (S, R, M) is one

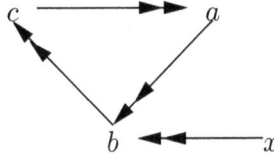

Figure 10

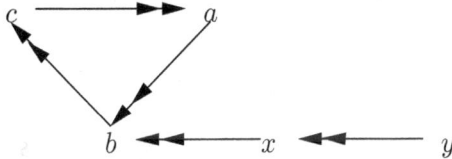

Figure 11

- *Since $M^*(c) = 1$, we get $M^*(a) = M(a) - M^*(c) = 1 - 1 = 0$.*

Example 2.13. *We continue Example 2.12 by modifying Figure 10 into Figure 11:*

First consider the graph of this figure as a Dung argumentation network. From that point of view we have $y = $ in and $x = $ out and therefore the loop $\{a, b, c\}$ stands alone as a three loop with the only extension for the loop is "all undecided". The fact that there is $x = $ out attacking b does not help or make any difference.

Let us now view the element of Figure 11 as having many-lives M, each having one life, i.e. we have $M(a) = M(b) = M(c) = M(x) = (y) = 1$.

Let us now calculate M^* from M for this network using Definition 2.10.

We get

$$M(y) = 1,$$
$$M^*(x) = 0,$$
$$M^*(b), M(b) - M^*(a) - M^*(x),$$
$$M^*(a) = M(a) - M^*(b),$$
$$M^*(c) = M(c) - M^*(b).$$

233

Substituting known values we get the equations:

$$M^*(b) = 1 - M^*(a)$$
$$M^*(a) = 1 - M^*(b)$$
$$M^*(c) = 1 - M^*(b)$$

which yields

$$M^*(c) = M^*(a)$$
$$M^*(b) = 1 - M^*(a)$$

If we allow $M^*(a) = M^*(c) = 1$ we get an anomaly since $c \twoheadrightarrow a$. So we must have (M^* is a $\{0,1\}$ function giving stable extension)

$$M^*(a) = M^*(c) = 0 \text{ and } M^*(b) = 1$$

Remark 2.14. *Example 2.13 raises several questions which require our answers:*

1. *We introduce the idea of solving Dung loops by using many-lives stable semantics. Namely:*

   (a) *Give all points in the loop single life, i.e., let $M(x) = 1$ for all x in the loop.*

   (b) *Choose a point in the loop. Fix this point (call it b).*

   (c) *Calculate M^* from M and get equations.*

   (d) *The equations in (c) might allow for several solutions. Do not allow for any solution which gives $M^*(x) = M^*(y) = 1$ when $x \twoheadrightarrow y$.*

We need to check under what conditions of the loop (geometry of the graph) we always get solutions.

Note that the calculation of M^* is not Dung like. We may have all attackers y of x all have $M^*(y) = 0$ but yet also $M^*(x) = 0$.

This is because of item *2 of Definition 2.10.

Remark 2.15. *In Definition 2.10 we mentioned that the function $M*$ is defined for each $x \in S$ using structural induction on the finite acyclic (S, R) network as the function derived from M. Let us explain how this is done.*

Define the Rank of $x \in S$ as follows:

- x is of rank 1 if $Attack(x)$ is empty.

- x is of rank 2 if all members of $Attack(x)$ are of rank 1

- x is of rank $n+1$ if all members of $Attack(x)$ are of rank $< n+1$ and at least one member of $Attack(x)$ is of rank n.

The structural induction is on the Rank of points x

Remark 2.16 (Case of Loops). *Our starting point is the definition of M^* from M in Definition 2.10, and Remark 2.15. The assumption there is that the network (S, R, M) is acyclic, and we use structural induction on the notion of Rank to define M^*.*

If we have loops we need to define the structural induction differently to be able to define M^* from M.

We proceed as follows:

1. *By a backward chain from point y to point x we mean a sequence of points $z_1, z_2, ..., z_n$ such that for each $i(i = 1, 2, ..., n - 1)$ we have that yRz_1, z_iRz_{i+1} and z_nRx.*

2. *The length of the chain in (1) is n.*

3. *If xRx we say the length of the chain in this case is 0.*

4. *Let $S1$ be a subset of S. We say $S1$ is a loop if for every x, y in $S1$ there exists a backward chain from y to x built up all of points of $S1$.*

   $S1$ is a maximal loop if it is not properly contained in a bigger loop.

5. *We say that a maximal loop $S1$ is a top loop if for every y and x in $S1$ such that there is a backward chain from y to x we have that y is also in $S1$.*

6. *Let $S1$ and $S2$ be two maximal loops Define a relation $\mathbb{R}$ on the set $\mathbb{S}$ of maximal loops by:*

   $S1\mathbb{R}S2$ iff for some y in $S1$ and x in $S2$ there exists a backward chain from y to x.

   Then $(\mathbb{S}, \mathbb{R})$ is finite acyclic.

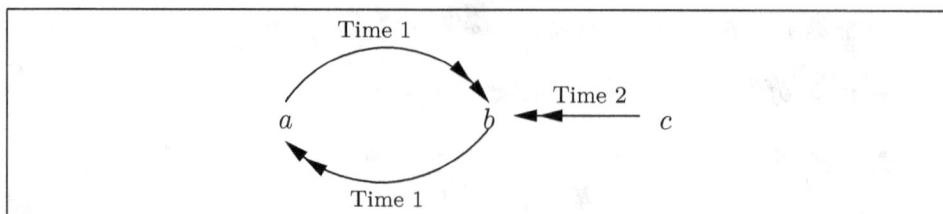

Figure 12

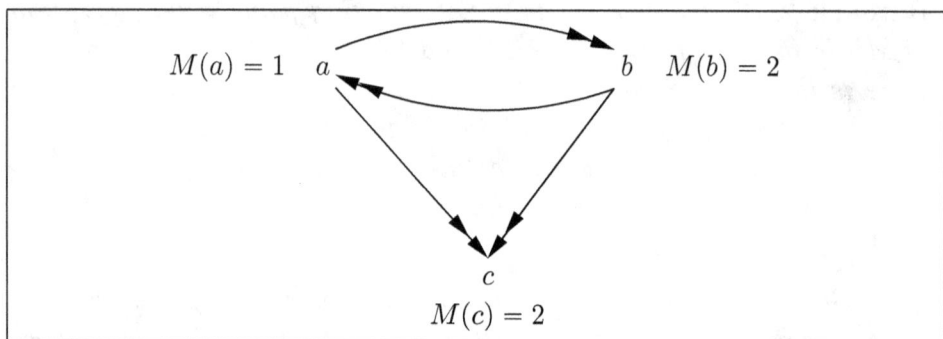

Figure 13

7. *For a maximal loop $\mathbb{C}$ in $\mathbb{S}$ let $\mathbb{M}(\mathbb{C})$ be defined in some reasonable way as the number of lives given to the loop as a unit, taking into account the lives of the members of the loop. For example define it as $\mathbb{M}(\mathbb{C}) = \min\{M(y)|y \text{ in } \mathbb{C}\}$.*

8. *Let $\mathbb{M}^*$ be calculated out of $\mathbb{M}$ as in Definition 2.10 for the system $(\mathbb{S}, \mathbb{R}, \mathbb{M})$.*

9. *Let M^* for the system (S, R, M) be finally defined for each y in a reasonable way from the values $M(y)$ and $\mathbb{M}^*$ (the max loop containing y), for example as $M^*(y) = \max(0, M(y) - \mathbb{M}^*(\mathbb{C})$, where $\mathbb{C}$ is the the unique loop containing y.*

Example 2.17. *To illustrate the ideas of Remark 2.16, consider Figure 13:*

In this figure the top loop is $\{a, b\}$. The minimum life of the top loop is 1 and therefore the M^* for loop members if $M^*(a) = 0, M^*(b) = 1$ and propagating to c we get $M^*(c) = 2$.

This calculation is consistent with activating the simultaneous attack of all elements of the loop $\{a, b\}$ on one another to get

$$M_1^*(a) = 0, M_2^*(b) = 1$$

and continuing attacking c, we get $M_1^*(c) = 2$.

It is possible to give other algorithms, for example, allow members of the top loop to attack all possible targets, not just only other members of the loop. In this case both a and b will attack c and we will end up with M_2^*, where

$$M_2^*(a) = 0, M_2^*(b) = 1, M_2^*(c) = 1.$$

3   Evolutionary temporal argumentation

In Subsection 3.1 we introduce evolutionary temporal logic and give some example from argumentation. In the next subsection we give many more examples.

3.1   Evolutionary propositional temporal logic for argumentation

Let us start with the classical propositional calculus with atoms Q and the classical connectives $\{\neg, \wedge, \vee, \rightarrow\}$. We have already said that we can turn any assignment to the atoms into a time dependent function by taking a flow of time $(T, <)$ and for each $t \in T$ look at a function $h_t(q) = h(t, q) \in \{0, 1\}$ for each $\in T, q \in Q$.

Example 3.1. *Now consider the flow of time $T = (1, 2, 3, \ldots)$ and the usual $<$ and assume for each t in T that we have a set $\mathbb{H}_t$ of assignments $\mathbb{H}_t = \{h_{t,i}, i = 1, 2, 3\}$ to choose from. So we can get a sequence $h_1, h_2, h_3, \ldots$ with h_n in $\mathbb{H}_n$. We can impose conditions on the choice of sequences. Examples of such conditions can be in a meta-language talking about the sequences, (not a temporal language but any other language). For example*

- *no change, $h_{n+1} = h_n$*

- *all $h \in \mathbb{H}_t$ must be obtained from some algorithms (e.g. be complete extensions of a varying argumentation network)*

- *For each t, h_t is generated from an argumentation network $\mathbb{A}_t$ as in Remark 2.6. Thus $\mathbb{H}_t$ is the set h_E, of all extensions E of A_t.*

- $\mathbb{H}$ *can be generated probabilistically*

- *and so on.*

Let us illustrate by defining on meta-level condition as an example, and so we choose the condition of continuity.

 We say that the sequence $h_1, h_2, h_3, \ldots$ preserves continuity if for each n, h_{n+1} is a minimal change from h_n. We have to define what we mean by minimal change, i.e. h_{n+1} is chosen from $\mathbb{H}_{n+1}$ representing a minimal change to h_n.

 (*) Given a set $\mathbb{H}^*$ of assignments $h \in \mathbb{H}^*$ and given $h_1 \in \mathbb{H}$, then $h_2 \in \mathbb{H}^*$ is a minimal change from h_1 according to a policy of change $\mathbf{P}$ of Hamming Distance defined in Definition 3.2.

 Note that we do not require that h_2 be unique, only that it be minimal.

Definition 3.2.

1. *Let $(T, <) = (\{1, 2, 3, \ldots\}, <)$*

2. *Let $Q = \{q_1, \ldots, q_n\}$ be a finite set of atoms.*

3. *Let $\mathbb{H}_t$, for each t in T be a set of assignments h*

$$h : Q \longmapsto \{0, 1\}$$

4. *We now define the Hamming distance policy $\mathbf{P}$ as follows*

 (a) *We can regard each $h \in \mathbb{H}_t$ as a vector $V_h = (h(q_1), \ldots, h(q_n))$ and for any two h_1, h_2 thus define $d(h_1, h_2) =$ the number of coordinates i for which $V_{h_1}(i)$ is different from $V_{h_2}(i)$.*

(b) Let $h_i \in \mathbb{H}_i$ be a sequence of assignment $i = 1, 2, 3, \ldots$. We say that this sequence preserve continuity according to policy $\mathbf{P}$, iff for each i, and each $h \in \mathbb{H}_{i+1}$ we have $d(h_i, h_{i+1}) \leqslant d(h_i, h)$.

An evolutionary temporal model for $(T, <)$, based on a sequence of sets of assignment $\mathbb{H}_t, t \in T$, is any sequence of assignments from $\mathbb{H}_t$ preserving continuity $\mathbf{P}$.

5. *An evolutionary temporal model for argumentation is any model bases on sets $\mathbb{H}_t$ obtained from respective argumentation networks $\mathbb{A}_t$, as defined in Example 3.1.*

The above definition is just for illustration, it is not suitable for the notion of continuity in the case of many-lives argumentation. The next section examines what happens in argumentation and what is needed.

Note that the notion of continuity is external (meta-level) to the temporal logic semantics. Some continuity policies $\mathbf{P}$ may be expressible as axioms on the temporal connectives (e.g. no change can be written as $G(A \rightarrow GA)$) but some may not. Such lines of research belong to pure traditional temporal logic and do not concern us here.

We now give examples to illustrate evolutionary temporal logic for argumentation. Compare with ($\sharp$1) and ($\sharp$2) of Subsection 1.2.

Example 3.3. *We illustrate evolutionary temporal logic for argumentation by two example networks, that of Figure 14 and that of Figure 15.*

1. *Analysis of Figure 14:*

   The two networks in this figure (network (i) and network (ii)) show the evolution of a network from network (i):

   $$S_1 = \{a, b\},$$
   $$R_1 = \{(a, b), (b, a)\}$$

   into network (ii)

   $$S_2 = \{a, b, c\}$$
   $$R_2 = \{(a, b), (b, a)(c, b)\}$$

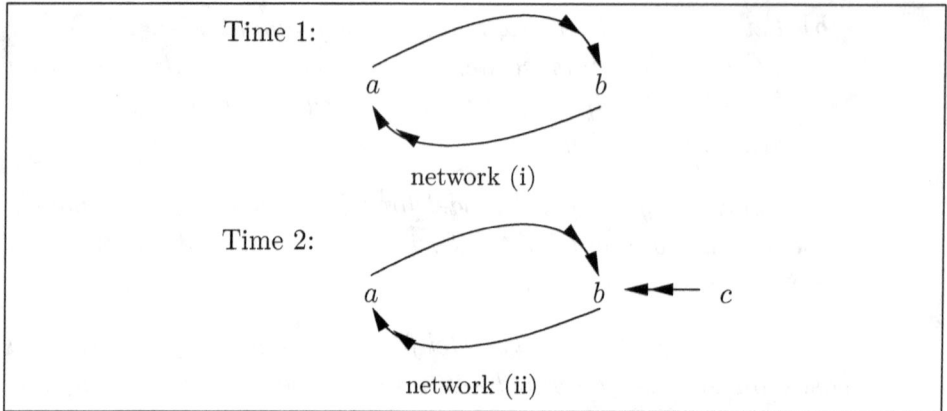

Figure 14

Evolutionary temporal logic can only talk about the change. It can only say that c showed up at Time 2 and that c attacks b.

This is not what we are interested in argumentation. We want to look at extensions. So what we want to say is either option 1 or option 2 or option 3.

Option 1. *At Time 1 there were three possible extensions. We chose at Time 1 the extension*

$$E_2^1 = \{a = \ in, \ b = \ out\}.$$

At Time 2 we got extra information of a new c attacking b and as a result we modified the chosen extension into

$$E_2^1 = \{a = \ in, b = \ out, c = \ in\}$$

Option 2. *At Time 1 there were three possible extensions. We chose extension*

$$E_1^2 = \{a = \ out, b = \ in\}$$

At Time 2 we got an extra c attacking b and so the only extension possible at Time 2 was

$$E_2^2 = \{a = \ out, b = \ in, c = \ in\}$$

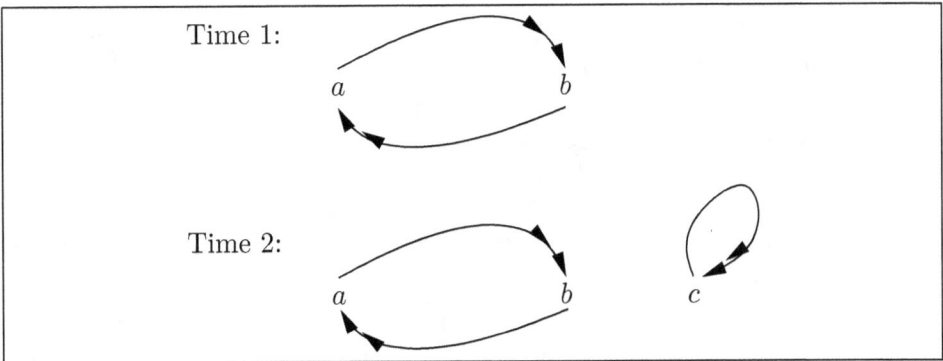

Figure 15

Option 3. *At Time 1 there were three possible extensions. We chose*

$$E_1^3 = \{a = \text{ und}, b = \text{ und}\}$$

At Time 2 we got an extra c attacking b and so the only possible extension was

$$E_2^3 = \{a = \text{ in}, b = \text{ out}, c = \text{ in}\}$$

We note that none of these options can be expressed in traditional temporal logic, because traditional temporal logic can only give the assignment generated by the chosen extension at time 1 and time 2, and, not how the extension was calculated, nor how the associated networks changed. So we have:

Problem 1. *What kind of temporal logic do we need? How do we extend temporal logic to suit our need?*
Answer. *We need what we describe in Definition 3.2, which we call* evolutionary temporal logic.

2. *Analysis of Figure 15:*

We make one more point. Consider Figure 15:

In this figure there are two independent parts, and there is no change in Time 2 on the {a, b} part of the network.

241

Therefore we should expect to say that the extension chosen in Time 1 remained unchanged in Time 2 as far as $\{a, b\}$ is concerned because the network did not change on $\{a, b\}$. What we do not want to say is Option 4.

Option 4 (we do not want this option). *At Time 1 the extension chosen was*

$$E^1 = \{a = \text{ in}, b = \text{ out}\}.$$

At Time 2 we changed our mid on the $\{a, b\}$ part and although the network did not change this part, we chose extension

$$E^2 = \{a = \text{ out}, b = \text{ in}, c = \text{ und}\}.$$

This presents us with a serious problem 2.

Problem 2. *Having chosen an extension E^1 at time 1, how do we continue modifying the same extension in future times without changing our minds like we did change in Option 4? In other words, how do we force/express continuity of our choice of extension, yielding only to necessary unavoidable change?*
Answer. *We can use the concept of continuity as item (c) of Definition 3.2. See also [30].*

Example 3.4. *Let $a =$ we cannot appoint Professor X.*
  $b =$ In the future Professor X can get big projects.
  We have that at Time 1 $b \twoheadrightarrow a$. This holds independently of the question of whether b is true or not. The reason being that we do not know the future (Time 2, 3,4,...), but we need to make a decision at Time 1 (take the extension $E^1 = \{b = \text{ in}, a = \text{ out}\}$.

Example 3.5. *We conclude with one more example showing that we may want the opposite of continuity. Consider networks (i) and (ii) of Figure 14 and assume that network (ii) comes temporally before network (i). Think that the cycle $\{a \twoheadrightarrow b, b \twoheadrightarrow a\}$ are two arguments*

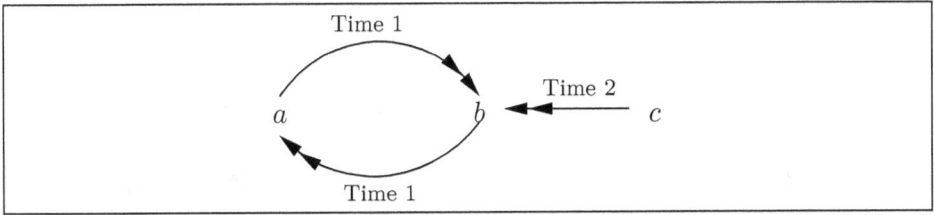

Figure 16: Combining the two parts of Figure 15, (Time 1 and Time 2) into this single figure by time stamping the arrows in it. The arrows $a \twoheadrightarrow b$ and $b \twoheadrightarrow a$ are timestamped "Time 1" and the arrow $c \twoheadrightarrow b$ is time stamped "Time 2". This is a much better notation because we have a single growing figure. See Remark 3.6.

attacking each other and that at Time 1 we have a witness c attacking b, so the only extension of the cycle is $\{a = in, b = out\}$. At Time 2 we have network (i), i.e. c withdraws, and we have the option of $\{a = out, b \ in\}$. We do not want continuity in this case. Our meta-level condition is to want $b = in$. In this case we take the option $\{a = out, b = in\}$.

We cannot express this condition in temporal logic, because the condition is Fb, which cannot always be true. But we can insist on an algorithm for computing the extension at any time t (this is meta-level for obtaining an extension h_t from (S_t, R_t)) which attempts to start with $b = in$ and checks if one can find an extension containing $b = in$.

Remark 3.6. *The perceptive reader might notice that the temporal progression described and discussed in Figure 14 can be represented in a single figure, where the arguments and attacks are time-stamped. See Figure 16, also compare with the general Figure 17.*

This perception is more than an alternative representation. It implies a criticism of what we are proposing here.

Criticism. *Why propose evolutionary temporal logic for argumentation, showing a sequence of temporal nodes t and argumentation networks (S_t, R_t) attached to t, why not put them all in one big argumentation network (S, R) with time stamping as in Figure 17. In this figure each node of the form z in the figure and each attack of the form*

(z_1, z_2) in the figure has the further annotation of a Time Stamp $T(z)$ and $T(z_1, z_2)$ respectively indicating the temporal moments in which the item exists. The annotation is a set of moments. In case there is persistence, that is an item which exists at a moment t continue to exist (and does not disappear after time t) we can use the annotation "$t+$".

So if $x, y \in S_t$ and $(x, y) \in R_t$ we put $(t, x), (t, y) \in S$ and $(t, x, y) \in R$.

We can retrieve (S_t, R_t) from (S, R).

Since the emphasis of "evolutionary temporal logic for argumentation" is on the argumentation part, it makes more sense to use (S, R).

Remark 3.7. Note that formally, from the point of view of formal argumentation, (S, R) of Example 3.5, looks like just another annotated argumentation network. It is in the meta- level that we interpret this annotation as leading to an evolutionary argumentation network and use it in our intended application. If we have a different application in mind, (see [29]) we might interpret the annotation differently and get different results. (See Example 4.3 in our discussion in the section Comparison with the Literature, in which [29] is discussed.)

Answer to criticism.

1. The notion of evolutionary temporal logic for arbitrary temporal sequences of systems is more general. We can have it for modal logic, for changing preferences, etc.

2. However, for some properties we are interested in argumentation, the big (S, R) with time stamping are more transparent in the evolutionary time stamping temporal logic approach. For example if some element x keeps attacking over time every element y other than himself, then the attack behaviour over time of x becomes an argument which can attack x. We can add that as a "temporal attack behaviour" which becomes an "argument".

3. Why not use both methods, depending on convenience?

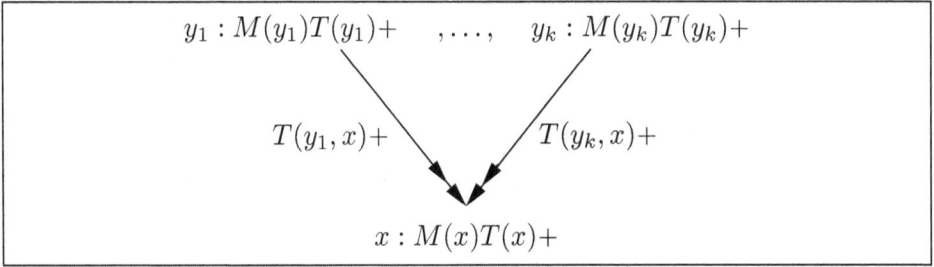

Figure 17: In this figure, x has $M(x)$ lives and exists at $T(x)$ and, afterwards, y_i, which exists at $T(y_i)$ and, afterwards, attacks x at time $T(y_i, x)$ and, continue to exist and attack afterwards, and has lives $M(y_i)$, for each $i = 1, ..., k$ and x exists from time $T(x)$ onwards. We have to assume that attacks (x, y) exist only at times where both x and y exist. However mathematically this is just a reasonable but not a necessary condition.

3.2   Further examples

This section examines examples from the application area of complaints about sex offenders. This area actually inspired the idea of many-lives argumentation networks. The temporal aspects come from the fact that the victims of a sex offender might complain at different times and so we need to time stamp the appearance of victims and their attacks. So the correct annotation is as in Figure 17.

The notation $T(y_1, x)$ is the time that y_i complained about x. It annotates the double arrow from y_1 to x. We need to explain in our notation the interaction between the many-lives of an argument x and the question of whether x is in or out or undecided.

- If the many-lives of x is positive then x is in (also we can say that x is alive).

- If the many-lives of x is 0, then we can say that x is out, or dead.

- If the number of lives of x is not known, or cannot be calculated because of loops, we can say that x is undecided or unknown.

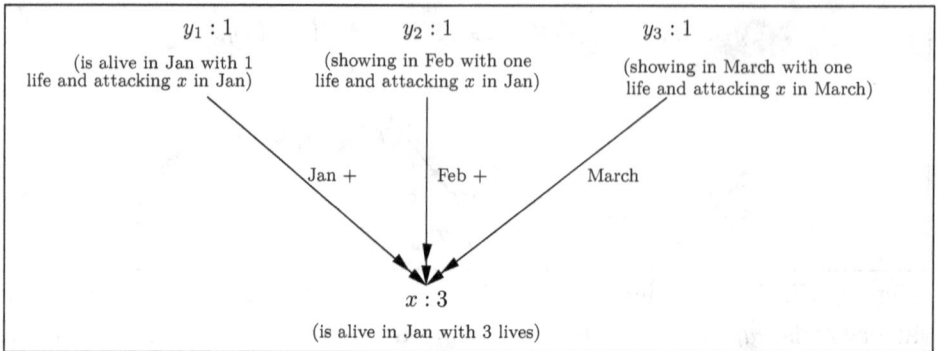

Figure 18

The question we ask is if we look at a time, say when there were only 2 complaints, we ask how many-lives does x have at that time? The answer is that x has $M(x) - 2$ lives, because of the fact that if $M(x) - 2$ more attackers come forward and complain then x will be "dead".

Figure 14 is a concrete example of this annotation (see Remark 3.6 for this temporal annotation):

x has only 2 lives remaining in January. In February he has 1 life left and when in March we have the third y_3 complain then x has 0 lives, i.e. x is dead.

We note that Figure 18 is a simplification of the temporal sequence. In January we had only the victim y_1 coming forward. We did not yet know about the victim y_2 who came forward in February, nor did we know of victim y_3, who came forward in March. So the January network should be the network of Figure 19 and not Figure 20.

We can use the convention that we always look at past figures from the point of view of the latest attack, in this case from the point of view of March, thus time-stamping all attacks.

So elements "show up" at the first time in which they attack others or are being attacked by others. So in Figure 18 we read that x and y_1 "showed up" in January, y_2 in February and y_3 in March.

To be consistent in using this notation/convention, we need the assumption of persistence, namely once there is an attack or an at-

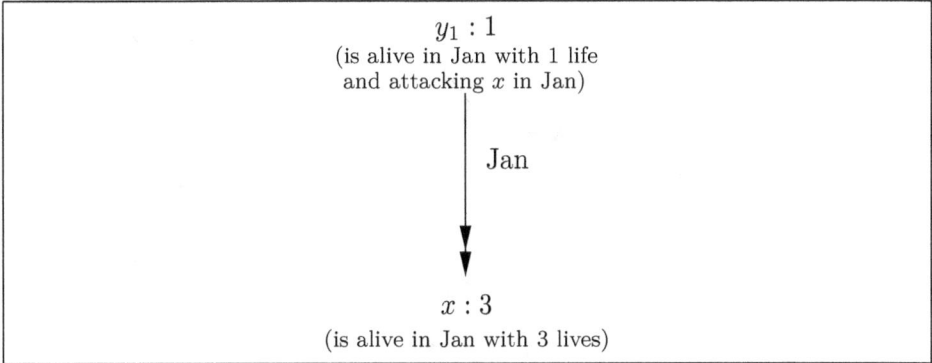

Figure 19

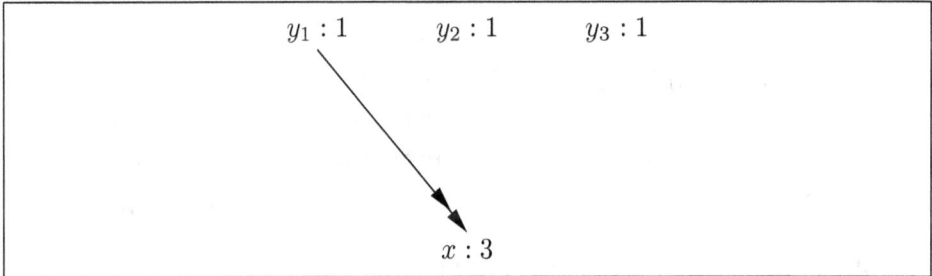

Figure 20

tacker or a target it does not disappear. Without this assumption we have to attach a set label to each element in $S \cup R$, stating all the time moments in which it exists. See Definition 1.14.

Obviously mathematically this is consistent but we need to consider applications where attacks can exist without an attacker and/or without a target. We further discuss this below following principle **PP4**.

We now state our third principle:[2]

---

[2]This paper continues our previous paper [2] entitled "Introducing Abstract Argumentation with many-lives" The previous principles **PP1** and **PP2** appear there, they are:

PP1: Every element x has a number $M(x)$ of lives (including possibly the value 0 in which case the element is out, or dead). To really kill x (reduce its many-lives to 0) you need to kill it $M(x)$ times (attack it by $M(x)$ different lives/in elements). In particular non-attacked elements retain all their many-lives intact and have the

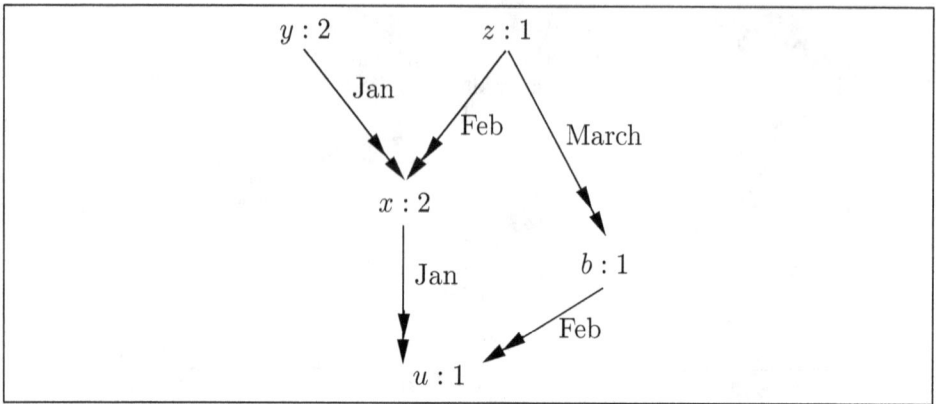

Figure 21

PP3: An attack from y to x is time stamped with a time $t = T(y, x)$ unique to y and x. Any algorithms governing the number of lives of any node in the system will take account of these time stamps.

Figure 21 describes the more general type of networks, which uses the time stamping mentioned in **PP3**.

In Figure 21, z attacks two targets at two different times. We need to calculate the situation (that is, the semantical extension, showing which element is in, or out or undecided and with how many-lives, also viewed as the survival situation, who is alive and who is dead) each month. In January the graph is very simple as illustrated in Figure 22. Note that the nodes z and b do not show up in the figure because in January they have not come forward yet.

The life of y is 2 because it is not attacked by anyone. Since y is alive it can attack x, reducing the life of x by 1. So x is still alive and the life of x is 1 even though it is attacked by y. x is alive so it can attack u and the life of u is 0. u is dead.

---

capability of attacking other elements (reducing the target's number of lives) if their value is not 0.

PP2: Although an element y may have $M(y)$ lives, when attacking any x it can kill/reduce only one of x lives.

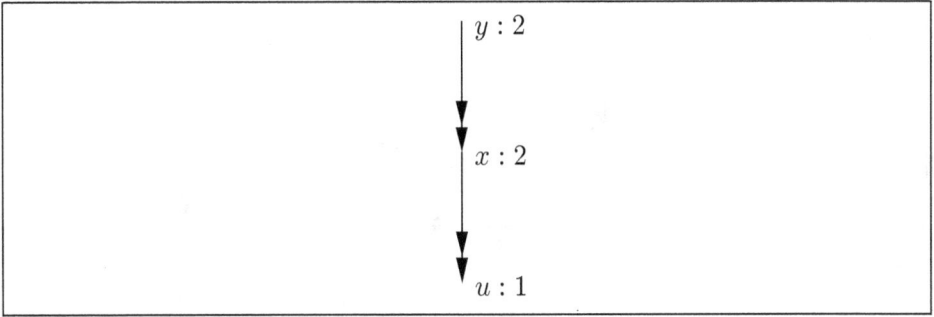

Figure 22

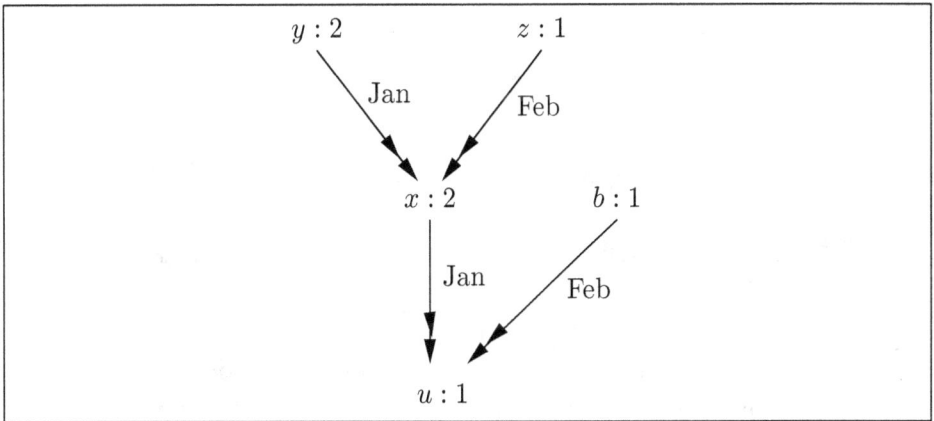

Figure 23

So the survival in January is

$$y : 2, x : 1, u : 0$$

Let us now move to February. The graph now is Figure 23 (recall our notation of Remark 2.5; where we say that alive means in and dead means out).

$y : 2$ and $z : 1$ are alive and attack $x : 2$. Thus $x : 2$ is dead and we write $x : 0$. However, $b : 1$ is alive and can attack u, and so $u : 1$ becomes $u : 0$.

We have the following survivals in February:

$$y : 2, z : 1, b : 1, x : 0, u : 0.$$

249

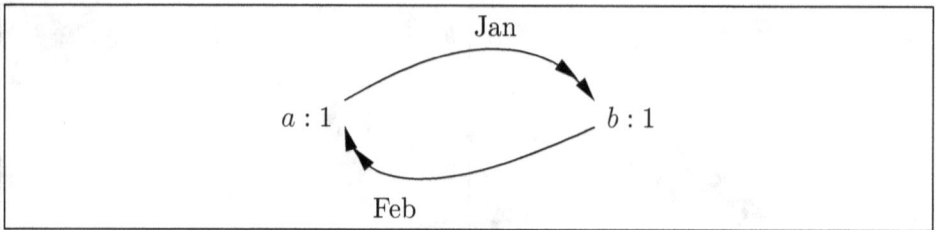

Figure 24

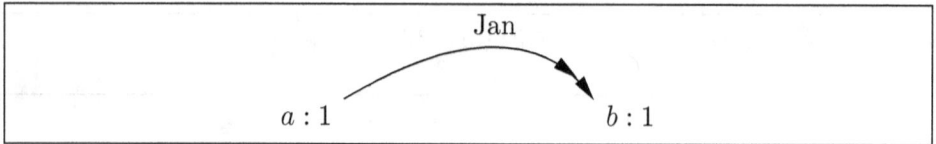

Figure 25

Let us now address another point of principle: we look at Figure 24.

The question we ask is what happens in January? Answer: we have Figure 25.

So the survival picture is $a : 1, b : 0$.

In February we have Figure 24. So we have a loop. We ask the question: what do we do with node b? Do we say that b was dead in January, and although we allow b to come back to life in February, b, once dead, can no longer attack?

So b cannot attack in February.

We now introduce a new principle,

PP4: In a system with time stamps, an element y may become dead at at time t but may come back to life at a later time s, with $t < s$. In such a case, we accept that y can be alive at time s but we do not allow y to attack any more at time s.

Let us refine better our understanding of principle **PP4**. Let us look again at Figure 23, and imagine that the node $b : 1$ is deleted from the figure. If $b : 1$ did not exist then $u : 1$ would have been alive in February, but it was dead in January. So do we consider u dead or alive? The answer is since its attacker x died in February then u would be alive in February if $b : 1$ was not there.

250

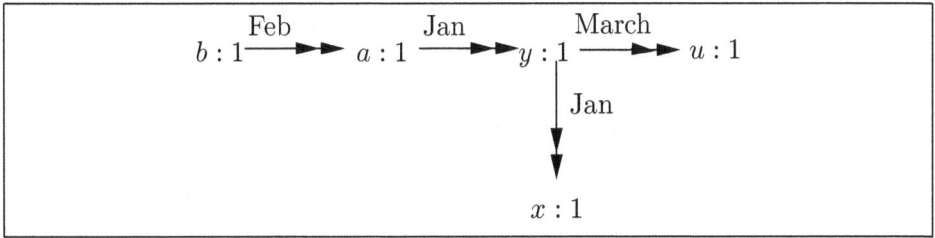

Figure 26

In March we have Figure 18 and the survivors are

$$y : 2, z : 1, x : 0, b : 0, u : 1.$$

If we apply this principle to Figure 24, the node b is dead in January. In February, the node cannot attack, having died in January and so is killed by node a. Without the principle **PP4**, the node b can counter-attack in February and we get the traditional network of two nodes attacking each other, which has three solutions, $\{a = 1, b = 0\}$, $\{a = 0, b = 1\}$, and $\{a = b = 0\}$.[3] See the next example 3.8.

Example 3.8. *Let us illustrate our computational options and offer possible refinements to the principle* **PP4**. *Consider Figure 26.*

Computation January. *At this time $b : 1$ has not come forward. Therefore $a : 1$ is alive/in, and since $a : 1$ attacks $y : 1$ we have y is dead/out (i.e. $M(y) = 0$) and hence $x : 1$ is alive/in and so is $u : 1$. The survival in January is therefore $\{a = y = x = u = in\}$.*

Computation February. *$b : 1$ comes forwards and attacks $a : 1$. So a is dead. So y becomes alive since it is no longer attacked by a. We have two options*

1. *If we apply* **PP4** *we do not allow y to attack x at February and so x is alive.*

---

[3]We use the simplified notation "$x = n$" for the expression "$M(x) = n$" or equivalently the expression "$x : n$" which we use in figures. We will be explicit when needed.

2. *Without* **PP4**, *y can attack x and x is dead. The two outcomes are therefore*

   (a) $b = in, a = out, y = in, x = in, u = in$

   (b) $b = in, a = out, y = out, x = out, u = in$

Computation March. In March we have the full Figure 26, including the March attack of $y : 1$ on $u : 1$, namely $y : 1 \twoheadrightarrow u : 1$. According to principle **PP4**, since $y : 1$ was dead in January then even though it came back to life in February and in march, it cannot attack any more and so the attack of $y : 1$ on $u : 1$ is to be discarded and ignored. It is at this point that we might fine-tune principle **PP4** into the more sensitive new principle **PP4***. The attack of $y : 1$ on $x : 1$ is a January attack and this attack was discarded because in January $y : 1$ was dead. Having come back to life in February, does not mean that we revive the January attack of $y : 1$ on $x : 1$. But the March attack of $y : 1$ on $u : 1$ is a new attack, newly executed in March when $y : 1$ is alive. So we can argue that it should be accepted and not discarded. To give a motivating example, suppose in January $y : 1$ complained that $x : 1$ sexually abused $y : 1$. A witness $a : 1$ came forward in January saying he heard clearly $y : 1$ boasting that $y : 1$ invented false accusations against $x : 1$. As a result of that testimony, $y : 1$'s complaint was declared false and rejected and $x : 1$ was declared innocent and the proceedings against $x : 1$ were terminated.

In February $b : 1$ attacked $a : 1$ saying that $a : 1$ was nowhere near $y : 1$ and could not have reported any boasting of $y : 1$. Thus the complaint of $y : 1$ against $x : 1$ is now (in February) credible. But the January proceedings against $x : 1$ are over and it stands to procedural reason that we adopt the view that "whatever is gone is gone". In February $y : 1$ credibility is reinstated and so $y : 1$ complaint against $u : 1$ is credible. There is no reason to reject it. We therefore could modify principle **PP4** into **PP4*** as follows:

PP4* In a system with timesteps an element y may become dead at time t. We thus declare dead at time t any attack emanating from y at any time $t' \leqslant t$. If at some later time s the element

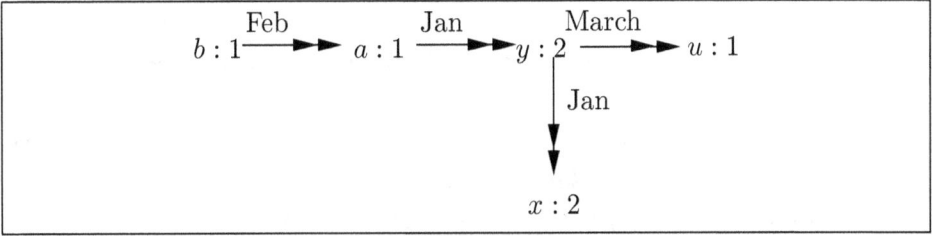

Figure 27

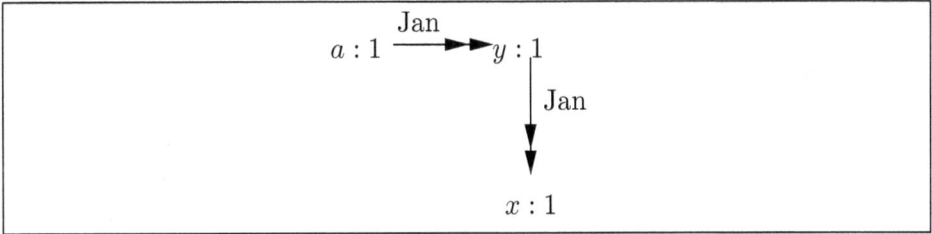

Figure 28

y comes back to life, then y coming back to life does not bring back to life any attack declared dead at any time $s' \leqslant s$.

According to **PP4***, in Figure 26, the March attack of $y : 1$ on $u : 1$ is alive and the survival picture in March is $b = \text{in}, a = \text{out}, y = \text{in}, x = \text{in}, u = \text{out}$.

Example 3.9. *Consider the situation of Figure 26 but let us give y and x the two lives. This is illustrated in Figure 27. We note that the network in the Figure is finite acyclic, allowing for the calculation which follows. If the figure contains loops or is one big loop itself, a specific algorithm for loops is required. See Remark 2.16.*

Let us calculate what happens in January. We get $a = 1$ attacks $y : 2$ and so y becomes $y : 1$. $y : 1$ is still alive and it attacks $x : 2$ and so we get $x : 1$.

The answer is

$$a = \text{in}, y = \text{in}, x = \text{in}.$$

Let us present the answer in Figure 28

Figure 28 is the same as the January part of Figure 26.

The problem is, if we look at this figure as the January part of Figure 26, we should execute the attacks indicated in the figure and get $a = 1, y = 0, x = 1$.

But if we look at this figure as the result of having already executed the attacks of Figure 27, then we do nothing and execute nothing.

This gives us ambiguity. We have two options:

1. *Do nothing and say that Figure 28 comes from Figure 27 after execution.*

2. *Continue the execution until the process is stable.*

3. *We note that the question of which option to use depends on our interpretation of the network. The sex offender interpretation requires option 1. Each complaint/attack reduces one life. We do not use the complaint again as if it were another victim complaining. In comparison, if we have a baby complaining/crying because it wants its nappy changed, then it will complain again and again until the parent cannot take it any longer and does the job (i.e. parent runs out of lives).*

PP5: *Let (S, R, M) be a network as in Definition 2.1 and Remark 2.16 and Remark 2.15. Consider M^* as defined from M in the above definitions and remarks. Continue the derivation of M^{**} from M^* etc., until we reach an $M^*...^*$ such that another application of the derivation does not give anything new. Call this M function $M(*)$. Principle PP5 says use $M(*)$ and not M^*. Compare with item 3 of Example 3.9.*

So to summarise, let us calculate the survival picture of the network in Figure 27 using principle **PP5**.

Computation January.

1. We start with Figure 29

    We make one pass of calculation as described already and get the network of Figure 28.

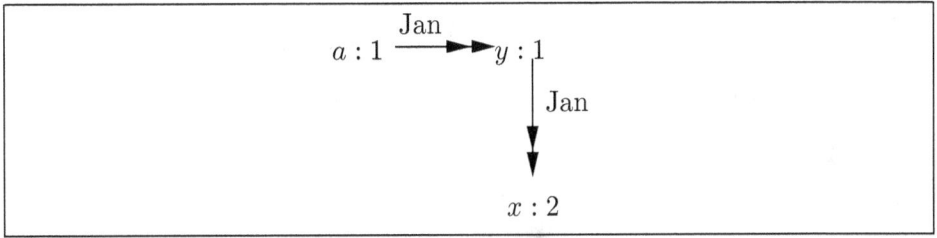

Figure 29

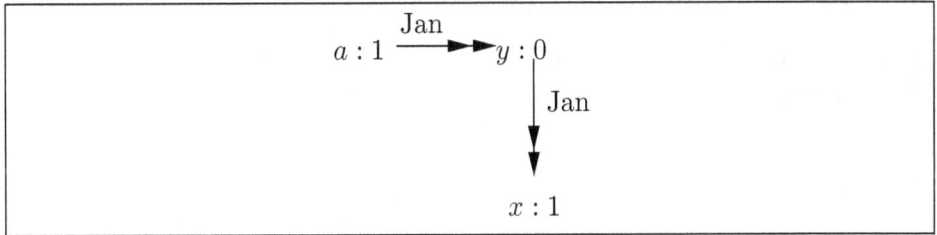

Figure 30

2. We make a second pass of calculation on Figure 28 and get Figure 30.

3. If we make another pass of calculation of Figure 30 we get the same figure. So we are stable and the survival picture for January is (using **PP5**) $a = $ in, $y = $ out, $x = $ in.

Computation February. The February network is Figure 31.

The attack $y : 2 \twoheadrightarrow^{\text{Jan}} x : 2$ is dead, but we left it in the figure for expositional reasons. y was dead in the January calculation (item

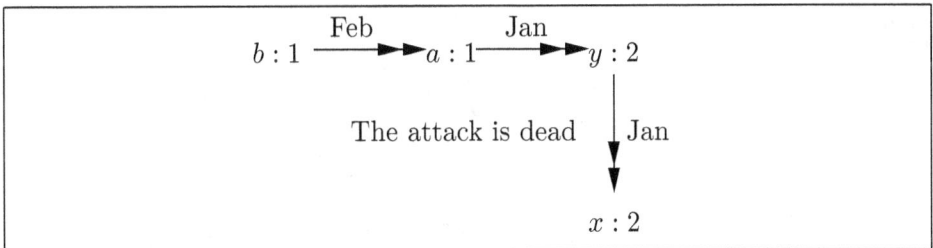

Figure 31

255

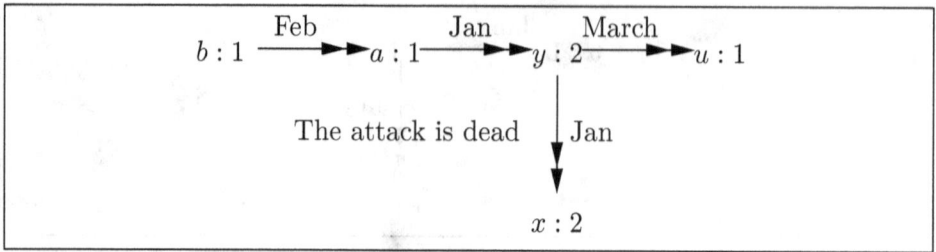

Figure 32

(2)) and so it cannot attack. $x : 2$ is not attacked in Figure 31. We get the survival picture $b =$ in, $a =$ out, $y =$ in, $x =$ in.

Computation March. This computation works on Figure 32.

The calculation is straightforward. The survival picture is $b = 1, a = 0, y = 2, x = 2, u = 0$.

Remark 3.10. *The temporal annotation aspects have no counterpart in the traditional Dung semantics, not even in any traditional modal temporal logic version of it. The main reason for this is because we use principle* **PP4*** *on the one hand and there may be more than one extension at a given time on the other hand. See Example 3.11.*

Example 3.11. *This example makes a methodological distinction which is useful at this point and also explains a possible problem/warning in using principle* **PP4***.

Consider the network of Figure 33. Assume all nodes have one life and all attacks are of strength one.

In January we have the network of Figure 34

We have three non-empty extensions in January, i.e. in the network of Figure 34

$$E_1^1 = \{a = in, b = out, x = in, y = in\}$$
$$E_2^1 = \{a = out, b = in, x = out, y = in\}$$
$$E_3^1 = \{a = b = x = \ undecided, y = in\}$$

We note the following

1. *The network of January (Figure 34) is a traditional network.*

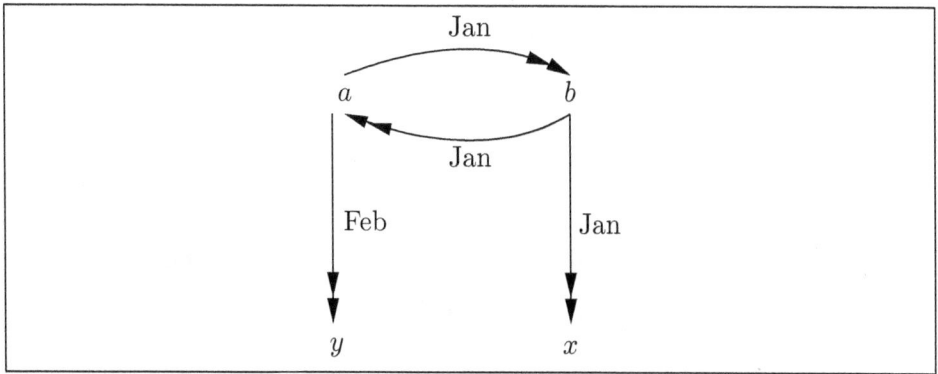

Figure 33

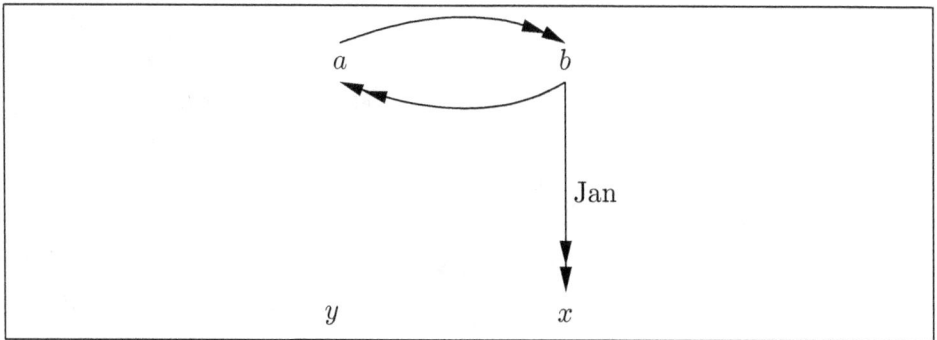

Figure 34

2. E_1^1 chooses $a = in$ and E_2^1 chooses $a = out$.

Let us now ask what is the network in February? It is the traditional Dung network of Figure 35.

The extensions are

$$E_1^2 = \{a = out, b = in, x = in, y = out\}$$
$$E_2^2 = \{a = out, b = in, x = out, y = in\}$$
$$E_3^2 = \{a = b = x = y = \text{ undecided}\}.$$

Our methodological point is the following: If in January we choose to resolve the top loop ($a \twoheadrightarrow b$ and $b \twoheadrightarrow a$) by letting $a = in$ can we now in February choose another extension for the top loop and take $a = out, b = in$?

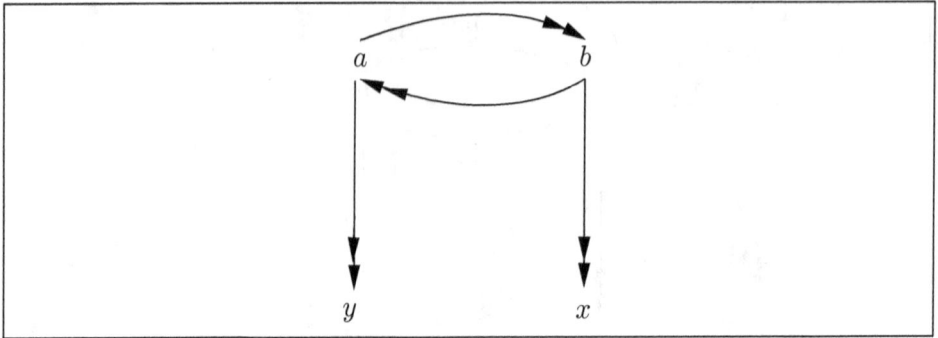

Figure 35

There is a consequence to this change of choice because if we take E_1^1 in January and E_2^2 in February we get that b was dead (b = out) in January and b came alive (b = in) in February. But then according to principle **PP**4*, the January attack $a \twoheadrightarrow x$ does not come back to life in February so we must have $x = in$ in February because x is not attacked in February.

We might argue that this is not acceptable because b came back to life owing to the administrative means (choice of extension) and not because of any substance.

Let us be clear about this point.

The top loop, namely ($a \twoheadrightarrow b$ and $b \twoheadrightarrow a$) is not internally affected between January and February. Therefore we might argue/expect that if in January we chose {a = in, b = out} then we chose the same in February and if in January we chose {a = out, b = in} then we choose the same in February. By switching choices between January and February, we activate principle **PP**4* generating possibly unwanted consequences. We therefore need to define/identify mathematically the circumstances under which we are making a change of choice and use this identification to modify principle **PP**4*. Let us call the yet to be defined principle **CPP**4* (**PP**4* with continuity).

The problem is how we formulate such a principle. If we use the declarative set theoretical definitions of extensions as in Subsection 2.1, how do we say that E_2^2 involves a change in choice and is not the correct February extension, which follows E_1^1 In other words, how do

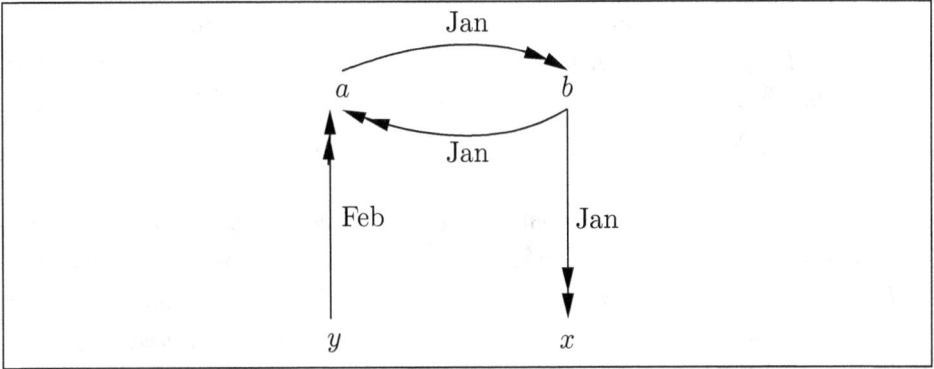

Figure 36

we define continuity in a set-theoretical way?

This is also a problem for traditional modal and temporal logics. Such logics do not deal with continuity in time.

To see the difficulty with the set theoretical instrument, consider the network of Figure 36

In this figure we are forced to move from E_1^1 to E_2^2 because y's attack on a forces it! However, it is not easy to tell the difference from the previous case. In fact, in February there is only one extension, so it does not matter what we did in January.

We postpone handling this question to a subsequent offshoot paper.

Remark 3.12. *There may be a way to maintain temporal continuity if the extension are chosen using an algorithm. We examine the algorithms used in previous Example 3.11. First we divide (S, R) into maximal loops (called SCC's, see [3]), then we choose points in the top SCC and propagate the attacks. We get a new (S', R') and repeat recursively. Each sequence of choices give an extension.*

If new points are added or deleted, we use the same algorithm but try to retain the same choices as much as possible.

Example 3.13 (Public Pressure). *Case 1: On February 11, 2021, the police announced that a young man named Yarin Sharaf, who was initially suspected of raping a 13-year-old girl at the Corona Hotel, was charged with the lesser offences of consensual, sexual harassment,* *threats and assault. Following the announcement, there was a wide public outcry and on 25.03.21 the prosecutor's office announced that it was filing an indictment for a more serious offence, which is rape. The change was following public criticism from the victim's family and women's organizations*

Case 2: A criminal known to the police was arrested on suspicion of murdering Yuri Volkov after detectives waited outside the house where he had been staying for hours. According to the suspicion, he stabbed the deceased after the deceased and his wife warned him that he almost hit them on the road. At first the police announced that he was charged with the relatively minor offence of manslaughter, but after public pressure the charge was changed to a charge of murder.

4 Comparison with the literature

We have already compared with the literature when we presented in Section 1, subsection 1.1 the Malkinson real example, and in subsection 1.2 the three views, Timed View, Modal View and Evolutionary View . Also relevant is Remark 3.7. In this section we discuss the differences between the views in more detail.

Example 4.1. *This example is to further illustrate the difference between the Timed View and the Evolutionary View.*

Consider legislation about driving licences. In many countries, traffic offence τ can give bad points on a driving licence D (of the offender). Usually when you accumulate 3 bad points your driving licence is revoked. If you continue to drive after your licence is revoked then the offence τ becomes more serious, say τ'.

The best way to view this is to say that D has 3 lives, and τ attacks D and takes one life from D.

Figure 37 shows an evolutionary sequence:

If we want to take the timed point of view, we have to write the following arguments (doing the evolution in the syntax) and we use $D0, D1, D2, D3$ where "Dx" is "D with x lives". We get a new table
38

D, x = Driving licence with x lives
τ = Offence
τ^* = Driving with no licence
$\neg D$ = Licence suspended

Time	Network
1	D with 3 lives
2	$D, \tau, \tau \twoheadrightarrow D$
	D two lives
3	$D, \tau \quad \tau \twoheadrightarrow D$
	D one life
4	$D, \tau \quad \tau \twoheadrightarrow D$
	D no lives
5	$\neg D, \tau, \tau^*$
	offence
6	$\neg D, \tau^*$
$\vdots$	$\neg D$
36	$\neg D$

Figure 37

Time 1:	$D3, \tau \twoheadrightarrow D3$
Time 2:	$D2, \tau \twoheadrightarrow D2$
Time 3:	$D1, \tau \twoheadrightarrow D1$
Time 4:	$D0, \tau \twoheadrightarrow D0$
Time 5:	$\neg D, \tau, \tau^*$
Time 6–36:	$\neg D$

Figure 38

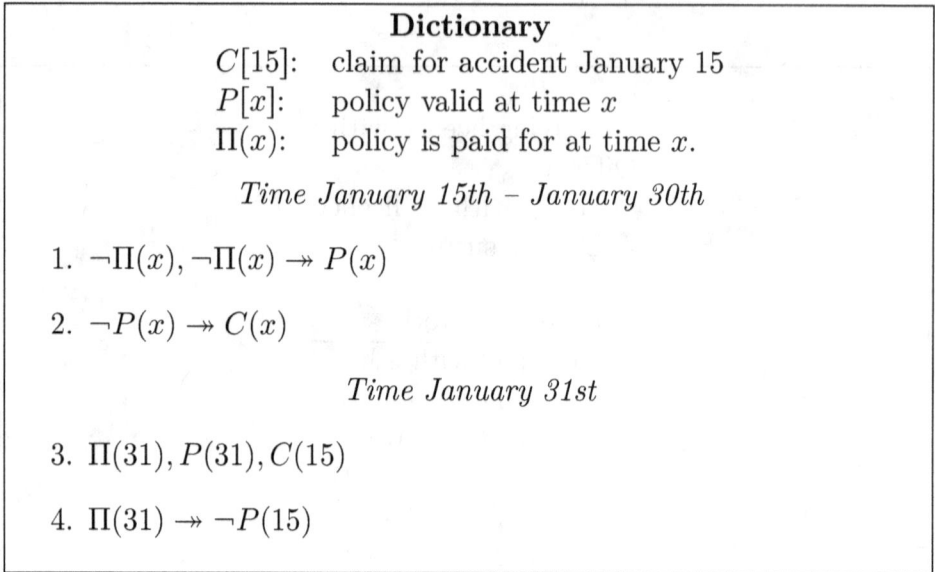

Dictionary

$C[15]$: claim for accident January 15

$P[x]$: policy valid at time x

$\Pi(x)$: policy is paid for at time x.

Time January 15th – January 30th

1. $\neg\Pi(x), \neg\Pi(x) \twoheadrightarrow P(x)$

2. $\neg P(x) \twoheadrightarrow C(x)$

Time January 31st

3. $\Pi(31), P(31), C(15)$

4. $\Pi(31) \twoheadrightarrow \neg P(15)$

Figure 39

In this example the timed view is forced to put the evolution in the syntax!

Example 4.2. *This example shows an application where the timed view does not work.*

In the UK if one has insurance paid for on direct debit then UK law says that if when it is time to renew the policy (say by December 31st) and something goes wrong and the direct debit does not work, then one is given one month (until January 31st) to renew (from December 31st).

So in this scenario the following can happen.

We have a claim C on, say, January 15th and the claim is rejected on the grounds that the policy is not valid, it not having been renewed! However, if payment is done by January 31st, then the policy is renewed retrospectively from December 31st and the January 15th claim is accepted.

The timed presentation of this scenario is as follows (Figure 39):

In the timed presentation we cannot avoid contradiction in the presentation at time January 31st.

We need to add to the timed presentation the attack $\Pi(31) \twoheadrightarrow \neg P(15)$ but to explain where it comes from we need the evolutionary representation of the insurance law.

The evolutionary reality allows for the consumer not to renew the insurance on December 31st and wait to see if a claim arises in the period January 1st to January 30th. If no claim arises the consumer can move to another new insurance company beginning a new insurance policy from January 31st. If asked why he is leaving the old company to the new one he can say he was hoping for a better deal.

Example 4.3 (Comparison with [29] Part 1). *This example compares directly with two important papers [32] from COMMA 2010 and paper [29] from 2015.*

We address directly the longer paper [29]] of 2015. The authors say in their abstract, and we quote:

> *Temporal Argumentation Frameworks (TAF) represent a recent extension of Dung's abstract argumentation frameworks that consider the temporal availability of arguments.*
>
> *In a TAF, arguments are valid during specific time intervals, called availability intervals, while the attack relation of the framework remains static and permanent in time; thus, in general, when identifying the set of acceptable arguments, the outcome associated with a TAF will vary in time.*
>
> *We introduce an extension of TAF, called Extended Temporal Argumentation Framework (E-TAF), adding the capability of modeling the temporal availability of attacks among arguments, thus modeling special features of arguments varying over time and the possibility that attacks are only available in a given time interval.*

1. *The first and second paragraphs of the above quotation declares that TAF is a temporal extension of the Dung approach. This means (and indeed is used in their paper) that they use the concepts of conflict free subsets and admissibility to form extensions*

and the the "Arguments Entities" they use and to which they apply the Dung machinery are "Temporally annotated arguments units".

This is not the case with our Chapter. Our Malkinson example and discussion in Section 1.1 does not conform to the basic dung machinery but we use time in evolutionary way.

2. *The third paragraph of the above quotation adds that their system E-TAF also temporally annotates the attack arrows. We also do that in our Chapter but we use all annotations in an evolutionary manner.*

3. *So what paradigm example application is compatible with the authors' machinery? Our answer is the consistency checking of legal laws that apply differently at different times and we want to verify the laws do not clash.*

For example taxation laws. The government may declare a new package of business tax increases spread forward over a period of 5 years and the author model may check whether any clashes arise. The key word is tax legislation into the future NOT LEGISTLATION INTO THE PAST.[4]

Let us now examine one of the authors examples which brings out the difference. We quote from their paper:

Begin quote 2, from [29] page 33, (I modified the notation):

The arguments are $\{A, B\}$

The Attack is $A \twoheadrightarrow B$

The temporal span is [0...60]

The temporal annotation for the arguments are

---

[4](The UK government does legislate into the past causing sometimes great resentment, and Evolutionary Temporal Argumentation is needed to model such legislation, but other EU countries do not do that and consider backwards Tax Legislation a TABOO!)

$E = \{(A, [0...40]), (B, [30...60])\}$, *and note that according to [29] E is the set of "Temporally annotated arguments units" for which the Dung Conflict free concept is applied.*

The attack of A on B (i.e. the double arrow $A \twoheadrightarrow B$) is annotated by $\{((A, B), [30...35])\}$

The authors say, and I quote

> *"Indeed, E is not a conflict-free collection of t-profiles, since the argument A attacks the arguments B in the time interval [30—35]"*

Going back to our proposed interpretation of consistency of legislation, a reasonable "consistency- conflict free" is the set

$$E - Con = \{(A, [0...29], (36...40)), (B, [36...60])\}$$

of arguments from time 0 to time 60.

The period [30...35] contains a conflict between A and B. We can decide in the Meta-level that A is dominant or we may not.

From the point of view of our Chapter, we look at the evolution from time 0 to time 60. We see that at time 30 there is an attack from A to B. Depending on the meaning of A and B we could extend the attack into the future up to time 60. This is forward looking.

Let us give a tax interpretation to A and B:

Assume A is a new tax on Builders B of luxury apartments. If the contract B starts at the time 30 when the tax A is instituted then we can adopt the view that it will continue to be valid until time 60 when the contract B terminates, despite the fact that the tax law A was canceled at time 36. The **PP** rationale could be that the contract B activity is still ongoing until time 60.

Example 4.4 (Comparison with [29] Part 2). *We continue our comparison with [29] by giving our own very simple algorithm which does the same job as the machinery in Sections 1–4 of [29].*

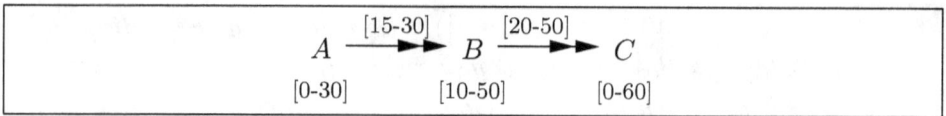

$$A \xrightarrow{\text{[15-30]}} B \xrightarrow{\text{[20-50]}} C$$

[0-30]   [10-50]   [0-60]

Figure 40

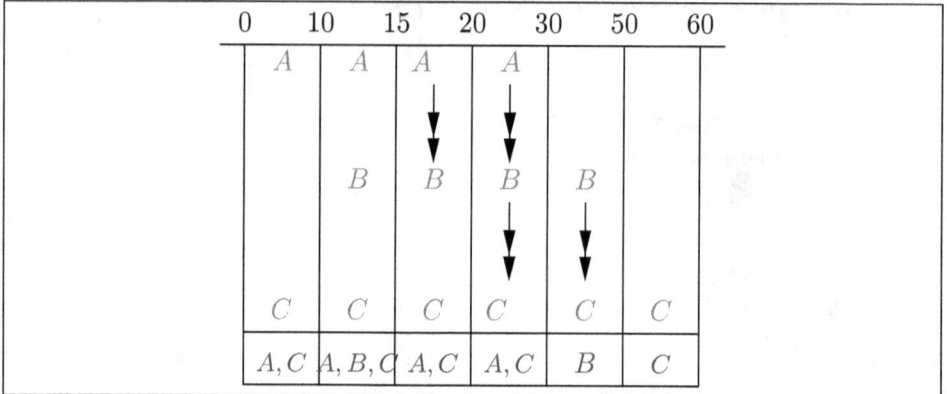

0	10	15	20	30	50	60
A	A	A	A			
	B	B	B	B		
C	C	C	C	C	C	
A, C	A, B, C	A, C	A, C	B	C	

Figure 41

Let us take the example of Figure 4 of [29], page 30.

This example contains a time-stamped network (E-TAF) containing several independent parts. We use one of the parts to illustrate our algorithm. This illustration will make clear what [29] is doing, and how [29] is different from our Chapter. We concentrate on the $\{A, B, C\}$ part of Figure 4 of [29]. We represent it in our own Figure 40.

Note that all figures in [29] are finite. The temporal annotations of all components, arguments and attacks, are a finite list of intervals. We now give the algorithm.

Step 1. *Construct vertical lines, ordered according to time of all starting points and end point of each interval appearing in the Figure.*

Executing step 1 will yield the following figure 41.

Step 2. *For each minimal box of the form $[a, b]$ in the figure, we include the units (arguments and attacks) which are valid in the interval of the box. This is within Figure 41 in colour red.*

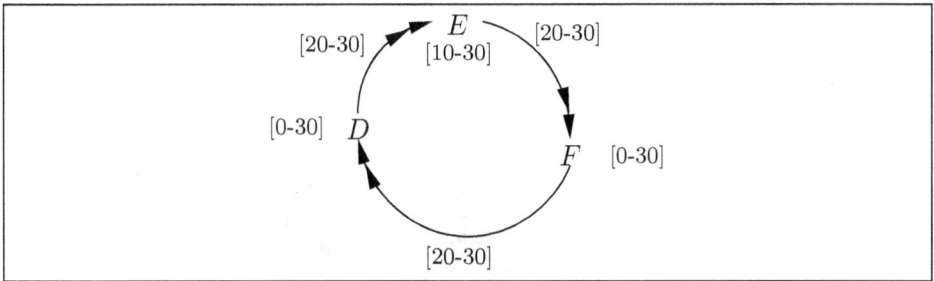

Figure 42

Step 3. *For each box compute all possible complete extensions. In our example there is only one. It is possible in general that there might be more. In general for each box $[a, b]$, let $E^1_{[a,b]}, E^2_{[a,b]} \ldots$ be all extensions. In Figure 41 at the bottom row of the figure we indicate the extensions for each in colour blue.*

Step 4. *The extensions according to [29] for the annotated Figure 40 can be obtained from the blue bottom box of Figure 41.*
   We get

$$C \; [0 - 30], [50 - 60]$$
$$B \; [10 - 15], [30 - 50]$$
$$A \; [0 - 30]$$

Indeed, in [29, page 31]. Example 5 (of [29]), this is exactly what is declared.

Example 4.5 (Comparison with [29], Part 3). *We continue our analysis of parts of Figure 4 of [29, p. 30].*
   We use our algorithm on two more subfigures of Figure 4 of [29, p. 30]. These are the loop of arguments $\{D, E, F\}$ and the loop of arguments $\{H, I, J, K\}$.
   These are presented here in Figures 42 and 43.
   Applying our algorithm to the network of Figure 42, we get the box Figure 44.
   Applying our algorithm to the network of Figure 43, we get the box Figure 45.
   The colour coding is as in the previous example 4.4.

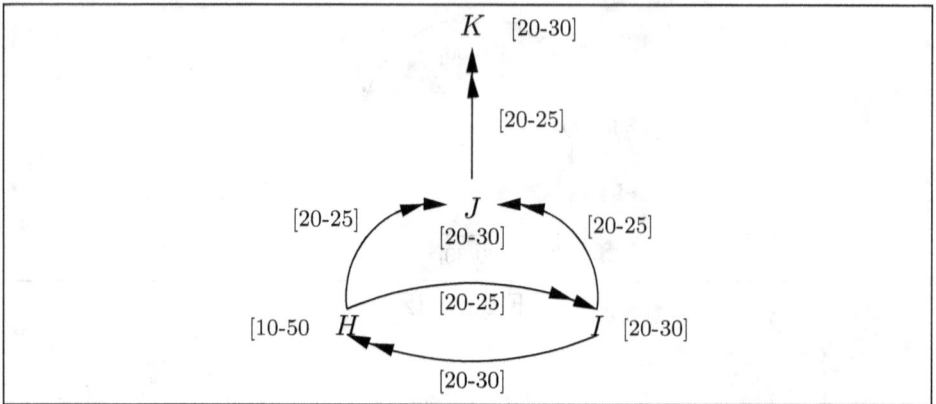

Figure 43

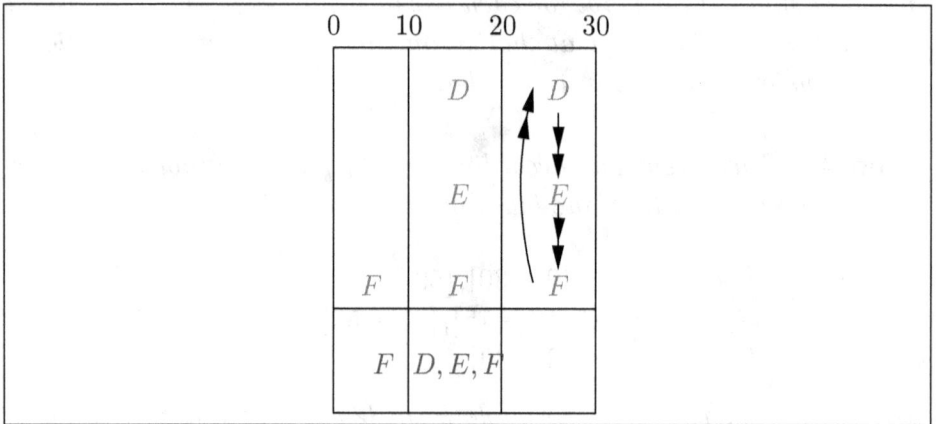

Figure 44

We make the following observations.

1. The results of the extensions we get agree fully with the extensions in [29, Example 5, bottom page 31, and top of page 32].

2. The authors of [29] use Dung machinery. For this reason, in Figure 44 the three cycle in box [20-30] has the empty extension. The way our algorithm works it allows us to choose the extensions in any box according to how we want. For example, we can choose CF2 semantics [3] for cycles in one box and Dung

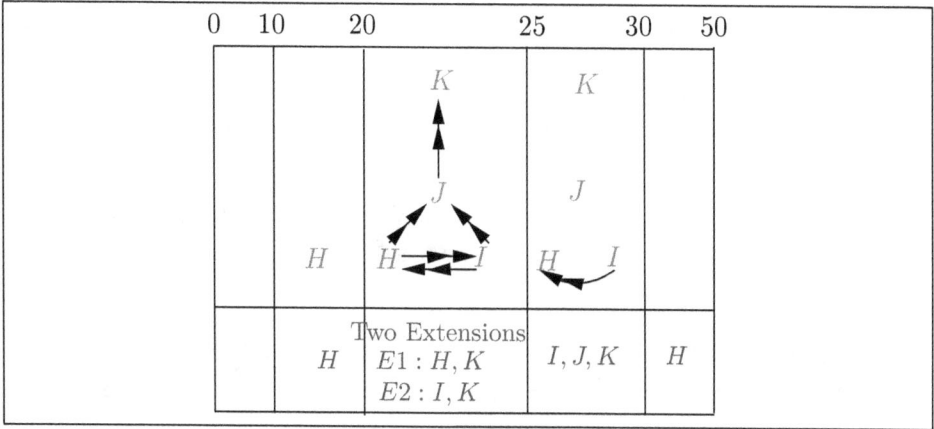

Figure 45

in another, all depending on the application area involved.

3. Our algorithm is simple conceptually and effective computationally. It can be easily generalised to an infinite number of intervals annotations. We just get an infinite number of boxes.

4. In [29, Sections 5–end] the authors investigate structured argumentation with time stamping. For the purpose of comparing with our evolutionary approach there is nothing new to compare. The temporal approach of [29] remains the same when structure is added.

5   Discussion, future research and conclusion

The many-lives approach is new (see [1, 2]), the idea of adding the many-lives function to abstract argumentation network. It does not fall under numerical argumentation. The way it is handled is inspired by the sex offenders case studies area.

There is a need for further research, investigating the place of many-lives in the general abstract argumentation landscape. There are many questions to be answered , among them the following:

1. What kind of semantics we should offer for systems with many-lives?

2. Can many-lives semantics simulate known semantics for traditional single life?

   For example can we simulate CF2 semantics by giving the elements of maximal conflict free sets more lives? (See Remark 2.14.)

3. How to handle support in the context of many-lives? Does support add lives? Is support (in the context of sex offender's many-lives) a higher level attack (on attack)?

4. How to define reinstatement? How many-lives to reinstate?

5. What is the best view of temporal change of a many-lives network?

6. What is the variation of the many-lives concepts across different application areas which use many-lives? (See Example 4.1.)

7. We are currently also looking at many-lives case studies in Nutrition. The liver for example can be attacked by a variety of foods, such as Alcohol, Sugar, Gluten and more. Such attacks combine in different ways, requiring/motivating new types of higher level attacks. In fact we require 3 dimensional argumentation networks for proper modelling.

8. We can regard many-lives as a resource (say M gives American \$). An attack destroys resources of the target but also costs resources of the attacker. We need to develop the evolutionary temporal logic of resource attack and defence.

9. The algorithm presented in the detailed analysis and comparison or our Chapter with the important 2015 paper [29] suggests we can write a new paper extending and simplifying their results also to the infinite case. We shall use Neibourhood Ultrafilter Semantics on the temporal line.

We leave these and other questions for follow up papers

Acknowledgements

This Chapter is a revised version of our paper from 2017, [6], submitted to a special issue of the *Journal of Applied Logics*. The referees did a very thorough job, but we did not revise in time. We thank the anonymous referees of that paper as well as Matthias Thimm and Leon van der Torre and the referees for most valuable comments.

References

[1] D. Gabbay, G. Rozenberg and L. Rivlin. Reasoning under the influence of universal distortion. *Ifcolog Journal of Logic and Their Applications*, 4(6), 1769–1900, July 2017.

[2] D. Gabbay, G. Rozenberg and Students of CS Ashkelon Introducing Abstract Argumentation with many-lives, In *Journal of Applied Logics*, Volume 7, Number 2, June 2020, pp 295-336.

[3] P. Baroni, M. Giacomin, G. Guida. SCC-recursiveness: a general schema for argumentation semantics. *J. Artificial Intelligence*, 168(1):162–210, 2005.

[4] Maximiliano C.D.Budán , Maria LauraCobo Diego C.Martinez Guillermo R.Simari. Bipolarity in temporal argumentation frameworks, *International Journal of Approximate Reasoning* 84 (2017) 1–22

[5] H. Barringer, D.M. Gabbay and J.Woods Modal and temporal argumentation networks. *Argument and Computation*, Vol. 3, Nos. 2–3, June–September 2012, 203–227

[6] D. Gabbay, G. Rozenberg, Temporal Aspects of many-lives Argumentation Networks, Submitted to *JAL special issue for Isralog'2017* (rejected, needs revision)

[7] D. Gabbay. Meta-logical Investigations in Argumentation Networks. College Publications, 2013, 770pp.

[8] D Gabbay, I Shapirovsky and V Shehtman. Products of Modal Logics and Tensor Products of Modal Algebras, *Journal of Journal of Applied Logic* 12 (2014), pp. 570-583.

[9] H. Prakken and G. Vreeswijk. Logics for Defeasible Argumentation, in Handbook of philosophical logic vol 4, D. Gabbay and F. Guenthner, editors, Springer 2002, Pages 219-318

[10] P M Dung, On the acceptability of arguments and its fundamental role in nonmonotonic reasoning, logic programming and n-person games Artificial Intelligence Volume 77, Issue 2, September 1995, Pages 321–357

[11] M. Caminada and D. Gabbay. A logical account of formal argumentation. Studia Logica, 93(2-3): 109-145, 2009.

[12] D. Grossi and S. Modgil. On the graded acceptability of arguments. In *Proceedings of IJCAI'15*, pp. 868–874, 2015.

[13] D. M. Gabbay Equational approach to argumentation networks *Argument and Computation* 3 (2-3):87 - 142 (2012)

[14] D. Gabbay, 2017, Abstract Generic Argumentation (AGA). In preparation

[15] C. Cayrol and M.-Ch. Lagasquie-Schiex. Graduality in argumentation. *Journal of Artificial Intelligence Research*, 23:245–297, 2005. [Davey and Priestley, 1990]

[16] S. Egilmez, J. Martins, and J. Leite. Extending social abstract argumentation with votes on attacks. In *Proc. 2nd Int. Workshop on Theory and Applications of Formal Argumentation*, pages 16–31, 2013.

[17] C. Cayrol and M.-C. Lagasquie-Schiex. Gradual acceptability in argumentation systems. In *Proc 3rd CMN (International workshop on computational models of natural argument)*, Acapulco, Mexique, pages 55-58, 2003.

[18] Gerhard Brewka and Stefan Woltran. Abstract dialectical frameworks. In *Proc. KR'10*, pages 102–111. AAAI Press, 2010

[19] G. Brewka, S. Ellmauthaler, H. Strass, J. P. Wallner, and S. Woltran. Abstract dialectical frameworks revisited. In *Proceedings of the Twenty-Third international joint conference on Artificial Intelligence* (pp. 803-809). AAAI Press. (2013, August).

[20] S. Polberg.Extension based semantics of abstract dialectical frameworks. In STAIRS 2014: *Proceedings of the 7th European Starting AI Researcher Symposium* (Vol. 264, p. 240). IOS Press. (2014, August).

[21] Claudette Cayrol, Marie-Christine Lagasquie-Schiex. Graduality in Argumentation. *J. Artif. Intell. Res.*, 2005

[22] Nielsen, SH and Parsons, S 2006, A generalization of Dung's Abstract Framework for Argumentation: Arguing with Sets of Attacking Arguments. in N Maudet, S Parsons and I Rahwan (eds), *Argumentation in Multi-Agent Systems (ArgMAS)*, 2006, pp 54-73

[23] D Gabbay and M Gabbay. Disjunctive attacks in argumentation

networks, *Logic journal of IGPL* Volume 24, Issue 2, Pp. 186-218. `http://jigpal.oxfordjournals.org/content/early/2015/09/10/jigpal.jzv032.full.pdf`

[24] D. Gabbay, I. Hodkinson and M. Reynolds. *Temporal Logic: Mathematical Foundations and Computational Aspects, vol 1: Mathematical Foundations* (Monograph) Oxford University Press, 1994, 671 pp.

[25] D. Gabbay, M. Reynolds and M. Finger. *Temporal Logic: Mathematical Foundations and Computational Aspects, Vol. 2: Computational Aspects* (Monograph) Oxford University Press, 2000. 600 pages.

[26] H. Barringer and Dov Gabbay. Modal and Temporal Argumentation Networks old version, in the Amir Pnueli Memorial Volume, *Time for Verification*, Springer LNCS, Doron Peled and Zohar Manna, Editors, 2010, pp 1-25.

[27] H. Barringer D. Gabbay J. Woods Modal and Temporal Argumentation Networks, expanded version, *Argument and Computation*, 2012, vol 3 issues (2-3), pp 203-227.

[28] Maria Laura Cobo, Diego C. Martínez, Guillermo Ricardo Simari: On Admissibility in Timed Abstract Argumentation Frameworks. ECAI 2010: 1007-1008.

[29] Maximiliano Celmo Budán, Mauro Javier Gómez Lucero, Carlos Iván Chesñevar, Guillermo Ricardo Simari: Modeling time and valuation in structured argumentation frameworks. *Inf. Sci.* 290: 22-44(2015).

[30] Timotheus Kampik and Dov Gabbay. The Degrees of Monotony-Dilemma in Abstract Argumentation published in J. Vejnarová and N. Wilson (Eds.): *ECSQARU 2021*, LNAI 12897, pp. 89–102, 2021.

[31] D. M. Gabbay. Compromise Update and Revision A position paper. In *Dynamic Worlds*, B. Fronhoffer and R. Pareschi, eds. pp. 111—148. Kluwer, 1999.

[32] N.D. Rotstein, M.O. Moguillansky, A.J. García, G.R. Simari, A dynamic argumentation framework, in: P. Baroni, F. Cerutti, M. Giacomin, G.R. Simari (Eds.), *Computational Models of Argument: Proceedings of COMMA 2010, Frontiers in Artificial Intelligence and Applications*, vol. 216, IOS Press, Italy, 2010. pp. 427–438.

CHAPTER 6
REASONING UNDER THE INFLUENCE OF UNIVERSAL DISTORTION. SEX OFFENDERS CASE STUDY

1 Background and orientation

We define and study the phenomenon of a universal distortion into a reasoning system or an argumentation network. Such distortions can happen for various reasons, for instance under the influence of alcohol or a fundamentalist religion, or as the result of a behavioural disorder such as paedophilia. We define the notion theoretically in the framework of abstract argumentation and present an actual case study of a sex offender. We then present a formal logical model.

This Chapter is a conceptual follow up to our paper [1], in which we modelled Reasoning Schemes, Expert Opinion and Critical Questions on the risk involved in the release from custody of a sex offender. Dealing with sex offenders is a high profile area of activity in any society. Once a sex offender is convicted and given a prison sentence, to apply for remission for good behaviour, the sex offender is expected to express regret and remorse and is offered the opportunity to join a therapy group in prison. This will enable the sex offender to apply for good behaviour and reduce the prison sentence by a third (in Israel). Of course it is not surprising that many sex offenders join a therapy group. What is more surprising, is that the sex offender therapists community uses logic and argumentation to treat these sex offenders. The community is not explicitly aware of this connection with the logic and argumentation community. They regard the sex offender as suffering from reasoning distortions (caused possibly by physical drives) and proceed to actually use argumentation to try to correct such distortions and reduce the temptation to re-offend. Once we, the authors, realised this, we were motivated to write the current Chapter and study reasoning distortions in general. Actually when you think about it, it is of great value to the logic community to have essentially a very high profile medical community using logic and argumentation. If the argumentation community could observe and model case studies from the therapy practice, this could immensely benefit both communities, as well as society in general. We envisage the argumentation community helping to improve the therapy methods of the sex offender community. Currently the therapy success rate is that out of the 100% set of sex offenders participating in therapy, 30% show significant improvement. Perhaps this success rate can be improved.

We would like to explain and make it clear to the perceptive reader of the argumentation community what to expect from this Chapter. The authors have three possible policy options

Policy Option E1. Observe the practical use of Logic and Argumentation in the sex offender therapist community, get new ideas for new theories of argumentation and write theoretical papers catering for advancing the research front of Argumentation and Computation. This is a safe bet and has serious value. To give an example from Applied Mathematics, we can observe how engineers push fluids through large pipes, miles long, and develop new theories of Turbulence. The advantage for us is that we do not have to model the application correctly or even mention it, it is enough to be inspired by observing the application, develop new theoretical logics and connect to other works of our theoretical colleagues. The disadvantage is that we give no help or better models to the sex offender therapist or the fluid engineer.

Policy Option E2. The other option is to try to model the case studies and practice using our knowledge and tool box of logic and argumentation, and be of more immediate use to the practitioners and to society. Unfortunately, in the case of the argumentation community and the sex offenders therapist community, we envisage two major problems.

1. The theoretical COMMA (Conference on Computational Models of Argument) community does not have sufficient experience in this type of modelling.

2. The sex offender community is medically minded. If you offer them a model, it is like offering them a new medicine. They would test it for years before finally accepting it.

So before we model practice, the two communities must understand each other much better and this will take time and effort.

Policy Option E3. The authors have decided on a middle option. Start with option 1 but at the same time try a first approximation model, to show the argumentation community that it is worth while to move later to option 2. The advantages of this approach are obvious, but there is also the risk of misunderstanding. Readers will criticise the partial model. We ask the reader to recognise an opportunity for further research and we will try to point out, as we develop this Chapter, any simplifications and shortcuts we employ.

So going back to the business of developing option 3 for this Chapter, we considered the influence of a rise of the sex drive on the offender's reasoning processes.

Our plan for this Chapter is as follows:

First recall the formal background material from Argumentation mentioned in Chapter 1. In Section 2 we say a bit more about universal distortion and then continue giving a general abstract view of distortion. This will prepare us for Section 3, where we discuss distortion in argumentation networks. We offer two main possibilities for distortion in the first two Subsections, these are

- Annihilator types of distortion. This means distorting by deleting (annihilating) some key argument that significantly changes the system, or in a system where arguments have strength, distorting by weakening this strength.

- Non-monotonic distortions. This means distorting by adding true or fake background additional information which gives a completely new perspective to the case.

We continue with Section 4 Subsection 4.1 by listing as examples various distortions by sex offenders. This prepares the connection with a sex offender's reasoning and in Subsection 4.2 we examine a real case study of a real offender.

The case study, presented in Section 4 is purely descriptive, relating reality as it is. It requires analysis and this we do in the next Section 5. The first Subsection, 5.1 presents the context (in the sex offender therapy community) of the case study and Subsection 5.2 analyses the arguments used in the case study. We are now in a position to understand, from the argumentation point of view, the nature and context of the sex offender reasoning distortions. We need however, before starting with our formal modelling, to understand comparatively the workings of reasoning distortions in general (not just that of the case of sex offenders).This we discuss in Subsection 5.3 and in fact Subsection 5.4 reveals that the Literary pragmatics community also deals with reasoning distortions in literary narratives.

We are now ready to start modelling universal distortions. Section 6 presents a first attempt at modelling. We begin in Subsection 6.1 with an intuitive semi- formal discussion of possible ways to define models, leading to Subsection 6.2, where we focus on the use of valuations, (the distortion being lowering the value and relevance/importance of some arguments, thus distorting the network). Subsection 6.3 summarises our formal initial options and Subsection 6.4 gives an initial valuation model, where each argument x is given a number $V(x)$ saying how many successful attacks are required for x to be out. This model is studied in detail and is an example of "export" from the sex offender area into formal argumentation. The approach in Subsection 6.4 is set theoretical, there is also an algorithmic approach which is discussed in Subsection 6.5. Section 7 gives better models, comprising of an initial discussion of how to do better, leading to the better model of Abstract Valuation Frameworks (AVF). The AVF model is presented generically, in accordance with our following of

Option E3 above. Section 8 compares with the literature and Section 9 concludes.

2 Abstract view of distortion

Reasoning distortions are common to every human being, but sex offenders have unusual and exceptional cognitive distortion. This motivated us to look at what happens when there is a major distortion of a reasoning network. The purpose of the current Chapter is to model the possible effects of such disturbances.

Let us list some familiar examples of universal disturbances.

1. A group of scouts equipped with compasses and maps, dropped on a hill in a national park is instructed to find their way to a meeting point. Nobody realises that there is a high concentration of iron ore in the area which distorts the compass readings and the group ends up moving in circles.

2. A man drinks a bit too much at a party and does not realise that the influence of alcohol is altering his perception of the reality.

3. A boy on a date is carried away by his hormones and does not respond to his girlfriend's objections to his sexual attentions.

4. A con man suddenly gets religion and changes his lifestyle.

5. A society is struck by an overwhelming natural disaster, such as an earthquake or hurricane and becomes subject to emergency laws.

6. A computer overheats and starts acting erratically.

7. A vital component fails in a complex system, affecting performance.

8. A cyber hacker maliciously penetrates a system and changes it.

9. Any small child has a major reasoning distortion in that the child does not have clear boundaries between what is real and what is imaginary. This creates problems for example when the child is a witness (being a sex victim).

10. Advertising distortions. These are hard advertising campaigns intended to create reasoning distortions favouring sales. One such method is to associate a product (e.g. fast cars) with basic instincts such as Macho attitudes in men.

> Psychologists Megan Vokey, Bruce Tefft and Chris Tysiaczny at the University of Manitoba (See [45]) analyzed advertisements in men's magazines to see what messages they were sending about what it means to be a man. They found that a significant number of the advertisements portrayed or promoted one or more of the following beliefs:
>
> - Danger is exciting.
> - Toughness is a form of emotional self-control.
> - Violence is manly.
> - ItÕs fine to be callous about women and sex.

Remark 2.1. *We note that some of the distortions are mistakes which happen as a result of reasonable reactions based on incorrect assumptions. No-one would call into question — at least to start with — the sanity of the lost scouts in trusting their compasses.*

Another example would be the drunk man crossing the road using the following assumptions:

Assumption 1.   the car is moving slower than it is

Assumption 2.   he can move quicker than he can.

In the second example he may incorrectly assume that he is much fitter than he is, or he may simply be calculating his agility based on what he knows about his speed of movement when sober.

To take this to the example of the sex offender, men (even very reasonable ones) often make a lot of incorrect assumptions about women. Can it be that the sex offender is simply someone who has low or non-existent empathetic ability and who assumes everyone in the world thinks like him or is simply an instrument of his own urges? This basic assumption would result in inconsiderate behaviour at the very least and of violence if this view of the world is contradicted in some way. (This is in fact a major distortion and it will be addressed in our second model in Section 3.)

The example of sex offenders making the wrong assumptions and resorting to violence when contradicted can take an extreme form. There are many examples of sex offenders killing themselves when caught, see [62]. This would be a reasonable and logical thing to do if the offender is a person who believes the rest of humanity is his toy, and then he discovers that the "toy" has turned back on him.[1]

We now consider our options for modelling universal distortions. Let us forget for a moment about practice and sex offenders etc. Let

---

[1]There has been very little work on the reasons behind the suicide figures, beyond the four categories set out by Emile Durkheim in the nineteenth century (see Wikipedia [63]) but at the time Durkheim was writing there was very little study, or even recognition of, sex crimes as a separate category especially those relating to children.

There is no doubt that convicted sex offenders have a higher suicide rate than in the general population (see [53]) but we are somewhat thrown into speculation about why pedophiliac sex offenders have a higher risk of suicide and why this risk is especially high for those who have used violence. One theory we have discussed is based on the general recognition that almost all sex crimes of any sort are to do more with the exercise of power than with sexual desire and following from this proposition, that the sexual exploitation of children is the ultimate exercise of power for an inadequate individual. So, when such an individual is caught and all power is taken away from him, he can see no more meaning in life. Also, if an individual has a pattern of violence it would make sense that he would find a solution to his problems in violence — i.e. self harm — rather than in introspection. Whether the individual resorts to suicide because he is unable to cope with the feeling of powerlessness inherent in judicial confinement, or whether he is simply rejecting life itself as a disappointment too heavy to bear is a matter which might be worthy of some consideration. Unfortunately, finding out the deeper reasons will always be problematic, given that by the time the pattern has played out, the determined Suicide is beyond human analysis.

us just take the idea of a theoretical approach to universal distortion in a reasoning system. We have two problems here:

1. Formally define what is a universal distortion in formal logical system such as classical logic or abstract argumentation.

2. When we model a practical area where there is a possibility of a practical universal distortion, we try to model the area using a formal system with compatible formal distortions.

There are however problems. The first problem is that to model a distortion in a system we first have to model the system itself and then consider how the system may be distorted, which is not as simple as one might think.

Remark 2.2. *We are aware that there are related papers on Argumentation Dynamics and revision which need to compared. The difference is in scale and intention. If we delete or add an argument or cancel an arrow of attack or arrow of support, then this is more of a local interference than a large scale distortion. With a universal distortion we make a big global change/interference. Of course if we take out an argument which attacks many other arguments, or take out a large set of arguments, then the effect could be global, but the intention may be local but with global consequences. The global consequence are done, however, as a side effect, without any general principle involved. In the case of argumentation dynamics, on the whole we deal with is local interferences while the case of distortion on the whole is global via some general principles making a global change.*

Consider the schema in Figure 1. In this figure, practical system P_1 is distorted by disturbance D to become the system P_2. The system P_1 is modelled by formal logical model $\mathbf{M}_1$. $\mathbf{M}_1$ is a natural best model for P_1. We would like to model the distortion by $\mathbf{M}(D)$ and formally apply it to $\mathbf{M}_1$ and get $\mathbf{M}_2$, modelling P_2.

It is important to note that the distortions, as modelled in this Chapter, do not change the underline logics, but distort the logical modellings of P_1/P_2. See, however, Remark 2.3.

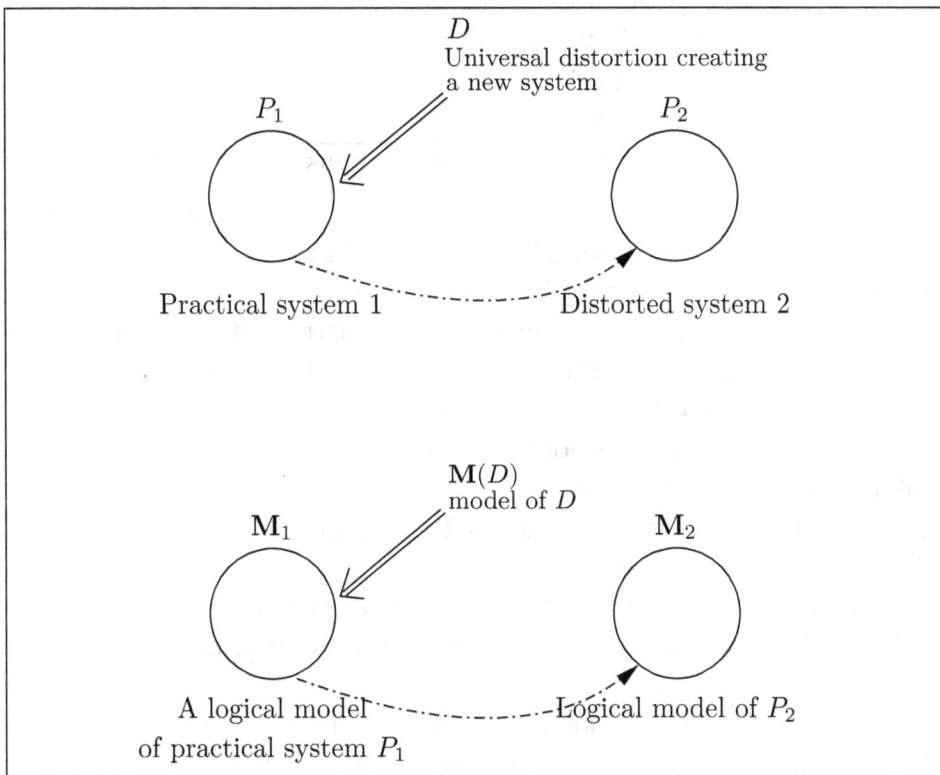

Figure 1

This schema looks reasonable but is problematic in its execution. Here is a list of some major problems. The emphasis is on "best natural modelling".

Problem 1. We assume system P_1, which is a practical system, is naturally modelled by the logical system $\mathbf{M}_1$. We know that P_1 is distorted into P_2, which we may naturally be able to model by $\mathbf{M}_2$. But we also need a reasonable transformation from $\mathbf{M}_1$ into $\mathbf{M}_2$, by the logical tool $\mathbf{M}(D)$. This may not be possible because it may be the case that $\mathbf{M}_1$ is not capable of distortion.

Let us be more specific. Many members of the argumentation and related communities use classical logic as a major modelling tool. Classical logic is not easy to distort. To see this, let us take the most basic deduction in classical logic.

1	A	assumption
2	$A \to B$	assumption
3	B	conclusion by modus ponens from (1) and (2)

This deduction is presented by a child to his mother. "Mummy, you said that if I was a good boy you would give me chocolate. I was a good boy yesterday, can I have my chocolate now?"

Unfortunately mother is harassed and distracted. So we expect a distortion. How can we interfere with the modus ponens deduction?

Classical logic can only add or delete assumptions and rules from a deduction. In our case, deletion of (1) or (2) or both can be the interference. Therefore the the only possible distortion in this case can be that the mother would say to her child "I don't remember". This is fine for this case.

Let us agree: interference/distortion in classical logic is mainly deletion/loss of data. So to clarify, if we give a proof in classical logic based on assumptions, then the proof can be distorted if some assumptions are deleted or lost. Adding assumptions will not invalidate the proof. On the other hand if a top executive is seen pinching his secretary's bottom, common sense will immediately see the executive as a sex offender. However, adding the information that his secretary is his wife might add a new perspective on the case and might mitigate the offence.

Problem 2. Let us now look at a different scenario. The drunk man at the party mentioned above wants to cross the road. A car is coming. He estimates the car is far away. He crosses the road and is hit by the car. The problem is that no matter how you model this in classical logic, the only formal distortion available is deletion, but in this case the real distortion is is not a matter of deletion. The man does not believe that there is no car coming. We can of course try to be smart and take a more complex model with data about car speed etc and delete the car speed as a distortion. But this model is too complex and violates the principle of simplicity. Another option to model this scenario could be for the drunk man to think: "A car is coming. But I have time to cross the road". In this case, the distortion could be represented by a deletion of the belief "I have time to cross the road". Again, we violate here the principle of simplicity. The fact is that a sober man sees the car far away and crosses the road and all is well. When drunk you do the same, except your reaction time is slow. The sober man does not calculate time, if this were the case, the drunk man would also calculate time and by the time he had finished his calculation, the car would have passed.

Let us now draw conclusions from the above discussion. To model distortion we need the following tools:

1. Understand formally how distortion can work in known logics, especially those logics which are used extensively in modelling reasoning. Let us call this "formal theory of distortion".

2. Identify those logics which are amenable to modelling distortions (of themselves) and try to use them to model those practical systems which are in practice prone to distortions. Such systems may unfortunately happen to be modelled by logics which are not capable of much formal distortion.

3. Study distortions in practical systems and try to understand how they work.

4. Use (1) and (2) to model (3).

We now give examples.

Example 2.1. *We discussed modus ponens in classical logic. Let us write a slightly different deduction: $A, A \to (A \to B) \vdash B$. There is not much that a distracted or drunk person can distort here except deletion.*

Suppose we work in a resource logics, say in linear logic. In linear logic the deduction above is not valid. You need two copies of A to get B, i.e. $A, A, A \to (A \to B) \vdash B$.

A drunk person has more scope for distortion in this logic, he/she may see double. So "A" becomes "A, A" and the deduction goes through.

Going back to the child, he may have needed to be a good boy for two days in a row, but he asked of his chocolate after the first day. The harassed mother did not notice.

On the other hand, crossing the road after the party, the man might think that there are two cars coming and might not attempt to cross the road at all.

Remark 2.3. *This is an opportunity to make a remark for readers familiar with the instantiated approaches to argumentation known as ASPIC or ABA, [7, 8]. These systems use arguments instantiated through proofs in classical logic, each in his own respective way. So it is quite possible to have several different arguments/proofs attacking another argument/proof, all using the same basic fact as part of their respective proofs. A universal distortion can be affected by rejecting classical logic in favour of linear logic, which allows the use of facts only once. Thus many attacks will be disqualified. In fact, the use of linear logic makes intuitive sense. Our perception of it in day to day reasoning is manifested in statements like "everything seems to depend on a certain key fact x".*

The idea of linear logic is that we can use an assumption only once, after that the assumption cannot be used again. A favourite example is if you have a dollar you can spend it only once and after that you do not have it any more. Many arguments might use the same assumption/fact in several contexts and so saying let us use linear logic would invalidate such arguments. No sex offender would create a distortion by saying "I use linear logic" but the sex offender might say "you rely in all your accusations on this one witness, this is wrong".

Consider the following argument
Assumptions:

1. *If a dollar can buy you a cup of coffee then get a dollar*

2. *a dollar can buy you a cup of coffee*

Conclusion

3. *I have a cup of coffee*

Assumption 2 need to be used twice in Modus Ponens.

3   Distortions in formal argumentation (towards modelling sex offenders)

This Section models the distortion schema of Figure 1 for the case of argumentation. We get our inspiration from practice in dealing with sex offenders. We present two models. One we call the annihilator model and one we call the non-monotonic model. Let us give a brief intuitive explanation first and then we define and discuss the models in Subsections 3.1 and 3.2 respectively. Let us offer the reader two images which illustrate the two possible models. Let us start with a normal normative person with a reasonable normative reasoning system. Let us forcefully inject this person with hormones that permanently enhance his sex drive. From then on his reasoning and behaviour become distorted. This is the first model. A change due to one single disturbance. We can obtain the second model if we assume the person say survives miraculously some accident and becomes a born again believer. This is the second model. A sudden injection of a set of truths which changes his reasoning and behavioural patterns.

We note that the two models can be combined into a single model containing all features from both models. However, for exposition purposes and also for export to cases other than sex offenders, it is good to identify two separate models.

Let (S, R) be an argumentation network (for background definitions see Chapter 1). Let $\mathbf{I}$ be a new node. Let $\mathbf{d} \in S$. Form the new network $(S \cup \{\mathbf{I}\}, R \cup \{(\mathbf{I}, \mathbf{d})\})$. What we have done is added an external node $\mathbf{I} \notin S$ and let it attack $\mathbf{d} \in S$. This causes $\mathbf{d}$ to be out.

This is why we call $\mathbf{I}$ an annihilator. We now look at $\mathbf{d}$. We say that $\mathbf{d}$ causes a distortion, if from the point of view of (S, R), a change of (the $\{$in, out$\}$) value of $\mathbf{d}$ can cause large scale changes in the extensions of the network. For example, if $\mathbf{d}$ attacks a large number of other elements of S, then a change in the value of $\mathbf{d}$ can cause a distortion.

We now explain the nature of non-monotonic attacks. In our 2009

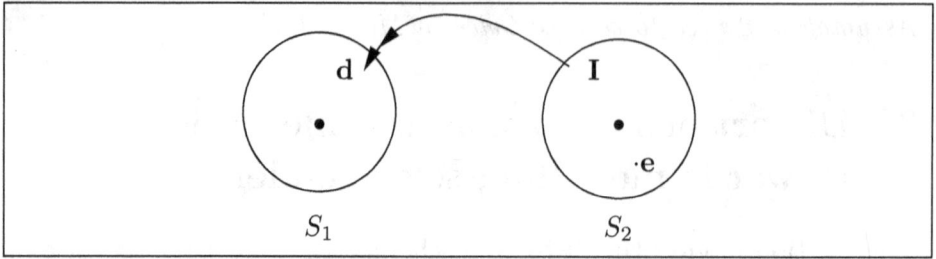

Figure 2: Schema for annihilator distortion

paper [5], we introduced the notion of non-monotonic attack. The nodes of the network (S, R) are non-monotonic theories. An attack from a theory Δ_1 to a theory Δ_2 is executed by forming $\Delta_1 \cup \Delta_2$. In such a context, a major distortion can arise if the underlying non-monotonic system is changed. For example we may inject into each theory of S the additional information, the theory Θ. Such a change may cause a large scale change in the nature of all attacks in the system.

3.1   Models for annihilator type of distortion

Figure 2 explains the model schematically. Our network (S, R) can be decomposed into the union of two networks (S_1, R_1) and (S_2, R_2).

S_1 is the undistorted system of argumentation. It allows for certain possible complete extensions which are considered acceptable and normative. The element $\mathbf{d} \in S_1$ is an argument which is out because of a related system S_2 in which a key factor $\mathbf{I}$ is in. $\mathbf{I}$ is an annihilator node. This $\mathbf{I}$ attacks $\mathbf{d}$ and so $\mathbf{d}$ is out. If $\mathbf{d}$ were in it would cause distortion. We see $\mathbf{I}$ as an annihilator for $\mathbf{d}$. The intuitive meaning of the concept of distortion is a large scale change in the extensions of S_1. Intuitively not every change of an element $x \in S_1$ from out to in will cause a large scale change. For example, if $\mathbf{d}$ attacks many elements of S_1 then if $\mathbf{d}$ changes from out to in, it may induce a large scale change in S_1. To explain our notation, we use $\mathbf{d}$ (for distortion) and we use $\mathbf{I}$ for the inhibitor. Changes occur in S_2 which force $\mathbf{I}$ to be out and so $\mathbf{d}$ becomes in and so $\mathbf{d} =$ in causes distortion. So what

$$e \xrightarrow{\hspace{3cm}} \mathbf{I}$$

value of   value of

e is $V(e)$ $\mathbf{I}$ is $V(\mathbf{I})$

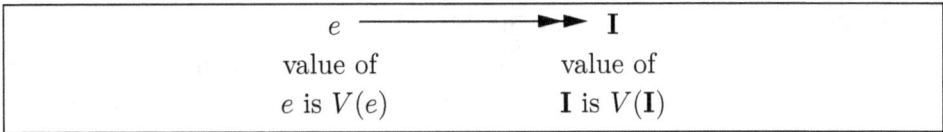

Figure 3: Attack with values

kind of changes can occur in S_2? For S_2 we use a seriously generalised and modified model, based on the technical instrument of a valuation function V employed in value based argumentation, see [3, 4] and for background material, see Chapter 1. The idea can be illustrated in Figure 3.

Our network has the form (S_2, R_2, V). Each node x in the S_2 has a value $V(x)$ attached to it, say a number $[0, 1]$. Assume $(x, y) \in R_2$, then if the value $V(x)$ is less than $V(y)$, then x cannot attack y. So if our S_2 is exactly the network described in Figure 3 and we have $V(e) < V(\mathbf{I})$, then we disregard the attack arrow $e \twoheadrightarrow \mathbf{I}$ and the complete extension of this network is $e = $ in, $\mathbf{I} = $ in. Similarly if $V(e) \geqslant V(\mathbf{I})$ then we do not disregard the attack arrow $e \twoheadrightarrow \mathbf{I}$. Now assume that it is the case that $V(e) < V(\mathbf{I})$. A distortion occurs when a change in V occurs to V' and $V'(e) \geqslant V'(\mathbf{I})$. In this case the distortion is generated because the extension now is $e = $ in, $\mathbf{I} = $ out.[2]

In Figure 2, the critical argument e changes value and as a result,

---

[2]We shall discuss in Subsection 6.1 the modification we need to the numerical comparison we have presented here. In Figure 3 we generalise as follows:

1. We allow the element e also to attack the value $V(\mathbf{I})$ and/or allow $V(e)$ to modify $V(\mathbf{I})$.

2. We allow for values to be transmitted (appropriately according to some algorithm) along attack (i.e. R) lines

3. These modifications will require us to work within the Equational Framework of [5].

We remark that one can also possibly use preference argumentation as our starting point, see Modgil [23]. This may be technically possible but we think the Bench-Capon valuation approach is more compatible with the sex offenders way of thinking. The therapists use numerical strengths in their tools. So distortions will change the numerical evaluations. In networks with preferences, a universal distortion can change the preferences.

I becomes out, and no longer inhibits **d**. The choice of single factors **d**, **I** and *e* are examples only and in practice there may be several of them. The choice of the Bench-Capon model (see Definition 4.7 in Chapter 1) is based on sex offenders practice and therapy. The therapy used in the sex offenders area changes the relative value of various factors in S_2 to eliminate the distortion.

3.2   Non-monotonic distortions

We begin with an explanation of how the non-monotonic mechanism works. Consider Figure 4.

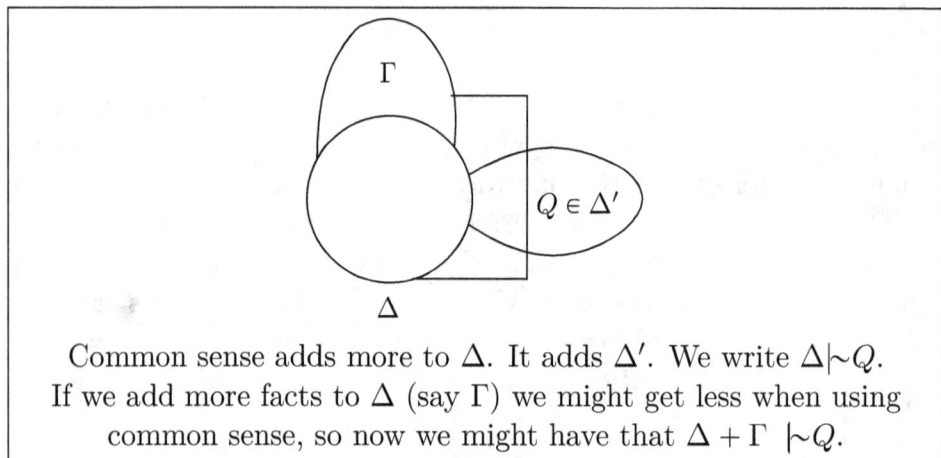

Common sense adds more to Δ. It adds Δ'. We write $\Delta \vdash\!\sim Q$.
If we add more facts to Δ (say Γ) we might get less when using common sense, so now we might have that $\Delta + \Gamma \not\vdash\!\sim Q$.

Figure 4

We start with given recorded data Δ. The non-monotonic commonsense mechanism adds more data to Δ. Call it Δ'. Δ' is not recorded, it may be wrong, but commonsense dictates it. For example if the data is that John is on a strict diet and he is offered a huge ice cream cone, we can add using commonsense that he did not eat it all. Maybe he had a taste. But who knows, maybe it was so tempting that he rejected his diet program then and there!

Now suppose we add the additional data Γ that this ice cream was offered on John's 60th birthday and that he had been exercising extra hard in anticipation of this event. Now it is not clear whether or not

John rejected the ice cream.

The perceptive reader might ask: what is the connection with sex offenders? We answer by an example showing how a normative teenager can end up accused of rape. Let us examine normal common-sense behaviour of a teenager asking a girl on a date. We can safely assume that if the girl says "no", then the boy should not force himself upon her. But in current conditions, with boys accessing pornography on the internet, this boy might have drawn the conclusion that girls never say "no" and mean it.[3] He might interpret the "no" as "yes — but try a bit harder to get me".

This is a distortion. Instead of

$$\Delta \vdash \text{"no" means no, respect it!}$$

we get

$$\Delta + \text{ porn movies } \vdash \text{"no" means "yes, but let us play the}$$
game of refusal and conquest".

So a universal distortion $\mathbf{D}$ in this context is a hidden set of assumptions such that every commonsense query $\Delta \vdash ?Q$ becomes distorted into $\Delta + \mathbf{D} \vdash ?Q$.

Remark 3.1 (Summary of Sections 3.1 and 3.2). *Before we move to the next Section, let us summarise what we have so far from the current Section. We have two possible models for universal disturbance.*

We start with a traditional network $(S, R)^4$ with R the attack relation on S and the universal disturbance is modelled as a change of R into R'. The situation of Figure 2 can be accommodated buy letting $S = S_1 \cup S_2$. Note that just changes R to R' connects us with many papers on argumentation dynamics existing in the literature. Are such existing Argumentation Dynamics models suitable for our purpose of

---

[3]No porno film will end in the first 2 minutes because the girl said no. Similarly no action films ends in the first 10 minutes because the hero gets killed, etc., etc.

[4]Consult Chapter 1 for some background material for the reader who is not from the argumentation community (hopefully from the sex offender therapist community).

*addressing sex offender therapy? We think they are not. We say, how-
ever, after looking at the list of arguments presented in this Section,
that we need to be more specific and look at value based networks of the
form (S, R, V), and generate the (argumentation dynamics) change of
the attack relation by changing V to V'.*

*As for the second non monotonic model, this is different. It has a
different form. The nodes are logical theories and the attack relation
is information input.*

*Fortunately, we have a new paper, entitled "the attack as informa-
tion input" which can unify both approaches. The paper shows that the
second model can simulate the first model. In short, more information
can kill an argument. See [11].*

*So at this stage our models have the form (S, R, V), where S is
a set of pieces of information, R is information input, V is a value
function on S and the attack machinery, extensions, etc., etc., is a
modification of some options from [11].*

*To be able to proceed we need to learn more from the sex offender
therapists about how they use logic and argumentation.*

*So we look in the next Section at some real sex offender arguments
and proceed to further look in the next Section at an actual case study
of a real offender.*

4   Distortions in the sex offender case

This Section presents the data about distortion in the thinking of sex
offenders. This is the area we want to model.

4.1   List of sex offender's arguments

We now show that our model is reasonably motivated by sex offenders' arguments. One of the main differences between sex offenders and
offenders in general is the distortions in the sex offenders' reasoning
process. These distortions can be characterised by the following features (as accepted in the professional community). Before we list the
features, we repeat our words of caution to our perceptive readers:

1. The list below is has been recognised and compiled by the community of sex offenders therapists.

2. We the authors looked at this list and as a result decided to start our theoretical modelling with the Bench-Capon valuation approach.[5]

3. We are not claiming that we are going to model or can model the features in the list below. Our first look at the list below gave us the impression that if we can model the features suggested by the list below, then we should start from the Bench-Capon valuation approach.

4. Our view shall be further refined as we go along and we shall offer more refined models until we discuss a much better model in Section 7.

So let us start the list:

1. Exaggeration. A simple insult can become a major attack which requires a serious counter-measure.

2. Generalisation. One girl rejected me and so I have no chance with girls and my only option is to take one by force.

3. Misinterpretation of facts. My wife smiles at someone and I am sure she is having an affair.

4. Unfounded deduction. A woman accepts my invitation for coffee, which means she agrees to have sex with me.

---

[5]The Bench Capon approach has two aspects:

(a) The Technical aspect, given an argumentation network (S, R), we can associate with each argument x a value $V(x)$.

(b) The qualitative aspect, the meaning we give to these values and what we do with them.

We use the technical aspect of Bench Capon, but give the meaning used in the sex offender case, which is some semi numerical strength/relevance/importance value to the arguments.

5. Extreme opinions.

- My wife says she wants a divorce, but if I force sex on her she will stay mine.
- Children love sex with grownups.
- I must have sex with this woman, otherwise my life is not worth living

The sex offender distorts the system in order to feel more comfortable with what he is doing. In the annihilator model he would change V in a way which puts anything having to do with himself in the highest V value.

The following list gives samples of sex offenders' rationalisations:

1. Kindheartedness.

- I was not attacking, I was only trying to help.
- I did not do anything.
- I exhibited myself in order to teach the children about sex, or the child was sad and I only amused him.

2. Helplessness.

- I cannot stop myself. My drive controls me.

3. Projection-blaming.

- She made me do it.
- I was drunk.
- My friends started it, I was just swept along by them.

4. I have the right to ...

- I spent money on her, she owes me.
- She is my wife, I have the right.
- She is my daughter, I created her.
- My wife denies me sex, so her daughter takes her place.

- It is ridiculous. A man cannot be accused of raping his wife any more than he can be accused of stealing his own radio.

5. Minimalization.

   - It did not bother her.
   - Other people do worse.

6. Justification.

   - She annoyed me. She deserves it.
   - Youngsters nowadays know more about sex than grown ups. They want sex. So what if she is only 12 years old?
   - I had a hard day and was a long time without sex.
   - She sleeps with everybody, why pick on me?

7. Self importance.

   - I am beyond the law.
   - All women adore me. I thought she was just playing hard to get.
   - I know what women think. I know she wanted me.
   - She contacted the police only because I stopped having sex with her and she just can't give up on me.

To put all the above arguments in some perspective, consider a recent incident reported in the BBC news, of a "proxy" sex offender (ex-girlfriend's dog beater), see [61]:

Speke man jailed for pouring boiling water on girlfriend's dog, by Andy Gill BBC North West Tonight.

Here we have an interesting case of a man who beat up his girl-friend's dog.

He was obviously taking out his frustrations on the animal, but the interesting thing is that when speaking about it in court, although acknowledging that he did it, he denies that he is capable of such a thing.

That is, he is holding two mutually exclusive views of reality.

1. He did it.

2. He is not a bad person and so could not have done it.

4.2   Case Study supervised by Dr Gadi Rozenberg

We describe here an example of distortion in an actual sex offender. The therapy has emotional, cognitive and behavioural aspects, [47] but this Chapter focusses on only the cognitive point of view. It includes therapy for changing the distortion [48] and also recommendations for therapy that are not modelled in our Chapter, as we are dealing only with the distortions themselves. The sex offender builds a view (a non-monotonic distortion) which makes him more comfortable with his actions. The therapy is to challenge the offender's view using logic. The sex offender's reasoning is distorted only in connection with his offences. His reasoning is sound in other contexts. The therapist shows analogy between the distorted context and a sound context and points out that the sex offender's reasoning is not consistent. Furthermore the therapy is conducted in groups. It it surprising but it seems that although one sex offender'd reasoning may be distorted for his own case, when faced with the identical distorted reasoning from another offender, the original offender can spot it as a distortion. So the sex offenders can see what is wrong in other offenders but not in themselves. This makes logic group therapy quite effective.[6]

---

[6]Yes, it is significant that this international community of experts dealing with sex offenders actually use argumentation extensively! It offers more opportunities for further research for both communities. We stress again that our purpose in this Section is to present the case study as it is, with a view to determining further what kind of theoretical tools we need if we want to model it. So we are sort of "casing the joint", to further refine our understanding of what kinds of tools we need. We are not yet ready for an initial model. We shall use footnotes to remark on items that require special modelling attention. The next Section will analyse the case study further, but still not yet model it. We shall summarise and offer an initial model in Section 7. We need to properly lead up to this model.

Once the therapy is successfully completed,[7] statistics show that 30% of the successful candidates do not re-offend in comparison with offenders who did not participate in the therapy, see [49].

The following is a description of this particular case [34]. Note the arguments and counter arguments between the offender and Dr Rozenberg. The time scale is 1 year and 8 months.

The case is of a native Jewish Israeli ultra-Orthodox man of about 39, with four children aged 7, 10, 11 and 13.

He would normally spend half of every working day in a religious study group and the other half working part-time as an estate agent, but he is serving a second prison sentence for committing two child sex offences. His first offence was for sexual contact with a child aged about 9, and when he met Dr Rozenberg he was serving a second sentence for sexual activity with two of his neighbours' children, boys aged 9 and 11.

Before the therapy, he denied the offences, stating that the children had falsely accused him but without being able to suggest a reason for their allegations. Following talks with a clinical criminologist, he gradually began to open up about the details of his behaviour. It was decided to accept him for group treatment in the prison Forensic Psychiatric Division for sexual predators. Treatment is about psychological introspection coupled with experience to lead to a change of pattern of thought and behaviour. During treatment all patients experienced a variety of interventions in their distortions. Group therapy lasted one year and eight months, during which the patient expressed a variety of distortions.

Presented below are some of the distortions of thinking and a summary of comments that arose during treatment. The goal was to try to enter his conceptual world and change the mindset and habits

---

[7]"Successful candidate" means the candidate realised his reasoning was wrong and expresses regret about his actions. Successful candidates can get a third of their prison sentence reduced. Of course with such an incentive some (but not all) sex offenders join therapy and express regret but for some of them it becomes genuine. The group therapy (10-14 sex offenders) takes 20 months. In short: Success means the candidate managed to finish the therapy and the therapists believe that his risk to sex offences recidivism is reduced.

of the patient.

When the subject was asked whether he was willing to participate in treatment, he said he would concentrate on the study of Torah to take his mind off thinking "prohibited" thoughts which would prevent him from re-offending. He was told that Torah study for a religious person is very important and can help, but he was asked how he coped with his aberrant drives after his first offence. The subject admitted that he decided to get married as soon as he could in the expectation that his sexual needs would be met by his wife.

It was explained to him that the two solutions he described are external solutions which, although important, are probably not sufficient. Marrying does not shut down a strong attraction to children, which is the central problem. Many victims of this drive honestly repent and believe that the repenting will "save them", but they repent every day, and fail and re-offend every day. Obviously repenting is not enough to break the cycle. In the same way, punishment is generally ineffective in preventing repetition. Instead there is a need for internal adjustment, usually guided by therapy.

At the beginning of the treatment the subject frequently referred to offences in the third person in an attempt to distance himself and so avoid responsibility. For example, he described the offences with the word "occurrence" and described each "occurrence" as a "mistake". It was then pointed out to the patient that he had committed actions of his own volition and, as he had elected to participate in treatment, so he had to regard the offences not just as accidents but as the result of his own thought and planning. To do this he had to start by describing his actions accurately and admitting their significance so that he could take responsibility for them. Review of his actions would extend to his description of even small things. For example, it would not be acceptable to say "the cup fell," it would be necessary to say "I dropped the cup".[8]

---

[8]The formal logician might ask, how are we going to model the subtle difference between "the cup fell," and "I dropped the cup"? After all, modelling the passive in classical logic (and bringing the difference of the passive as compared with the active) is not easy. We say there is no need to micro-model here. We simply annotate one with responsibility and the other without it. So we read the meaning

After we dedicate time and effort in reviewing relatively minor details, it is then easier to set things right in more significant areas. For the purpose of illustration, we asked a group of patients the following question:

Raise your hand if you have ever said to yourself "I'm going to rape now!" or "I'm going to commit a sexual offence".

To date, none of Dr. Rozenberg's patients has raised his hand. All have justified their actions with a variety of explanations and rationales. Such explanations have been along the lines of "I am going to have fun", "it is not offensive", "it is just a game", "the child will love it" and so on. It is therefore critical in our therapy to make clear the importance of correct and precise definitions. We also correct common statements like: "I must say something" and explain that the word "must" implies that there is no choice, and we always have a choice. Even if someone puts a gun to our head and demands we do something, we can still choose not to do so. In pointing this out we make our patients more aware that there is always a choice and therefore it is more accurate to say "I want to say."

The patient made an attempt to transfer responsibility by blaming his inclinations on the sexual attentions of a teacher to which he was subject at about 11 years old. He related that he still had harrowing nightmares and maintained that these memories filled him with "prohibited" sexual impulses. He had not told anyone about it until treatment but he believed that because he was a victim of sexual assault he had become an offender.

After expressing sympathy for the patient, the therapist asked if the patient believed that every victim of sexual abuse became a sex offender. Various studies were presented to him and he had to admit, most importantly to himself, that despite the trauma he nevertheless

---

of "I dropped the cup" as saying "I dropped the cup and I am responsible for this" and we read "the cup fell" as "the cup fell and I am not responsible for it". The problem with the sex offender reasoning is that the sex offender does not want to take responsibility for his/her actions. For this reason saying the "the cup fell" in the passive attaches less responsibility than saying "I dropped the cup". Similarly lowering the valuation V on the descriptions of the offences is designed to distance the responsibility for the offences from the offender.

still had a choice. Dr Rozenberg tried to show him that even though he had suffered harm, he had managed his studies, started a family and functioned as a father.

Another attempt to reduce personal responsibility was the patient's emphasis on feelings of helplessness. He spoke of a strong sexual attraction to children and a lack of ability to resist his impulses, while telling himself that this was his inescapable destiny.

In this case we used examples from religious texts which emphasise the personal responsibility given to each human being by divine power, with righteous behaviour bringing appropriate rewards and wicked behaviour bringing just punishment.

With regard to the patient's inability to overcome his sexual urges we presented him with a strategy we call "The Policeman Test".

In The Policeman Test, the patient is asked if he believes he would commit the offence if there was a policeman standing nearby. Of course if the individual is not suffering from a mental illness the answer is "no" which again proves the existence of choice.

We pointed out to the patient that he had controlled his impulses in the case of his own children, because of the paternal love he felt towards them – a connection which he did not feel towards the other victims.

The patient initially tried to minimise the severity of his crime by emphasising that he had not raped but had "only" committed indecent acts. We explained to him that illicit sexual activity is not a competition and that he cannot compare one offence with another but has to recognize that any harm to a victim is a severe blow. We pointed out that if a man is severely beaten, he will suffer physical pain and psychological damage and that it is not reasonable then to tell him that he should not worry about it because some people have been stabbed to death. Also, we emphasised that an injury can disrupt the entire world of the victim, who will internalise it and carry a lifelong trauma that will affect all future actions and relationships. In this way an indecent act is never trivial but is in fact a severe injury. We also used his own argument that he himself was abused, an experience which has left him with painful and ineradicable memories which have distorted his social interactions and which set him on the path to becoming a sex offender himself. This was given as an example of why he should consider the outcome of his actions and the injury to his own victims.

In the same context the patient argued that his offences could not have been too serious as the children did not object immediately but waited some time before complaining about what had happened to them. We countered that he himself remained silent about the abuse that he had suffered, even though he was well aware of how badly damaged he was by the behaviour of his teacher.

One more argument that the subject used was that the children agreed to his actions and did not protest. We then asked the subject whether he would be willing to commit seppuku. When he found himself unable to answer that question, on the grounds that he did not understand the word, we explained to him that "seppuku" is a Japanese word referring to ritual suicide, which was an act expected of an honourable Samurai in certain circumstances. Just as the subject did not know how to react to the word that he did not understand, so a child who does not understand sexual activity would not know how to consent to, or reject such activity.

5 Analysis of the case study

5.1 Initial analysis

Taking a first look at the case study of the previous Section, we find ourselves puzzling over the process. The first question we might ask is why it takes 20 months to put forward certain arguments to the sex offender. Granted there must be time taken for administration and it is group therapy but still 20 months is a long time. Is there some need for additional, time consuming steps forced by the logical nature of the therapy and the logical attack on the universal distortion of the sex offender?

Let us start by giving more details of the therapy process:

The goal of the therapy is prevention of further offences.

Treatment makes the following assumptions:

A There is no complete cure, but the offender can learn how to avoid abusive behaviour.

B Even if the offender continues to have deviant thoughts he can choose not to act upon them.

C SUD mode (Seemingly Unimportant Decision), see [50]. For example, a sex offender with a paedophilic disorder is asked to deliver packages on a regular basis to an office which happens to be next door to a kindergarten. The seemingly unimportant decision to accept the job may lead to abusive behaviour.

D The offence is planned rather than impulsive.

As is customary in the international sex therapist community, we are using the Relapse Prevention Model to reduce risk in sex offenders. We believe that correction of cognitive distortion is an important part of the therapy.

At the best of times, even a treated offender may continue to experience the urge for deviant sex. This treatment model seeks to help the offender in managing such urges but it is not necessarily a cure nor does it guarantee removal of the urge to re-offend. We focus our treatment on the identification of the offender's sexual offence chain and cycle, and the development of plans to prevent the offender from experiencing a total relapse. We develop the relapse prevention plan after extensive education on the sexual offence chain and cycle. This incorporates an examination of the progression from the initial urge through the stages that culminate in the sexual offence. We then help the offender to understand his own offensive chain and cycle, and to identify his specific pre-offence thoughts, feelings and behaviours. We identify the progressive and self-re-inforcing nature of the pre-offence components to help the offender to recognise that his offence is not a spontaneous event, but the product of a generally predictable series of thoughts, feelings and behaviours.

The relapse prevention plan then takes each step of an offender's chain and/or cycle and generates options [51], diversions and/or alternate behaviours that interrupt his sexual offence path.[9] All activities carried out in therapy relate directly or indirectly to interrupting his offence chain and cycle, and strengthening the relapse prevention plan. The offender is required to acknowledge all his sexual offences during therapy, whether they are known or unknown to other people. The goals of therapy include identification of the patient's chain and cycle of offending, a reduction of denial, working toward taking full responsibility, recognising the impact on victims and developing victim empathy, recognising the impact of victimisation on family members and friends, planning for regaining the trust of family members, self-management of deviant sexual arousal, and working toward implementing an effective relapse prevention plan.

Other issues, such as healthy attitudes toward sexuality, substance abuse and anger management are also addressed. The offender is expected to increase coping skills for all activities, especially when stress or gratification needs are present. Usually, the offender is expected to be in treatment for 12 to 18 months successfully to develop an effective, individualised relapse prevention plan

Treatment focuses on the emotional, cognitive and behavioural aspects and the steps upon which the Group focusses are:

1. Familiarity, working on a establishing contact with the candidates.

2. "I and the other" — patients are asked to draw a picture/image of themselves and draw significant figures in their lives and the work on such drawings allows focussing on the life history of the individual. Other methods at this stage are guided visualisation,

---

[9]The perceptive reader should observe that logically this is a sort of time action cyber protection model.

Let us rephrase this:

The Cyber Protection plan then takes each possible step of a Hackers chain and break in cycle and generates options, diversions and/or alternate behaviours that interrupt the possible Hackers break in plan.

therapy cards, and "Anibi"[10]

3. Empathy or identification of emotions — work on the different feelings and emotions of the patient and the others. This can be done using pictures with facial expressions, writing a letter to a victim and a letter from a victim, reading the testimonies of the victims and so on.

4. Sexual functioning — It is imperative that we convey to the patient not only what is prohibited, but also what are the alternatives.

5. Offence cycle — can be identified as a summary of most of the therapy progression.

   (a) trigger: accelerator of the cycle, which may be an event which is not necessarily sexual in nature

   (b) feelings, thoughts (cognitive distortions),

   (c) dis-inhibitors (like alcohol, drugs and pornography),

   (d) planning,

   (e) focus on the offence,

   (f) reconstruction.

Finally, we work on risk situations and ways to deal with them.

5.2   Further analysis

Continuing our discussion about how to model the sex offender's reasoning distortions and their therapy, we get our clue from the following part of our case study. We said in our description above of the therapy process that:

"In the same context the patient argued that his offences could not have been too serious as the children did not

---

[10]anibi — name of special Therapointing cards with pictures that the patient should choose and tell about himself in an indirect and non-threatening way.

complain immediately but waited some time before complaining about what had happened to them. We countered
that he himself remained silent about the abuse that he
had suffered, even though he was well aware of how badly
damaged he was by the behaviour of his teacher."

It is clear from the above description that there was an internal inconsistency in the patient's statements. The therapist detected it and
pointed it out to the patient and made him aware of the distortion in
his reasoning.

We understand this as typical of many cases of universal distortion.
The distortion affects only part of the system leaving some of the
system unaffected, thus exposing a detectable internal inconsistency.

Furthermore, in group therapy with other patients, although each
patient is unable to detect inconsistencies in his own narrative, each
patient does detect such (even identical) inconsistencies in the other
patients narrative. So the therapist can point out to each patient the
similarities of his own narrative to other narratives and thus making
him able to detect his own inconsistencies.

Therefore the therapy proceeds along the following lines:

1. (a) Take steps to gain the trust of the patient and let him
       disclose more and more of his reasoning network so that
       internal inconsistencies can be better detected

   (b) Put several patients in group therapy, let them give their
       own narrative to their fellow patients. The other patients
       will detect inconsistencies in others but not in themselves
       and then therapy can proceed to make them see the incon-
       sistencies in themselves.

   This process, as described above, takes time.

2. Find ways to encourage in the patient a desire to be helped.

3. Use logic and arguments to have the patient see his internal
   inconsistencies and minimise the distortion

5.3 Comparison with other kinds of distortions

There are cases—such as the universal distortion which results in victims joining fundamentalist (and sometimes murderous) religious movements such as ISIS— which might not be amenable to this type of treatment. Since part of the distortion involved in such cases is that no-one has the correct view of the world except the victim him/herself and other people subject to the same distortion, the victims are not looking for a way to change their attitudes and therefore will not be receptive to understanding their own illogic. Even if we gain the trust of the patient and point out some inconsistencies, the patient might even resort to re-enforcing his beliefs by increasing the distortion instead of decreasing it.[11]

One of our authors remembers what is now an amusing anecdote from her childhood in Yorkshire in the 1950s — a place hardly notable for its cosmopolitanism. As the only Jewish child in her school she was often challenged by her classmates on biblical matters to which they had been exposed during their Sunday school lessons. On one occasion she was cornered by a group who said "You killed Our Lord". Startled by the inconsistency of this argument, the girl responded along the following lines: "Didn't Jesus die to save you from your sins?" This fact was acknowledged. The girl continued: "If He did not die, you would not have been saved." This was also acknowledged. "So what are you complaining about? If the Jews did kill your Lord, we did you a favour." At this point, the whole logic of the argument descended into violent rejection. Whether that was because the attackers were unwilling to accept the truth of the argument, or whether they were simply annoyed with their victim for being a smart-aleck is now something we will never know but the main problem with the counter-argument was that it did not go back far enough to the basic assumption. The problem was not that the Jews had"killed our Lord" but that the Jews were not Christian. Finding and altering such a basic assumption would be extremely difficult and some might say

---

[11]The discussion about ISIS and religious fundamentalists is speculation/conjecture by the authors. We plan to ask for funding and get permission to form therapy groups and see what happens.

completely impossible.

The story of ISIS is not so amusing, but it follows similar lines. An orthodoxy is disseminated, based on certain Ôtruths' which the adherents take for granted and assign to these truths as much reality as we would assign to the existence of France. Once these rock-solid truths are established everything else flows from them with perfect logic. No amount of arguing can shift that unless the original assumption is destroyed. This does occasionally happen, particularly when the victim finds himself on the receiving end of the sort of cruelty which he may have inflicted upon others, having been convinced that what he has been told to do is necessary for the building of whichever ideological paradise is his particular poison.

With regard to point 2 above Ñ that a solution to the problem can be found only if the patient wants to be cured — in the clinical field we have not once met a sex offender who rejects any change at all and argument therapy is more successful than most in preventing recidivism. For example if the recidivist rate is 10 percent, offenders who take argument therapy will have a 7 percent recidivist rate.

Some of the patients have serious organic mental disturbance that would manifest itself whatever treatment was given but some of the patients are not curable because although they are aware that their actions are illegal, they do not want to be cured. That is, they have a low empathetic index and regard their own wishes and feelings as superceding those of any others. In their case any logical argument would hit the brick wall of the patient's own desires.

With reference not to the distortions of sex offenders, but to the distortions formed by religious fanaticism, we might take the example of those who have been ensnared by an organisation such as ISIS. Many column inches in British newspapers have been devoted to the question of why star students have forsaken a comfortable suburban life in the UK to become cold blooded murderers in Syria. There is also speculation about whether or not such people might be amenable to therapy to rid them of their lethal ambitions but it is likely that intervention will work only if the subjects themselves want to change. Since there have been some defections, we can see that a change of mind is possible but a Moslem who has espoused the most literal interpretation of his religion is absolutely convinced of the existence of Paradise and Hell and would not regard any killing for the sake of Islam as murder–even the killing of innocent co-religionists.

In 2014 a Taliban group stormed a school in Peshawar and murdered 141 people, all of them Moslem. See `http://www.bbc.com/news/world-asia-30491435` (accessed May 16, 2017, 1230p hours UK time). The justification for this was that the guilty adults were being punished while the innocent children were being fast-tracked to Paradise before they had been corrupted by their wrong-dealing parents and teachers. According to one survivor, "the terrorists shouted at the boys to say a 'kalma' (an especially holy prayer) before they shot them". It is obvious that the assassins regarded their actions as a kindness, saving the children from an eternity in Hell.

Such people go to their own executions with the certainty that martyrdom will not only ensure their immediate acceptance into Paradise but also the acceptance of their close family members.

It may be observed that although the effects are more extreme and result in more casualties, the rationalisations for the activities of fundamentalists of this type follow the same structure as the arguments used by sex offenders, such as "I have no choice" or "I was converted/corrupted/brainwashed in childhood by a teacher/parent/neighbour" or "I was trying to educate the unbeliever", "this is what I am, I cannot change", etc.

We now conclude our discussion. We note that the above observations allow us to formally model distorted systems via detecting its internal inconsistencies. This is done by internal analogies and isomorphisms. So we need to define such concepts in a plausible way. Our latest model from previous Sections had the form (S, R, V), Where S is a set of atomic arguments, R is a binary relation on S and V is a (qualitative and technical variation of) Bench-Capon valuation. The distortion is modelled as a change in V. We also mentioned that we can generalise and take an information input model as in [11]. Let us for the moment remain within the framework of general abstract set S. We need to introduce analogy and isomorphism. If S has no internal structure, then the isomorphism will be just an abstract auto-morphism of (S, R). This is useless for practical modelling. We must give S internal structure.

The second idea we got from this Section is the need to model the religious fundamentalist who perpetuates and expands his distortion. We leave the details of this modelling case for a subsequent paper. Religious fundamentalists cling to their distortions, even at the cost of denying obvious facts. Dealing with such an attitude is an entirely different ball game. The sex offender knows deep down that there is something wrong and so there is some hope for therapy with him.

Dr Gadi Rozenberg points out that therapy is more difficult with sex offenders who are lawyers or academics. They seem to be more resistant to his therapy logic arguments. We might be tempted to consider a different point of view regarding the role of therapy.

A lawyer believes he is cleverer than the therapist, and demonstrates his intelligence by using ever more devious arguments. One theory that may be advanced is that what is happening is a duel or a game of chess. Sex offenders know they are doing wrong — there are very few who are convinced they are right and they are the ones with psychoses or mental disabilities who cannot in any case keep up an argument — so it could be that when the therapist argues with an offender of normal mental capacity is that he or she is convincing the patient of the therapist's superior intelligence whose logical arguments should be taken on board as having validity.

In that particular model the it is lawyer who resists who is fighting to keep up this structure of his own power. Only when the therapist proves to be cleverer (by putting forward arguments he cannot counter) will he begin to cave in and perhaps finally see things as the therapist sees them.

Dr. Rozenberg refutes this theory. While agreeing that beginning therapy can be described as a fight, in his experience, "winning" increases resistance and cure starts only when the patient realises the therapist cares about him and that they both have a common goal.

We do think however that the "power" model might work with religious fanatics. They believe they have power (God, Allah) on their side and so they feel powerful. However much you argue with them, they have to keep up this illusion of their own power. Even being in prison will not shake them, because they can believe they still have all the power of Truth on their side. It is only if you have better arguments that you will overcome this resistance.

The argumentation would have to be of an extremely high (and possibly superhuman) quality but it might mean that what is needed in such cases is not psychologists but lawyer — and lawyers of a particularly high calibre.

5.4   Connection with the pragmatics community

There is a connection with the Pragmatics community which study text and testimonies to check for internal consistencies and distortions. The need for this arises especially in allegations of sex victims' testimony, such as children. Paper [25] is a sample. The pragmatics community studies text for consistency and coherence. We study arguments of sex offender for the same. To what extent the methods are similar and the mistakes and distortions are similar remains to be studied. The pragmatics community is vibrant and connects with other further away communities such as literary analysis and language and psychology. It is exciting for us to look forward to working with this community.

See also Chapter 7, Section 1 for the Israeli guidelines and Chapter 7, Section 3 for the UK approach. We quote its summary

> SUMMARY (quoted from [25])
>
> "In evaluating the truthfulness of children's allegations of (sexual) abuse, German forensic experts have focused on qualitative aspects of the content of a witness's statement. Within the overall credibility assessment of a witness's statement, known as statement validity analysis (SVA), they have developed a technique referred to as criterion-based content analysis (CBCA), which utilizes content criteria that supposedly are indicative of the truthfulness of

a statement. While first validation studies of CBCA criteria have been undertaken, a theoretical basis of why and under what circumstances deceptive and truthful accounts should differ with respect to these criteria has been wanting. The reality monitoring (RM) approach is proposed as a theoretical basis for discriminating between fabricated and self-experienced events. The present experiment links forensic CBCA credibility criteria to the reality monitoring approach and tests the relative validity of CBCA and RM criteria in discriminating between fabricated and self-experienced video recorded accounts of adult participants. Transcripts rated for the presence of CBCA and RM criteria by trained experts could be classified in an above-chance fashion. On the basis of a factor analysis of CBCA and RM criteria, commonalities and differences between the two approaches are noted."

Another sample paper is [24]. Again we quote from it

Abstract
"This study describes the linguistic differences between the discourse of truth and discourse whose objective is to mislead. The intention to mislead arouses cognitive and emotional functions in the speaker that affect his speech. An examination of the linguistic characteristics that distinguish between the discourse of truth and that of invention among 48 native Hebrew speakers who were asked to tell both true and invented stories found 13 criteria that differentiate between the two types of discourse. The criteria were classified according to the cognitive and emotional functions affecting the speaker, also addressing his level of awareness of these functions. The objective of this Chapter is to demonstrate the effectiveness of the linguistic examination in differentiating between truth and deception. This effectiveness is due to the uncontrollable psychological processes that cause differences between the

discourse of truth and invention. The results may enable
us to construct an instrument for linguistic examination
to differentiate between the two types of discourse."

Paper [26] studies the way an offender views his actions in a way
that makes the offender more comfortable with himself. This is directly related to the way we analyse our case study. Again we quote
from [26]:

Abstract
"This article deals with the strategies the storyteller uses
to influence the listener's perception and thinking. It is
based on qualitative research, which examined the narratives of 12 men who killed their female partners. After
entering prison, the murderer attempts to salvage some
part of his social image. He does this using an assortment
of means in two areas: the content of the narrative and
its linguistic style. In terms of content, all the storytellers
present themselves as extremely positive and their wives
as very negative. With respect to language, the killers use
verbs that distance them from responsibility, they hedge,
repeat words and phrases to persuade, and use figures of
speech they expect will impress their listeners. This artificial discourse is cunningly interwoven in terms of content
and story art to recreate an alternative reality of a man
who is normative and whom society can accept."

6 Formal models of distortion, a first attempt towards sex offender case

In Sections 2 and 3 we discussed modelling distortion in logic and argumentation. In Sections 4 and 5 we discussed distortion in the minds
of sex offenders . We are now ready to connect the two discussions
and attempt to model the distortions we see in the thinking of sex
offenders using the theory of distortion in logic and argumentation.
First we identify the principal distortion features we have observed

in sex offender thinking (sections 4 and 5). This will help us decide what formal logic features we need to model them.

Sex offenders like to feel comfortable with themselves. They are doing no wrong. They are doing good. They educate small children. Women find them irresistible and in fact to the extent that they can be blamed, it is not their fault. They are the victims.

So when presented with arguments and facts to the contrary they distort the evidence and the reasoning using more or less the following:

1. any argument against them is valued as insignificant (lowering its V value) and adding various other qualitative V values that make it less important.

2. add more arguments to show that they are victims.

So from the above it is clear that we need to use formal argumentation systems of the $(S, R, V_1, V_2, ...)$ and the distortion is achieved by lowering /changing the V values.

Furthermore, the distortion the sex offender offers is only partial, affected only in those parts of the system containing arguments attacking his integrity. Other similar parts of the system, not directly related to him, remain intact. This creates structural inconsistency in the sex offender narrative. The therapy, as we have seen in the case study, makes use of this inconsistency. So we need to be able to show formally that parts of S look like/are structurally isomorphic to other parts of S, and further show that the structure of one part is valued differently from the structure of the other part (thus showing the inconsistency). This necessitates that we give internal structure to the elements of S so that we can use the internal structure to define the similarity.

We now have an initial idea of what we need, so let us proceed with our modelling, first with informal discussion and then with a more formal one.

The reader is warned that formal machinery can acquire a life of its own. We have identified that we need to develop formal models of the form $(S, R, V_1, V_2, ...)$, where the Vs are valuation and the elements of S have additional internal structure. Once we look at such structures formally, we have to deal with them in the context of formal argumentation which could mean that we deal with features beyond what is needed directly to model the distortions of sex offenders. Look at it as export of new ideas from the sex offender field into argumentation, pushing argumentation in new directions, not necessarily fully correlated with sex offenders reasoning. This is what is happening in Sections 6.2 to 6.4 and later in Section 7.

This is not surprising, it happens all the time in Science. For example the bouncing of a ball striking a wall necessitated the Dirac δ function, which in turn motivated the development of the the the mathematical theory of distributions. For the formal argumentation reader, we recommend to view Sections 6 and 7 as new theories of argumentation arising from the application area of sex offender reasoning.

Let us now discuss the correlation between features of the sex offender case study and corresponding formal properties required to model these features.

Feature 1. Many lives annotation to arguments

In recent years (especially in the last 2 years), there have been many cases in the press of senior politicians and celebrities both in Israel and the UK who have been accused by victims of sex offending. The patterns are all very similar. A victim from years past accuses the person x of sex offence. x denies it all and gives some explanation pointing out that if it were that serious, why wait so many years to complain? A short while after the first complaint more victims come forward and complain. The number rises to k victim complaints. At which point either x resigns and/or the police investigates and/or x is condemned in the social media, etc., etc. We call k the numbers of lives of x, notation $k = V(x)$.

For example an Israeli minister x resigned after four complaints. An Israeli general y was prosecuted after two complaints, etc.

So the obvious formal addition to the formal Dung theory of abstract argumentation is to add the many lives annotation function V. Networks have the form (S, R, V), where S is the set of arguments, $R \subseteq S \times S$ is the attack relation and $V : S \mapsto \{1, 2, 3, \ldots\}$ is the many lives function. The meaning and role of V is very clear.

- $x \in S$ is considered "out" if the number n of "live/in" attackers of x is $\geqslant V(x)$.

Although the idea behind the addition of the many lives function V is very intuitive, its formal mathematics is not as simple, see [44] and Subsection 6.5. It requires a special additional research paper. There are conceptual problems to be resolved and the new concepts must agree with the old concepts as special cases. This is in addition to showing how to model the sex offender case study. Here is a partial list of questions.

First we note that the obvious distortion in (S, R, V) is to change V. The sex offender will increase V to suit himself.

$\mathbf{Q}_1$:   Suppose we have two nodes x and y, each having three lives with x attacking y and y attacking x. Clearly none can 'kill" the other (assuming the attack takes only one life). So $E1 = \{x, y\}$ is "conflict free" and is indeed a "complete extension".

How many lives does x, y each have left? The obvious answer is 2 lives each. But note what we have here! The set $E1$ is a complete extension to the original network but it also a new network with two lives for each member. So extensions are not just sets of points but are networks with different, derived/calculated new V.

Repeat the process on this new network $E1$ and get a further new network $E2$, this time with one life for each member. $E3$ is a traditional Dung network. So how do we continue? If we continue our calculations as Dung would do then we should have 3 traditional extensions, but if we uniformly repeat our own process we get only $\emptyset$.

So we lose uniformity or we lose compatibility. We need an algorithm which will do a compatible job uniformly and unambiguously. We now describe this algorithm through our example. More details in Subsection 6.5.

We start with a and b having 3 lives each. Choose either a or b from the network. Say choose a. The case of the choice of b is symmetrical.

*   Attack along the arrow, i.e. a attacks b. b now has 2 lives.

* b is still alive so b attacks a, now a has 2 lives.

First cycle complete
Start second cycle

** a continues and attacks b. Now b has one life.

** b attacks a, now a has one life.

Second cycle complete.
Start third cycle

*** a attacks b, now b has 0 life.

***   carry on, b attacks a but b is dead and so now a still has 1 life.

Third cycle complete
Start fourth cycle

**** a attacks b, b has 0 life.

**** continue to a, a has 1 life.

Fourth cycle complete and is equal to third cycle

Stop because nothing is new, (fourth cycle equals third cycle).

If we start with b we end up with b having one life and a having 0 life

If we start with a and b together we end up with both $a = b = 0$ life.

This is a uniform process yielding all options.

The mathematical implementation of it in the general case is complex, as we shall see in Subsection 6.5. See also [44]. See also the next question.

Q2: How do we find all "extensions" in the general case using say the cycling algorithm?

The reader should note that we still need to develop and investigate our cycle algorithm. Our Policy Option 3, which we follow in this Chapter, is just to explain the formal properties and what they entail in principle but not necessarily develop them in detail in this Chapter. So we tell you that the cycling algorithm as described in the previous question, needs to be modified a bit. Take for example a traditional 3 cycle network (traditional means one life only). That is, $S = \{a, b, c\}$ and with aRb, bRc and cRa. This has as extension all undecided. If we cycle through starting with a, we get

- $a = 1$, attacks b, making $b = 0$.
- $b = 0$ attacks c, leaving $c = 1$.
- $c = 1$ attacks a making $a = 0$.

We now have stability and so the extension $\{c = 1, b = a = 0\}$.

We do not have the rule that if all attackers y of a node x which is dead ($x = 0$) are all dead (all $y = 0$) then the said node x comes back to life (x becomes $x = 1$)!

Q3: Can we offer an equational approach to (S, R, V)? What do the solutions mean?

Q4: Can we view V as a special case of weighted/numerical annotation and compare with existing numerical/fuzzy argumentation papers? See Example 6.8.

The reader can see that these questions are questions of the integration of the new ideas within the old framework ideas. Actually [44] sees the formal model as a survival game. The complete extensions being survival groups unable to completely kill one another and containing all others which they can protect.

Feature 2. The attack as information input

Many of the claims of the sex-offender add more information. For example, if y complains about x that he raped her, then a very common answer is that there was consent. This adds more information and changes the nature of the attack/offence. There are two ways to see this:

1. x sends information I to y and so $y + I$ (the information y with I added to it) no longer attacks x.

2. x adds an evaluation $V_I(y) = $ "there was consent".

This new V_I lowers the value of y as an attacker.

The (1) interpretation is wide ranging and requires a new research papers. See papers [11] and [1]. This is the non-monotonic approach.

For our purposes, we use the (2) interpretation. It is simpler and more uniform with the numerical many lives V. This forces us, however, to consider the network of the form $(S, R, V_1, V_2, \ldots)$, where $V_i(x)$ are general formulas of predicate logic which can also be numerical. These mixed possibilities, however, push us to adopt an algebra of the labels $\mathcal{A}$ and the modelling of Section 7.3. Section 7.3 can be very general, see [55].

Feature 3. Internal isomorphism

We note the sex offender case study, where the subject claimed that the child y was not seriously abused and as evidence the offender put forward $V(y)$, that the child has not complained for 20 years. Furthermore, the offender claimed that he x himself is actually a victim. He put forward $V(x)$, that he himself was abused by his teacher 20 years ago and that he, x, was made to be like he is by his teacher abuser. The therapist pointed out that x is inconsistent. He cannot use $V(z)$ to suit himself.

The case of x is similar to the case of y. We cannot express this if x and y are atomic. We need to put content into x and y. We do not need to add much, just

$$y = \text{offender} \ \bigcirc \ \text{child}$$
$$x = \text{teacher} \ \bigcirc \ \text{offender}.$$
$$V(x) = 20 \text{ years passed}$$
$$V(y) = 20 \text{ years passed}.$$

So now we can show the similarity between the two cases, but this means that our model puts structure into the argument. This is discussed in Section 6.1.

Feature 4. Bipolar networks

Sex offenders bring character support. This requires using bipolar systems with attack and support. Support increases $V(x)$ = number of lives and decreases any numerical values $V_{\text{risk}}(x)$, saying how strong is a risk to let x loose in society. We have not addressed bipolarity in this Chapter. We treat this aspect in [44], see also [1].

6.1 Informal discussion

The previous Section showed us that there are two ways to detect distortion in a patient.

The first is to detect internal inconsistencies in his own reasoning system, and the second is to compare his reasoning system with other similar systems. Both ways can be used simultaneously. This is what group therapy does. Put together several sex offenders, let them describe their systems to each other and point out and detect inconsistencies in each other. This means that in order to model distortions we need to use a family of argumentation networks with values, of the form (S, R, V), where the audience for each network in the family are all the other networks. To do this successfully we need a similarity mapping and a good definition of the values function V. We first discuss the similarity mapping and then we discuss the values V. To model a similarity mapping we need to instantiate the atomic arguments in each network S. This instantiation gives the arguments internal structure which can be used to define a similarity mapping. Without the internal structure, if we just leave the arguments as atomic, any similarity mapping would have to be an arbitrary function from S into S and this is too abstract. All we need is some reasonable minimal instantiation. We need not go as far as ASPIC or ABA [7, 8] but it is sufficient to regard diagrams of finite predicate models.

The diagram idea is very simple. Consider the statement a offended b. This statement is atomic. In classical propositional calculus we can only denote it by an atomic letter say $q =$ "a offended b". The internal structure is lost. So if we also have $q' =$ "a' offended b'", we cannot point out the similarity between q and q'. However if we allow the letter "O" for "offend" in the language then we can say that "aOb" is similar to "$a'Ob'$".

Imagine for example a set of elements, say $\{t, a, b\}$ and predicates $\{O \text{ (binary) and } C \text{ unary})\}$. We can form the following atomic statements U (the universe from which we form our arguments) using the diagram of this language:

$$U = \begin{aligned} &\{C(a), C(b), C(t), tOb, tOa, tOt, bOb, bOa, bOt, aOa, aOb, aOt, \\ &\neg C(a), \neg C(b), \neg C(t), \neg tOb, \neg tOa, \neg tOt, \neg bOb, \neg bOa, \neg bOt, \\ &\neg aOa, \neg aOb, \neg aOt\}. \end{aligned}$$

The unary and binary predicates and the list of elements allow us to define the similarity mappings.

The valuation V is defined on elements of U. Let us at this stage take a simple two valued $V(X) \in \{0, 1\}$, as we first want to illustrate the instantiation.

Let us form an example of an (S, R, V):

$$S = \{aOb, tOa, \neg C(b), C(b), \neg C(a), C(a)\}$$

Where the meaning is:
$t =$ teacher; $a =$ patient; $b =$ child
$aOb = a$ sex-offends b
$C(b) = b$ complains

$V(X) =$ "X" is a serious matter
$V(aOb) =$ "a sex-offending b" is serious.
 The reasoning goes as follows:

$$\begin{array}{ll} \text{Assumption 1.} & aOb \\ \text{Assumption 2.} & \neg C(b) \\ \hline \text{3.} \quad \text{Conclusion} & \neg V(aOb). \end{array}$$

a offends b, b has not complained, so the offence is not serious.[12]
The analogous argument is:

4. tOa teacher offended patient

5. $\neg C(a)$ patient did not complain

6. $V(tOa)$ but nevertheless the patient thinks it is serious.

We can therefore point out the analogy function α and detect a distortion.

The function α is:
$$\alpha : a \mapsto t$$
$$\alpha : b \mapsto a$$
$$\alpha : O \mapsto O.$$

We point out to the patient that:

For aOb you said $\neg V(aOb)$
but for the analogous tOa you said $V(tOa)$.

In case we have other networks by other patients we can also have inconsistency detected by other patients. Our patient says:

$$\begin{array}{c} aOb \\ \neg C(b) \\ \hline \text{Conclusion} \ \neg V(aOb). \end{array}$$

---

[12]We are modelling the following argument from the case study:

"In the same context the patient argued that his offences could not have been too serious as the children did not complain immediately but waited some time before telling other adults what had happened to them."

Note that "$\neg C(b)$" attacks the valuation "$V(aOb)$" and not the argument "aOb". This is not allowed in the Bench-Capon model.

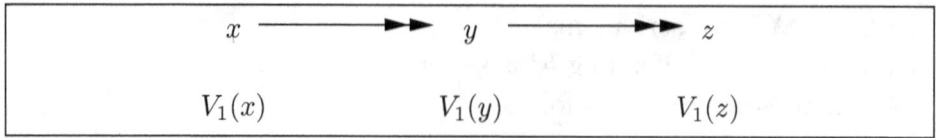

$$x \longrightarrow y \longrightarrow z$$

$$V_1(x) \qquad V_1(y) \qquad V_1(z)$$

Figure 5

Other patients (for example the abusing teacher) will point out that the reasoning is wrong! It should be $V(aOb)$, even though other patients themselves will say in their own respective networks

$$
\begin{array}{ll}
1. & tOa \\
2. & \neg C(a) \\
\hline
3. & \text{Conclusion } \neg V(tOa)
\end{array}
$$

In fact, if the teacher who abused is also present at the therapy it is likely that he will recognise the inconsistency but will exempt himself.

6.2   Discussing valuations

Let us now turn to examine what kind of valuation function V we need to use. Consider the network (S, R, V_1) in Figure 5.

Assume that V gives values in $[0,1]$.

Assume that $V_1(y) = V_1(z)$ and that $V_1(x) < V_1(y)$.[13]

Let us see what we need to do and what the Bench-Capon model does.

1. According to Bench-Capon, since $V_1(x) < V_1(y)$, x cannot attack y and so we have one extension E.

$$E_1 = \{x = \text{ in}, y = \text{ in}, z = \text{ out}\}$$

To model the sex-offender's argument we must also allow x and $V_1(x)$ to attack $V_1(y)$ and so we need sound procedures and

---

[13]We really want relative strength. Taking values in $[0,1]$ gives us relative strength for any finite set of arguments. We take the smallest number as strength 1 and present all the others as multiples of it. We shall insist on the multiples to be natural numbers, which is a restriction on V.

definitions of how to do that in a way that generalises the Bench-Capon model as a special case, as well as being able to model the sex-offender's application. So, for example, if we allow $V_1(x)$ to reduce $V_1(y)$ to a new $V_1'(y)$, say $V_1'(y) = V_1(y) - V_1(x)$, then we get the $V_1'(y) < V_1(z)$ and so y with its new value cannot attack z and the extension will be

$$E_2 = \{x = \text{ in}, y = \text{ in}, z = \text{ in}\}$$

2. We also need to be able to transmit the V value from x to z somehow. This will be addressed later in Section 7 Example 7.4, but we can already use Figure 5 to show the difference in approach.

The Bench-Capon model says that we get the extension

$$E_1 = \{x = \text{ in}, y = \text{ in}, z = \text{ out}\}$$

We ask, what is the V of this extension? The obvious answer is that it is the same as before. In fact, Bench-Capon does not worry about this question. He just uses V to get the extensions. We can however generalise and say, for example, that we get a new $V = V_2$, with

$$\begin{aligned}
V_2(x) &= V_1(x) \\
V_2(y) &= V_1(y) - V_1(x) \\
V_2(z) &= V_1(z) + V_1(x).
\end{aligned}$$

The above is an arbitrary illustration. It does not necessarily fit the sex-offender's application area, but it shows what kind of options we have. The network of Figure 6 is analysed in Section 7, Example 7.5, using the Equational Approach.

When we consider the above directions we would need to generalise, we find that we might consider using the equational framework of [7]. Put differently, since many lives is numerical, we need to augment the equational framework into an equational system (S, R, V) with valuations. We have to say how to generate equations for such a

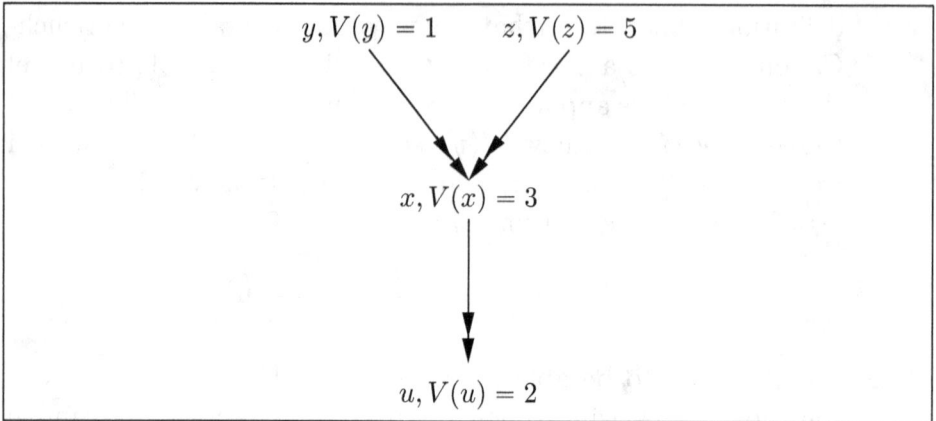

$$y, V(y) = 1 \qquad z, V(z) = 5$$

$$x, V(x) = 3$$

$$u, V(u) = 2$$

Figure 6

system. We should do this in a general way, as a theoretical endeavour and mention the connection to the case of sex-offender's modelling. At this point we are not committing to the equational approach. We might prefer to use the traditional Dung set theoretical fixed point approach, because the many lives although numerical, are natural numbers and solving equations can yield rational or real numbers. We shall address this in Section 7. See [10].

Let us now give a more specific comparison between our view of value based argumentation and the Bench-Capon view. Consider the network of Figure 6. We have arguments $\{x, y, z, u\}$ and value function V.

The values V are relative strength. It is a small generalisation over Bench-Capon (he would only write the order $V(y) < V(u) < V(x) < V(z)$) but we need to quantify V, in order to better represent our view. Bench-Capon (see Definition 4.7 of Chapter 1) will say y cannot attack x but z can. So according to him, Figure 6 is equivalent to Figure 7 (without values).

Our view is different. Let us call it the HML-view (How Many Lives-view). We know the saying that cats have nine lives, so to make sure a cat is dead, you have to kill it 9 times.

So we interpret $V(a)$ as saying how many lives a has. There are still several options for us in interpreting the number $V(a)$, and its

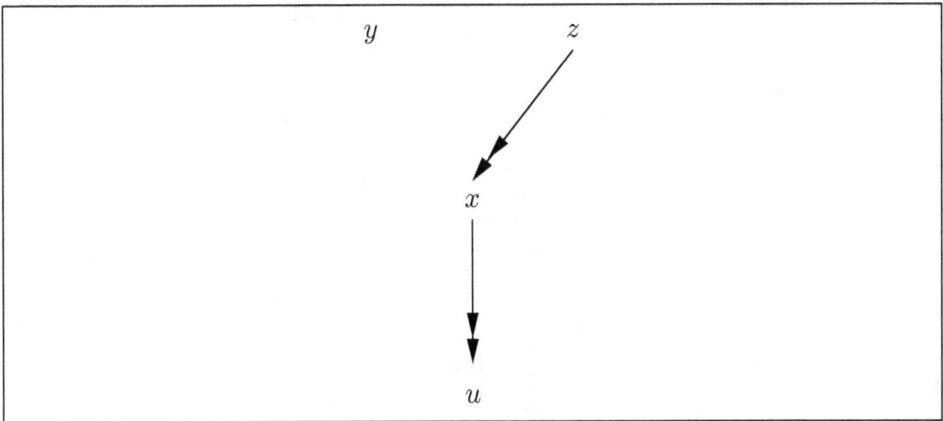

Figure 7

relationship with the numbers V of the attackers of a.

Let us look at Figure 6 and see what are our options in reading it.

1. For any x, $V(x) = 0$ means x is out/dead and cannot attack.

2. $V(x) > 0$ means that x is alive/in and can attack. There is still the question of how does it attack? In what manner? The options for the manner of the attack are dealt in 3., 4. and 5. below.

3. We can understand the number $V(x)$ as indicating how many different in/live attackers are needed to have $x = $ out/dead.

   This understanding requires that we count the number of y such that y attacks x and $V(y) > 0$. The attack of y on x is not influenced by the number $V(y)$. All we need is that $V(y) > 0$. So if for example $V(y) = 7$, y counts as attacking x only once.

4. Another option is to take into account the number $V(y)$ in the consideration of y attacking x. If we want to do that we need to give V a completely different interpretation. Think of $V(x)$ as saying how many missiles does x have to attack or protect itself. So in Figure 6, y has one missile, z has 5 missiles and since they have missiles to shoot, they are in. x has 3 missiles

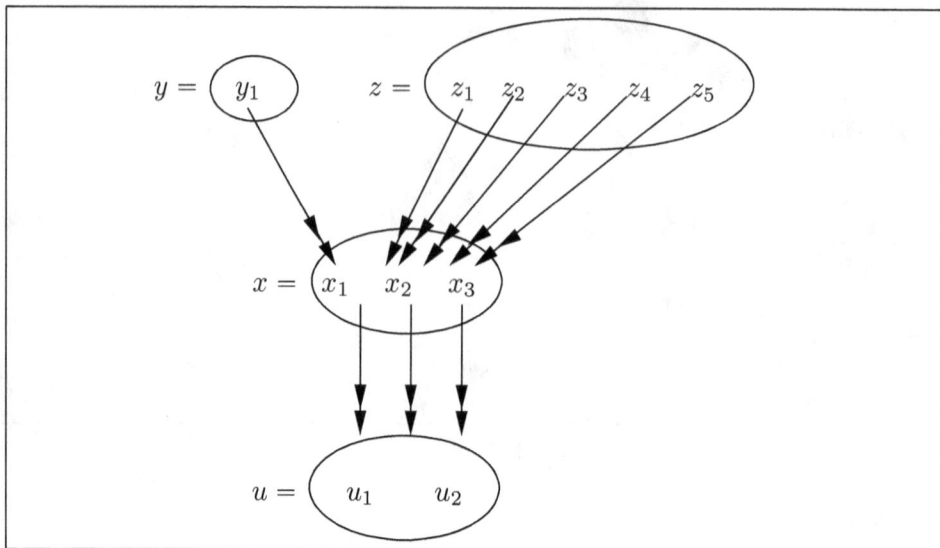

Figure 8

. So if y and z shoot their combined missiles at x, x can only counter with 3 missiles, so x will be dead/out.

This is the cowardly approach, everyone shoots. We can adopt the brave approach, (5) below.

5. Only the attackers y with $V(y)$ greater or equal $V(x)$ can shoot/ attack x.

6. We can use any other criterion to define which nodes y can attack x.

So Figure 6 becomes according to (4) above Figure 8

Perhaps a better way to think of the circles in Figure 8 as people with guns and the value V giving how many bullets they have. The people can be live/active/in or dead/not active/out or unknown/un-decided.

We can follow for example a variation of (5) and require that from among the active people, we select one with maximal bullets and let him attack. So in Figure 8, it is the z which attacks the x. The result is that x is dead because z has more bullets than x has lives. If z were not alive, then the attacker chosen would have been y and y has only one bullet and so x would have survived with two lives left.

The reader should note that this simple idea of HML (How Many Lives) uses the traditional Dung point to point attack idea to define a new type of simultaneous attack on both the argument a and its value $V(a)$.

We need not work out the details in this Subsection, since we just want to explain the idea here.

Let us further remark that in Figure 5, for the case of

$$V_1(x) < V_1(y) = V_1(z),$$

if we were to be more specific and have $V_1(x) = 1$ and $V_1(y) = V_1(z) = 2$, then according to our HML model the extension would have been $\{x = $ in, with $V(x) = 1, y = $ in with $V(y) = 1$ and $z = $ in with $V(z) = 1\}$.

6.3   Intermediate summary of options

We need to pause for a moment and summarise our options for the new concept of argumentation networks with values. We list points of difference with the Bench-Capon approach.

Option 1. Geometrical values.   The Bench-Capon approach is essentially geometrical. Let (S, R) be an argumentation network. Bench-Capon essentially defines a function $\beta(x), x \in S$, telling us which points in S are not allowed to attack x. That is, $\beta(x) \subseteq S$. This is done externally at the meta-level. For example we can give colours to each node and list which colours are stronger than which colour and forbid a weaker colour from attacking a stronger colour. We call this approach geometrical because we can work with R_β instead of with R, where

$$x R_\beta y \text{ iff } (x \notin \beta(y)) \wedge x R y.$$

So for example Figure 6 describing (S, R) is transformed into Figure 7 describing (S, R_β), where $\beta(y) = \beta(z) = \beta(u) = \varnothing$ and $\beta(x) = \{y\}$.

Option 2. Geometrical partition values. The Bench-Capon approach essentially divides S for each $x \in S$ into two subsets, $\beta(x) =$ set of nodes which cannot attack x and $S - \beta(x) =$ the set of nodes which can attack x. The obvious way to generalise this partition is to look at several disjoint subsets of S, forming a partition of S,

$$S_1, S_2, \ldots, S_k, k \geqslant 1$$

and require that any attack on S to be represented by nodes from some combination of S_i. Thus for each $x, \beta(x) \subseteq \{1, \ldots, k\}$. We allow x to be attacked by $\{y | yRx\}$ only if for each $j \in \beta(x)$ we have

$$\{y | yRx\} \bigcap S_j \neq \varnothing.$$

A further generalisation is to say that $\beta(x)$ simply lists the subsets of S which can attack x. That is $\beta(x) \subseteq 2^S$, and we have that x can be attacked only if $\{y | yRx\} \in \beta(x)$.

The above generalisations are still geometrical. Their meaning is that we want attacks on x coming from different audiences and $\beta(x)$ gives the possible acceptable mixtures of audiences.

To connect with the numerical $V(x)$ discussed in Section 5.1, let for example $V(x) = 2$. We can take: $\beta(x) =$ all subsets of S containing at least two elements.

The above generalisation is still geometrical. We can still define

$$xR_\beta y = xRy \text{ and } \{z | zRy\} \in \beta(y).$$

Option 3. Non-geometrical partitions. Option 2 becomes non-geometrical when we connect the geometry with the notion of extension. In other words we require of any acceptable complete extension to satisfy the following:

- x is "out" iff $\{y | yRx \text{ and } y = \text{"in"}\} \in \beta(x)$.

To see the difference, consider Figure 9

Assume that $\beta(z) = \{\{u\}\}$. We do not care about the other values of β. In the pure geometrical interpretation, (Option 2) since the

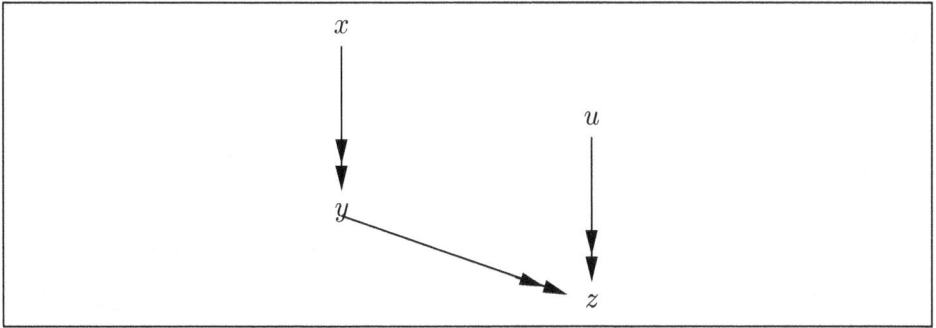

Figure 9

attackers of z is the set $\{y, u\} \notin \beta(z)$, we delete the attacks $y \twoheadrightarrow z$ and $u \twoheadrightarrow z$ and end up with $R_\beta = \{x \twoheadrightarrow y\}$ and the only extension we get is $\{x, z, u\}$. However, if we use Option 3, since $x = $ in and $y = $ out, and $u = $ in, we have that the set of "in" attackers of z is $\{u\} \in \beta(z)$ and so the attack of u on z is accepted and the extension is $\{x, u\}$ only.

This can be expressed in Abstract Dialectical Framework (ADF) [27], see Chapter 1.[14] In this framework, a node x is "in" iff some Boolean formula B_x holds for $\{y|yRx\}$ under the truth values assignment to the ys generated by $(y = $ "in"$)$. Consider for example Figure 9, in this figure z is attacked by y and u. We can consider the formula $B_z = (y \wedge \neg u) \vee (\neg y \wedge u)$. According to this formula, z is "in" if exactly one of its attackers is "in". Compare with Example 4 of Brewka and Woltran's paper [27]

Option 4. Dynamic non-geometrical. This option is where the function β of Option 3 changes dynamically with the construction of the extension. Such a possibility was hinted at in Section 5.1. We need the equational approach to implement it. We cannot say more

---

[14]ADF is a recent powerful framework which we considered modifying and using for our purposes. This option is still on the table. ADFs are defined for all of "Dung standard-semantics" as well as for some other semantics (like stage, semi-stable, ...). We shall propose in Section 7 our own Abstract Valuation Networks, AVFs, which we will discuss and compare with ADFs in the light of the complete semantics.

in this summary Section.

6.4   The formal many lives valuation models

We now develop several formal models to reflect our discussion in the previous Subsection.

Our first model has the form (S, R, V) where S is the set of arguments, $R \subseteq S \times S$ is the attack relation and $V : S \mapsto \{0, 1, 2, \ldots\}$ gives for each argument $x \in S$ the value $V(x)$ in the set of natural numbers indicating how many 'lives' x has.

For example we have the idea that a cat has nine lives. So to get it 'dead' you need to 'kill' it nine times. In argumentation terms, for x to be 'out', it needs to be attacked by at least $V(x)$ number of attackers which are 'in'.

To give an example, suppose the wife is considering taking the family with the children to a three week holiday to India. The husband wants to argue against it. To really "kill" this option he needs several arguments

1. The holiday is too expensive, we cannot afford it

2. The 3 weeks package group holiday is too long, we will have to take the children out of school

3. India is not the best place to go, we will have to be careful what we eat and drink. Westerners are not immune to local infections.

4. The flight is too long, the younger children cannot take it.

5. The real reason might be that the husband simply does not like travelling but he cannot say this to his wife.

Note that each attacker y can kill only one life of the cat x. The attacker himself may have say $V(y) = 2$ lives, but for the purpose of attacking x it can take out only one life of x. If there are only $m < V(x)$ such attackers, we have two ways to view this:

1. The attack fails and the number $V(x)$ remains unchanged

2. The attack fails but the value for x is reduced to $V'(x) = V(x) - m$.

We are now ready with our formal description of the model.

For the purpose of exposition and clarity of conceptual progression, we start with networks (S, R), that are finite acyclic.

Definition 6.1. *A network (S, R), with S non-empty and R a binary relation on S is said to be acyclic if there does not exist a finite sequence of nodes of the form $(s_1, \ldots, s_n)$ such that $n > 0$, and $s_n R s_1$, and for each $0 < i < n$ we have $s_i R s_{i+1}$.*

The next Definition 6.2 is a technical definition needed for later proofs. It recursively defines the distance of a point from the top nodes of acyclic networks. For example in Figure 6, nodes y and z are of level 1, x is of level 2 and u is of level 3.

It is placed here in the Chapter but its use is later.

Definition 6.2. *Let (S, R) be a finite acyclic network. We define the notion of a node x in S is of level n, $n = 1, 2, 3, \ldots$ as follows.*

- *x is of level 1 if there is no y such that yRx.*

- *Assume a subset S_n has been defined of nodes of level less than $n+1$. Assume each point of S_n has a unique level $k < n+1$ and that if any such point y is of level $k > 1$, then for some point x in S_n of level $k - 1$ we have xRy.*

Let z be any point of level n and assume u is a point such that zRu holds. We must have that u is different from x. If u is not in S_n declare u as a point of level $n + 1$.

Let S_{n+1} be $S_n \cup \{u : u \text{ is declared of level } n + 1\}$.

Since S is finite the process will terminate.

Proposition 6.1. *Let (S, R) be a finite acyclic network, then it has at least one point x such that there is no y such that yRx holds. In terms of Definition 6.2, the node x is of level 1.*

Proof. Assume otherwise, then for each x there is a y such that yRx. Choose any point x_0 in S. Then there is an x_1 such that x_1Rx_0. Continue and find x_2 such that x_2Rx_1. Continue by induction and get an infinite sequence $x_0, x_1, \ldots$. All points in the sequence must be different because we have no cycles. This contradicts the assumption that S is finite. $\qquad\square$

Definition 6.3 (*V*-network)**.**

1. *A finite acyclic argumentation network has the form* (S, R), *where S is a finite non-empty set of arguments and $R \subseteq S \times S$ is the acyclic (see Definition 6.1) attack relation.*

2. *A function V from S into the set of natural numbers $\{0, 1, 2, \ldots\}$ is called a many-lives (ML) valuation on S.*

   We require that if x is not R attacked $((\neg\exists y)(yRx))$, then $V(x) > 0$.

Remark 6.1. *This remark motivates Definition 6.4 of the notion of V-semantics.*

Our starting point is a network $(S, R), S \neq \varnothing, R \subseteq S \times S$. Let us choose $x \in S$. An attacker of x is any y such that yRx. This attacker y can "kill" x if y is "alive" (i.e., y is in) in which case we must have that x is out. To defend x against y we need an element d such that dRy.

This is the traditional semantics, where each element $z \in S$ has one life $(V(z) = 1)$.

When x has more than one life, say $V(x) = 2$, any geometrical attacker y on its own is not endangering x to be "dead", even if y is alive. We need two such live attackers at least. Therefore we have to think in terms of sets $Y \subseteq S$ of attackers of any node x. If the number of elements of Y is at least $V(x)$ and for all $y \in Y$ we have yRx, then Y is a potential threat to x. If all its elements are alive then Y can kill x.

Now let us look at defence. How can x be defended against Y? We simply need to reduce the number of live members of Y to be less

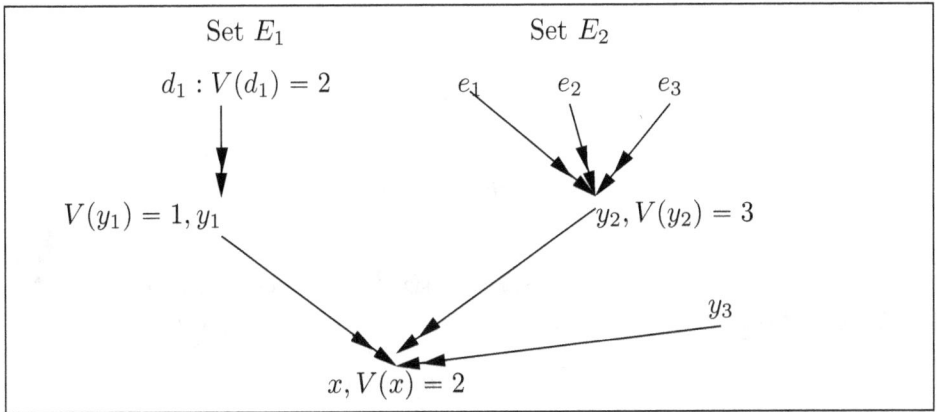

Figure 10

than $V(x)$. So (consider Figure 10 and) assume for example that $Y = \{y_1, y_2, y_3\}$ and $V(x) = 2$ and that $y_1 R x, y_2 R x$ and $y_3 R x$.

So to defend x against Y we need to be a threat to at least two of the elements of Y. Say we have attackers of y_1 and y_2. We already agreed that such attackers must be subsets of S, E_1, E_2 with enough elements in them to be more than $V(y_1)$ and $V(y_2)$ respectively. Say if for example $V(y_1) = 1$ and $V(y_2) = 3$, then E_1 must contain at least one element d_1 and E_2 must contain at least 3 elements e_1, e_2, e_3 such that $d_1 R y_1$ and $e_1 R y_2, e_2 R y_2$ and $e_3 R y_2$ all hold.

Figure 10 describes this situation geometrically (i.e., in terms of R).

Let us state this formally. Let E be a set containing $E_1 \cup E_2$. We can say that E defends the node x against the attack of Y on x because there exist $E_1 \subseteq E$ and $E_2 \subseteq E$ such that the situation of Figure 10 holds.

Namely E_i attacks with sufficient force (i.e., number of nodes z in E_i is at least $V(y_i)$ respectively) to reduce the number of unattacked nodes of Y to be less than $V(x)$.

Let us now consider the set $E' = E_1 \cup E_2 \cup \{x\}$. Is this set conflict free? The answer is yes. Although x attacks d_1, we have $V(d_1) = 2$ and so x cannot kill d_1 even if x were alive.

Does this set E' protect its members? The answer is yes. The potential attackers of x are the sets $\{y_1, y_2\}, \{y_1, y_3\}, \{y_2, y_3\}$ and $\{y_1, y_2, y_3\}$. The set E' has subsets E_1 attacking y_1 and E_2 attacking y_2 so that all attack sets are left with not enough attackers on x.

So E' is a complete extension in this Figure 10. It is conflict free, it protects its members and it contains all those it protects.

Definition 6.4 (V-semantics). *Let (S, R) be given and let V be an ML valuation on S. We define the V-semantics for (S, R).*

1. *A set $E \subseteq S$ is said to successfully V-attack another node $x \in S$, iff the following holds:*

   (a) *For all $y \in E, yRx$*

   (b) *$V(y) > 0$ for all $y \in E$*

   (c) *The number of elements of E is greater or equal than $V(x)$.[15]*

   Let $E \subseteq S$, we say that E is V-conflict free iff the following holds and such that $V(y_i) > 0$.

---

[15]To explain condition (c) assume for example that $\{y, z\}$ attack x (i.e. yRx and zRx holds) and that they are the only attackers of x. Assume further that $V(y) = 2, V(z) = 1$ and $V(x) = 3$. Condition (c) reflects the understanding that $V(y) = 2 > 0$ means that y is alive and can generate only a single attack on x. Since x has three lives and only two attackers it is not going to die and as a result of the attack will have only one life left.

We can change our assumptions and allow y to have two attacks on x, one attack for each of its lives. In this case x will die as it is attacked 3 times, twice by y and once by z. We can bring the difference mathematically and uniformly as follows:

Let us define functions $\delta_i(x)$ for $\in S$, as follows:

- $\delta_1(x) = 1$, if $V(x) > 0$ and $\delta_1(x) = 0$, if $V(x) = 0$.
- $\delta_2(x) = V(x)$.

Then condition c can be written as

$$\sum_{y \in E} \delta_1(y) \geq V(x).$$

If we use δ_2 in the above equation we get the alternative approach, as described above. There are other possibilities for the use of δ_1, for example we can sum only on y in E for which $V(y) > 1$ (instead of $V(y) > 0$). This means that we allow elements to attack only if they have at least 2 lives.

    (a) For no $x \in E$ and $E' \subseteq E$ do we have that E' successfully V-attacks x

    (b) For all $e \in E, V(e) > 0$.

2. *Let $E \subseteq S$ be a set of arguments and $x \in S$. We say that E V-protects x iff the following holds.*

    (a) For no subset E' of E do we have that E' successfully V-attack x

    (b) Let $E' = \{y_1, \ldots, y_k\}$ be all elements of S such that $y_i R x$ holds and such that $V(y_i) > 0$ and y_i is not successfully V-attacked by any subset E' of E. Then $k < V(x)$.[16]

We say E is a V-admissible if E is V-conflict free and V-protects its elements.

3. *We say that E is a V-complete extension if it is admissible and contains all the elements x such that $V(x) > 0$ and it V-protects x.*

4. *Let E be a V-complete extension. Define for each x in E the value $V_E(x)$ to be $V(x)$ — [the number of elements y in E such that yRx holds]. $V_E(x) > 0$, since E is conflict free*

Lemma 6.1. *Let E be V-admissible set an let x be an element which is V-protected by E and such that $V(x) > 0$.*
    Then $E \cup \{x\}$ is V-admissible.

Proof. First note that $E \cup \{x\}$ certainly V-protects its elements. The question is whether it is conflict free. Let $E' \subseteq E \cup \{x\}$ and $z \in E \cup \{x\}$ be such that E' successfully V-attacks z.
    We show that this is impossible.

---

[16]Using the δ function, we write

$$\sum_{y \in E'} \delta_1(y) < V(x).$$

Case 1. x does not appear in E' and z is not equal to x. Then this is impossible because E is V-conflict free.

Case 2. $z = x, x \notin E'$. Let $\{y_1, \ldots, y_k\} = E'$. So on the one hand we have the $y_i R x, i = 1, \ldots, k$ and $k \geqslant V(x)$ and on the other hand E V-protects x so for some y_i we must have that E successfully V-attacks y_i. The two options are impossible together since E is V-admissible.

Case 3. $x \in E'$ *and* $z = x$. In this case we have on the one hand that $\{y_1, \ldots, y_{k-1}, y_k = x\}$ successfully V-attacks x and so again $k \geqslant V(x)$ but also we must have that E must successfully V-attack some y_i. Again this is not possible because E cannot successfully V-attack any y_i nor x.

Case 4. $x \in E'$ *and* $z \neq x$. Similar to Case 3, since E V-protects z. $\qquad\square$

Lemma 6.2. *Let (S, R, V) be given. Then there exists the smallest V-complete extension.*

Proof. Start with $\varnothing$. This is a V-admissible set. If there are *no* points $x \in S$ such that $V(x) > 0$ and there are no attackers $y R x$ with $V(y) > 0$, then $\varnothing$ is a complete V-extension since it obtains all the elements it V-protects.

If there are $x \in S$ such that $V(x) > 0$ and for all $y, y R x$ implies $V(y) = 0$, then $\varnothing$ V-protects such x and so these xs can be added to $\varnothing$. Continue this process and get a V-complete extension. $\qquad\square$

Example 6.1. *We now illustrate the concepts of V-attack and V-admissibility in an example.*[17] *Consider Figure 11.*

Let us compute the complete extensions E in two ways:

1. Propagation along the tree:

(a) a, b are alive, since they are not attacked $V(a) = 1 = V(b)$.

---

[17] I thank one of the referees for giving this example to show that the original definition of V-attack needed to be corrected.

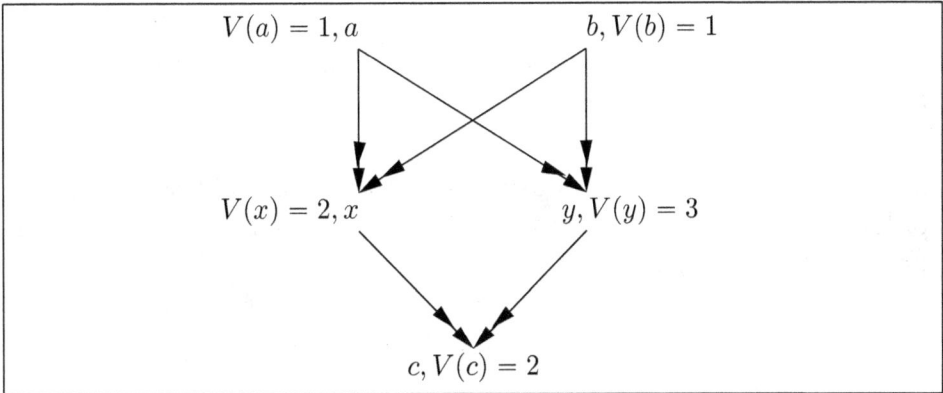

Figure 11

(b) x is dead, being attacked by $\{a, b\}$. So $V_E(x)$ becomes 0. y is alive but with reduced $V_E(y) = 1$.

(c) Since now x is dead and $V(c) = 2$, c remains alive with reduced $V_E(c) = 1$.

Therefore the complete extension when computed in this way is $\{a, b, c, y\}$ with $V_E(a) = V_E(b) = V_E(c) = V_E(y) = 1$.

2. Let us look at the set $E = \{a, b, c, y\}$ and check whether it is conflict free, protects itself and contains all elements it protects, as defined in Definition 6.4. Clearly the attacks of $\{a, b\}$ on y fails because $V(y) = 3$ and $|\{a, b\}| = 2$.

Similarly the attack of $\{y\}$ on c fails. So E is conflict free. Can E protect itself? the set $\{x, y\}$ attacks c. But a subset $\{a, b\} \subseteq E$ attacks x and kills it and $\{y\}$ on its own does not kill c. So c is protected. What would V_E be for the elements of E? According to item 4 of Definition 6.4, we get $V_E(a) = V_e(b) = 1$, $V_E(y) = 1$ and $V_E(c) = 1$. This is the same as calculated in (1) above.

Theorem 6.1. Let (S, R, V) be a finite acyclic network. Then there exists a unique V-complete extension E such that

$$S = E \cup \{y | E \text{ successfully } V\text{-attacks } y\}.$$

Proof. We define the sets E_n^+, E_n^- by induction on $n = 1, 2, \ldots$. E_n^+ is the set of elements that are certain to be "in" at step n and E_n^- are the elements that are certain to be "out" at step n.

Step 1. We know by Proposition 6.1 that S has at least one point of level 1, i.e. a point x such that $\neg \exists y(yRx)$. Let $E_1^+ =$ the set of all point in S of level 1. Let $E_1^- = \varnothing$.

Step 2. Let E_2^- be the set of all points y such that E_1^+ successfully V-attacks y Let $E_2^+ = E_1^+$.

Step 3. Let y be any new point not in $E_2^+ \cup E_2^-$. Let $\mathrm{Att}(y) = \{z | zRy\}$. Consider the set $A(y) = \mathrm{Att}(y) - E_2^-$. Let $E_3^+ = E_2^+ \cup \{y | \text{number of elements of } A(y) < V(y)\}$. (This means that there are not enough attackers or potential attackers to "kill" y. So y is for sure "in".)
  Let $E_3^- = E_2^-$.

Step 4. Let E_4^- be the set $E_4^- = E_3^- \cup \{y | y$ is successfully V-attacked by $E_3^+\}$.
  Let $E_4^+ = E_3^+$.

Steps $2k+1, 2k+2$. Continue by induction as done in steps 3 and 4 in terms of steps $2k-1, 2k$.
  Since the sets E_n^+, E_n^- can only increase and S is finite the process will become stable say at E_m^+, E_m^-.
  We now show that

(*) $S = E_m^+ \cup E_m^-$.

Assume in order to reach a contradiction that there exists a z_0 such that z_0 is not in E_m^+, nor in E_m^-. z_0 must have some attackers z (i.e. zRz_0) for otherwise $z_0 \in E_1^+$.
  Let $y_1, \ldots, y_k, u_1, \ldots, u_e, y_1', \ldots, y_r'$ be all R attackers of z_0. Assume $y_1, \ldots, y_k \in E_m^+$, $y_1', \ldots, y_r' \in E_m^-$ and that $u_1, \ldots, u_e$ are all the rest of the points which are neither in E_m^+ nor in E_m^-.

338

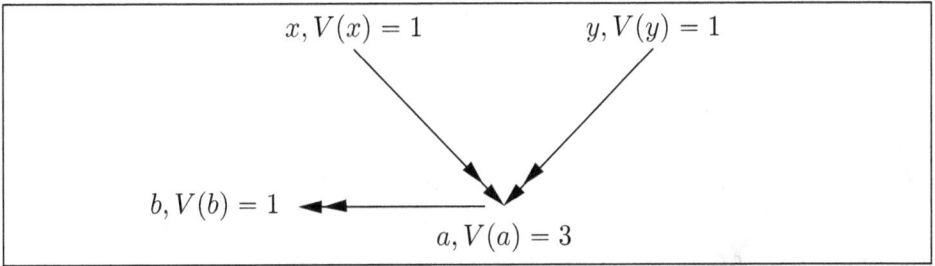

Figure 12

We bear in mind that there may not be any points y or u or y', in which case we write $k = 0, e = 0, r = 0$ respectively.

Since E_m^+ is not successful in V-attacking z_0, we have that $k < V(z_0)$. Since z_0 is not a member of E_m^-, we have that $k + e \geqslant V(z_0)$. This means that there exists at least one point z_1 (say $z_1 = u_1$) such that $z_1 R z_0$ and $z_1 \notin E_m^+ \cup E_m^-$.

We repeat the process for z_1 and get $z_2, z_3, \ldots$, an infinite sequence of pairwise different points (otherwise we get a cycle). This contradicts the finiteness of S and therefore (*) is proved.

Note that we get a unique "stable" extension.

$\square$

Lemma 6.3. *Let (S, R, V) be any network, with or without cycles, such that $V(x) = 1$ for all $x \in S$. Then the V-semantic notions coincide with the Dung traditional ones.*

Proof. Note that since $V(x) = 1$ for all $x \in S$, we have that:

$$x \text{ is } V\text{-attacked by } \{y\} \text{ iff } yRx.$$

$\square$

Example 6.2. *Consider Figure 12.*

This figure represents a network $M = (S, R, V)$, with $S = \{x, y, a, b\}$ and with $R = \{(x, a), (y, a), (a, b)\}$ and with $V(x) = V(y) = 1$ and $V(a) = 3$ and $V(b) = 1$. According to our definition, the set $E = \{x, y, a\}$ is a V-complete extension, because x and y are in since they are not attacked and a is in because to be out we need 3 attackers which are in and we have only 2. So the set E is V-conflict free which V-protects its elements, (its elements are not V-attacked). We cannot add to E the element b since it is attacked by a. The set $E1 = \{x, y, b\}$ although V-conflict free, does not V-protect it elements. $E1$ cannot V-protect b against a since it cannot V-attack a. What is missing from our theoretical considerations is the option to define a new function indicating how many lives we have for each of the elements of this V-extension. We mentioned in the beginning of this Subsection that we have two options. Option 1 is that the number $V(x)$ remains unchanged. If we adopt this option then we have no problems. Option 2 was to reduce the $V(x)$ number in view of the attacks on x. There may not be enough attacks to take x out but in this option the number of lives of x $(V(x))$ is reduced. This option is problematic as we shall now see.

We now need to be careful with our notation for otherwise we get confused. The V-extension is $E = \{x, y, a\}$. It is a set of nodes. We can ask how many lives do the elements of this set have according to Option 2? We know the answer that each of these elements have now one life. Let us write the function V_E to indicate how many lives each element of the extension has. So we have in this case

$$V_E(x) = V_E(y) = V_E(a) = 1.$$

So the notion of V-extension E must also include a function V_E for the V-extension E. So the set theoretic approach models the view that $V(x)$ simply says how many "in" attackers are required to force x to be "out", but if there are not enough such attackers, then x remains "in".

However, the number of lives of x is reduced. This presents us with a problem:

Consider the same network with the new V_E, namely the network $M' = (S, R, V_E)$. This network has values all 1 for $\{x, y, a\}$ and 0 for b (which therefore can be ignored) and therefore the complete extension for it is, say, E' with $x = y = $ "in" and $a = $ "out" (and b is already "out" and is ignored, so we are really looking at the network without b).

Let us look at what is happening here in the following way. We *apply our V semantics (Option 2) to M and get M'. If we apply our V-semantics (Option 2) again to M' we get M" with E'. On the other hand if we use Option 1, and apply the V semantics (Option 1) to M we get M and apply again we still get M. So for Option 1 the process stabilises after one application but in Option 2 it does not.*

What happens in traditional Dung semantics (see Chapter 1)? After one application we get an extension which is a set of conflict free elements. If we regard this set as a network with no attack relation and apply the semantics again we get the same set again.

So we ask should we look at any semantics, traditional or new, as an algorithm for generating a sequence of networks which can go on until it stabilises?

Put differently, we ask: do we make a connection between E and E' and say that E' is a second level extension for M?

This makes the semantics concept like proof theory; we keep proving from the data until we stabilise and can prove nothing new any more. This is what we do in Semantic Tableaux.[18]

This process is especially interesting in the case of finite acyclic

---

[18]The meaning in practice of the function V_E can be illustrated from Talmudic and Islamic law. The Talmudic legal system may require two independent witnesses to refute a claim. If I claim that I was standing on the sidewalk when the car hit me, then two independent witnesses are required to say that I was standing on the road. If I bring only one witness, the function V_E says we need to wait for one more.

In Islamic law we have the following example, and we quote from `http://www.islamhelpline.net/node/905` (visited on September 25, 2016):

> If the woman who is raped accuses that so and so specific person or people raped her, then there are only two ways an Islamic Court can convict the accused rapist/s: The accused rapist confesses to his heinous crime; or she produces four witnesses to justify her claim that so and so person raped her. If the accused rapist does not confess, and the woman is unable to produce the four witnesses; then the Court can levy upon her the case of kazaf or falsely accusing somebody. Under no circumstances can a woman who claims she was raped be charged, accused, convicted, or punished for zina (fornication or adultery) in an Islamic Court of Law. All she has to do is say that she was raped, and her word will be taken as the truth.

networks $M = (S, R, V)$. By Theorem 6.1, the network M' is unique, so we ask, if we continue the process and generate $M'', M''', \dots$ what do we get at the end? Do we get the traditional Dung ground extension of (S, R)? The answer is no. Consider the network $N = (S, R)$ of Figure 5. In this network $S = \{x, y, z\}$ and $R = \{(x, y), (y, z)\}$. Let V be the valuation with $V(x) = 3, V(y) = 2$ and $V(z) = 1$. Then in N', z is out and x and y are in and in N'', y is also out. Neither case equals the Dung extension for (S, R).

Actually this example of the network $N = (S, R, V)$ above also shows that there is no way that N could be translated into a traditional Dung network (S_N, R_N) with possibly additional points (i.e. with $S \subset S_N$), because any traditional extension of any network is stable after the first step. We can never get a sequence like N, N', N''.

6.5 Algorithmic semantics for abstract valuation frameworks with multiple lives valuations

Theorem 6.1 above in the previous Section, showed that for the case of acyclic networks, there is an algorithmic way of obtaining the grounded V-extension, which coincides with the V-extension as defined set theoretically in Lemma 6.2. The correspondence is problematic in the case of networks with loops, and needs to be investigated. This is the task of this Section. The algorithmic semantics we will define we shall call VW-semantics, and VW-extensions. The formal machinery will come later in the Subsections.

To proceed in this direction, we need to look at more revealing examples with loops. It is good to look at loop examples before we decide how to handle loops.

Example 6.3. Note that the propagation computation does not give the same results as the set-theoretic definition of Definition 6.4 and item 4 of this definition. This is because of the cyclicity of the network. We need the algorithmic VW-semantics

Consider Figure 13

1. If we propagate, we need starting points

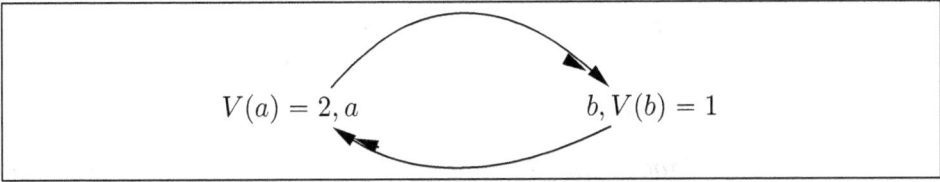

Figure 13

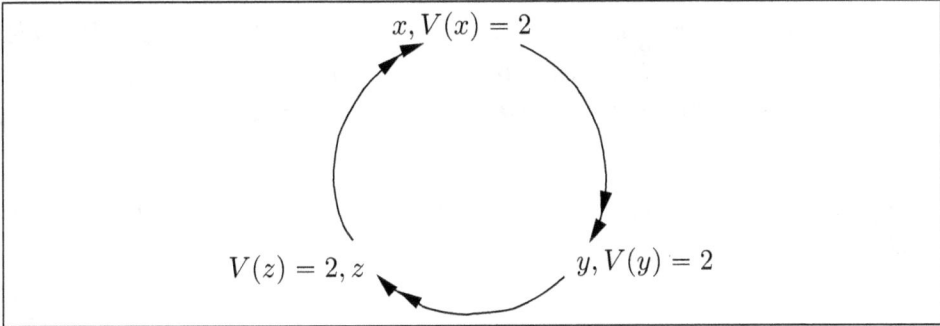

Figure 14

(a) *Start with a. a kills b and we emerge with $E = \{a\}$ and $V(a) = 2$.*

(b) *Start with b. b attacks a and reduces its life to $V_E(a) = 1$ and now a kills b. We emerge with $E = \{a\}$ and $V_E(a) = 1$.*

(c) *Both a and b attack simultaneously. We emerge with $E = \{a\}$ and $V_E(a) = 1$.*

2. *If we calculate extensions set theoretically following Lemma 6.2, the only complete extension is $E = \{a\}$ with $V_E(a) = 2$.*

Example 6.4.

1. *Consider the loop in Figure 14*

   The network is $M = (S, R, V)$, where $S = \{x, y, z\}$, $R = \{(x, y), (y, z), (z, x)\}$ and $V(x) = V(y) = V(z) = 2$. There is only one V-extension, (according to Definition 6.4) it being $E = \{x, y, z\}$ and we have $V_E(x) = V_E(y) = V_E(z) = 1$.

   Consider now $M_E = (S, R, V_E)$. This has only one V-extension, it being $\varnothing$.

2. *Let us now approach the extensions of Figure 14 differently, us-ing the fact that (S, R) is a cycle. We go in steps:*

Step 1. *Choose an element in S and let it attack. Say we choose $(x, V(x) = 2)$. Since we are cycling and attacking, the V will change. So we subscript the V by an increasing index. Let $V(x)$ be $V_1(x)$. We keep the index "1" until we go a full cycle and cycle back to x, in which case we increase the index to 2. x attacks y and so we get $(y, V_1(y) = 2 - 1 = 1)$.*

Step 2. *y can still attack. It has $V_1(y) = 1 > 0$. It attacks z and we get $V_1(z) = 1$.*

Step 3. *z can attack x and we get $V_2(x) = 1$. We have re-turned to x and thus got $V_2(x)$.*

Step 4. *x attacks y and we get $V_2(y) = 0$. We use the notation V_2 because it is the next loop.*

Now we have several ways of continuing:

$(W1)$: *Stop. y cannot attack, we have $V_2(y) = 0$.*

The extension is: $V_{w1}(x) = V_{w1}(z) = 1 . V_{w1}(y) = 0$.

$(W2)$: *Skip y and look at z, z has $V_1(z) = 1$. Strictly speaking, since we are now in the second cycle we should rename the V as $V_2(z) = V_1(z) = 1$. Anyway, z has 1 life so it can attack. So we let it attack and get $V_3(x) = 0$. It is $V_3(x)$ because we returned to x.*

The extension is $V_{w2}(x) = V_{w2}(y) = 0, V_{w2}(z) = 1$.

You see that the extension depends on where we start.

Let us call these VW-extensions. A formal definition will be given later in this Subsection. In the meantime, let us proceed with only an intuitive grasp of this concept, to be refined by looking at more examples, leading to the sequence of definitions beginning with Definition 6.5 below. So at the moment we know that the algorithmically (to be defined) VW-extensions depend on where we start. If we start with x, we get the extensions

$$V_{w1}^x(x) = V_{w1}^x(z) = 1, V_{w1}^x(y) = 0$$
$$V_{w2}^x(x) = V_{w2}^x(y) = 0, V_{w2}^x(z) = 1.$$

the other possible VW-extensions are obtained by symmetrical permutations. It is better to adopt the W2 computation option for the VW-semantics approach because mathematically we keep on going until nothing changes.

Example 6.5. *Let us check what VW-extensions look like for Figure 12.*

We have in mind the conjecture that for finite acyclic (S, R) the V-semantics option 2 and the VW-semantics are the same, if we start from all unattacked nodes in the graph, i.e. $\{x | \neg \exists y (y R x)\}$.
Figure 12 is acyclic. So let us check. We start the sequence of steps and define V_1.

Step 1. $V_1(x) = 1, V_1(y) = 1.$

Step 2. x, y attack a so $V_1(a) = 1.$

Step 3. a attacks b so $V_1(b) = 0.$
Stop.

What we get is the same as the level 2 extension E' of Example 6.2. See the proof of Theorem 6.1. This proof confirms that the conjecture is true.

We are now ready to define the VW-extensions for an arbitrary argumentation network (S, R, V), where we allow for loops.

The following sequence of definitions develops the semantics for general Abstract Valuation frameworks (AVF) for the case where the valuations are many lives (x is "out" if at least $V(x)$ attackers are "in"). For a further in depth analysis of the many lives option, see our new paper [44]. We are giving formal definitions to what we did in Examples 6.2, 6.5 and 6.10.

The level of writing is aimed at members of the COMMA (Conference on Computational Models of Argument) community

Given an AVF network (S, R, V), the semantics relies on resolving loop cycles in S through the use of step by step algorithm and then using the SCC ordering of the loop cycles. Thus every extension is stable (no undecided). This is reminiscent of the CF2 semantics (of [40]), and indeed we shall offer a comparison.

To start we the notion of an SCC, taken from [40].

Definition 6.5. *Let (S, R) be an argumentation network with $S \neq \varnothing$ and $R \subseteq S \times S$.*

1. *We say that a sequence $(x_1, \ldots, x_n)$ of elements of S is a cycle of length n if we have*

$$x_1 R x_2, x_2 R x_3, \ldots, x_{n-1} R x_n, x_n R x_1.$$

2. *Define a relation $x \approx y$ on S by setting $x \approx y$ iff $x = y$ or x and y share a cycle. That is, $x \approx y$ iff there is a cycle $(z_1, \ldots, z_n)$ of length n such that $x = z_j$ and $y = z_i$ for some $1 \leqslant i, j \leqslant n$.*

3. *Since $\approx$ is an equivalence relation on S (see [40]), let*
   - *$x^{\approx} = \{y | x \approx y\}$, for any $x \in S$.*
   - *$S \approx = \{x^{\approx} | s \in S\}$.*
   - *$R^{\approx} = \{(x^{\approx}, y^{\approx}) | aRb \text{ for some } a' in x^{\approx}, b \in y^{\approx}\}$.*

   $R^{\approx}$ is well defined and is an antisymmetric relation (see [40]).

4. *Let $<$ be the transitive-reflexive closure of $R^{\approx}$. That is, $x^{\approx} < y^{\approx}$ iff either $x^{\approx} = y^{\approx}$ or for some $z_1^{\approx}, \ldots, z_k^{\approx}, k \geqslant 1$ we have $x^{\approx} R^{\approx}, z_1^{\approx}, z_1^{\approx} R^{\approx} z_1^{\approx}, \ldots, z_{k-1}^{\approx} R^{\approx} z_k^{\approx}$ and $z_k^{\approx} = y^{\approx}$.*

Then $<$ is a partial acyclic ordering on $S^{\approx}$.

Our purpose is to define the VW-semantics for AVF of the form $(S, R, <, V)$ where $S \neq \varnothing$, S finite, $R \subseteq S \times S$ and V is a function on S giving natural numbers values in $\{0, 1, 2, 3, \ldots\}$ to elements of S. We added the value 0 for technical convenience. $V(x) = m$ means that x has $m \geqslant 0$ lives. To be "out", it needs to be attacked by at least $m \geqslant 0$ attackers y such that $V(y) > 0$. Of course if $V(x) = 0$, then x is already out.

We need to use the truncated subtraction symbol defined below:

$$x \,\dot{-}\, y = \text{ def.} \begin{cases} x - y, \text{ if } x \geqslant y \\ 0 \text{ if } x < y. \end{cases}$$

Definition 6.6 (VW semantics for a complete cycle with a front F).

1. *Let $M = (S, R, V)$ be an AVF network which is a complete cycle. This means that for any $x, y \in S$, there exists a sequence $t_1, \ldots, t_k$ in S such that*

$$x R t_1 \wedge t_1 R t_2 \wedge \ldots \wedge t_k R y.$$

   We allow $k = 0$, in which case the condition is $x R y$. We allow $x = y$ in which case the condition is $x R x$.

2. *A set $F \subseteq S, F \neq \varnothing$ is called a Front. Let us choose such an F. This choice determines what we are going to get in the next item 3.*

3. *We define a sequence of networks $M_i = (S, R, V_i, F_i), i = 0, 1, 2, \ldots$ by steps as follows:*

   (a) *Let $V_0 = V, F_0 = F$.*

   (b) *Assume that V_i, F_i have been defined and that $F_i \neq \varnothing$. We define V_{i+1}, F_{i+1}.*

      i. *Let F_{i+1} be the set of all y such that for some x in F_i we have $x R y$. This set is non-empty because (S, R) is a cycle.*

ii. Let $V_{i+1}(z) = V_i(z)$ if $z \notin F_{i+1}$.

iii. Let $V_{i+1}(z) = V_i(z) \dotminus \{$the number of $y \in F_i$ such that $V_i(y) > 0$ and $yRz\}$.

We stop the process when $V_{n+1} = V_n$. The process stops since at each step m some $V_m(z)$ is reduced.

4. When we stop at step n, we say that the Front F resolved (S, R, V, F) into (S, R, V_n, F_n).

5. We say that $M = (S, R, VF)$ was resolved into $M = (S', R', V', F') = (S, R, V_n, F_n)$.

Definition 6.7 (Front obtained externally).

1. Let (S_0, R_0, V_0) be a complete cycle and let S_1 be a set of nodes with $S_1 \cap S_0 = \emptyset$. Let $R_1 \subseteq S_1 \times S_0$. Thus the S_1 nodes attack nodes in S. Let V_1 be a valuation on S_1.

2. Define a cycle network $M = (S, R, V, F)$ as follows.

   (a) $S = S_0$

   (b) $R = R_0$

   (c) $F = \{x \in S_0 | (\exists y \in S_1)(yR_1x)\}$

   (d) V is defined by

   $$V(z) = \begin{cases} V_0(z), & \text{if } z \notin F \\ V_0(z) \dotminus \{\text{number of } y \in S_1 \text{ such that } yR_1z \\ \text{and } V_1(y) > 0\}, & \text{if } z \in F \end{cases}$$

3. We say that (V, F) was induced on (S_0, R_0) by the external attackers system (S_1, R_1, V_1).

4. We say that the network (S_0, R_0, V_0) with the external attackers (S_1, R_1, V_1) is resolved into $M' = (S', R', V', F')$ when this AVF (namely M') is what resolves M (of item 2 above) according to Definition 6.6.

Definition 6.8 (VW-semantics for general AVF, (S, R, V)).   *Let (S, R, V) be given. We look at the acyclic network of SCC's for (S, R) as defined in Definition 6.5.*

1. *Let X_i be the top nodes SCCs in the ordering of Definition 6.5. These X_i are actually sets S_i being equivalence classes of elements of S, according to the equivalence relation of Definition 6.5. Let $M_i = (S_i, R_i, V_i)$ be the cycles defined using these top SCC equivalence classes, with $S_i \subseteq S$ is the set of elements of the cycle and with*

$$R_i = R \restriction S_i$$
$$V_i = V \restriction S_i$$

   There are several possibilities for M_i because of Definition 6.5

   (a) *$S_i = \{x_i\}$, one node with $\neg(x_i R_i x_i)$.*

   (b) *S_i is a proper cycle as defined in item 1 in Definition 6.6.*

2. *We arbitrarily choose a subset $F_i \neq \varnothing, F_i \subseteq S_i$. This choice determines the VW-complete extension of the VW-semantics (for our (S, R, V)) which we are defining. Different choices of F_i will give us different extensions.*

3. *We now have systems $M_i = (S_i, R_i, V_i, F_i)$. We are going to resolve cycles according to Definitions 6.6 and 6.7 by going down the acyclic ordering $<$ of Definition 6.5 as applied to our initial (S, R, V). Note that we use an inductive step by step definition. It is important to note that even if a cycle attacks another cycle, we resolve completely the top cycle first and only afterwards propagate the values as external attackers (as in Definition 6.7, item 4) to the next lower cycles. (Example 6.6 below illustrates this point.)*

Step 1. *We are ready to resolve each M_i into $M_i' = (S_i, R_i, V_i', F_i')$ following Definition 6.6. We distinguish two cases as in (1) above.*

(a) $S_i = \{x_i\}$ with $\neg(x_i R x_i)$, let $M' = \{S_i, R_i, V_i, \{x_i\}\}$.

(b) Otherwise for this case (b) let M' be as resolved as we do in item 6 of Definition 6.6.

4. Inductive Step $k + 1$.
Let us assume that we have already resolved cycles

$$M'_i = (S'_i, R'_i, V'_i, F'_i)$$

at step k. We now resolve more cycles at step $k + 1$.

Let $M^*_j = (S^*_j, R^*_j, V^*_j)$ be cycles in the ordering of the SCC's (according to Definition 6.5) that come immediately below at least one of the M'_i cycles. Actually we have that R'_i is R restricted to S'_i and $R^*_j = R \upharpoonright S^*_j$ and $V^*_j = V \upharpoonright S^*_j$. We use notation with $*$ to differentiate between higher cycles (in the ordering) and lower cycles so that we will not be confused between cycles.

This means that for each S^*_j there exists at least one S'_i and $x \in S'_i$ and $y \in S^*_j$ such that xRy. (These xs are the external attackers (in the sense of item 4 of Definition 6.7). Also recall that our starting point was one big (S, R, V), which was divided into SCC cycles as in Definition 6.5.

Let F^*_j be the set of all $y \in S^*_j$ such that for some S'_j and some $x \in S'_j$ we have xRy. Define $V^{**}_j(y)$ to be $V^*_j(y) \dot- \{$the number of elements x in some S_i such that xRy and $V'_i(x) > 0\}$.

We now have a system of cycles $(S^*_j, R^*_j, V^*_j, F^*_j)$ which can be resolved according to Definitions 6.6 and 6.7.

This completes step $k + 1$.

5. The process terminates at step n, for some n, because the original network is finite.

6. The extension we get is $(S, R, V^\sharp)$ where $V^\sharp(x)$ is defined as follows:

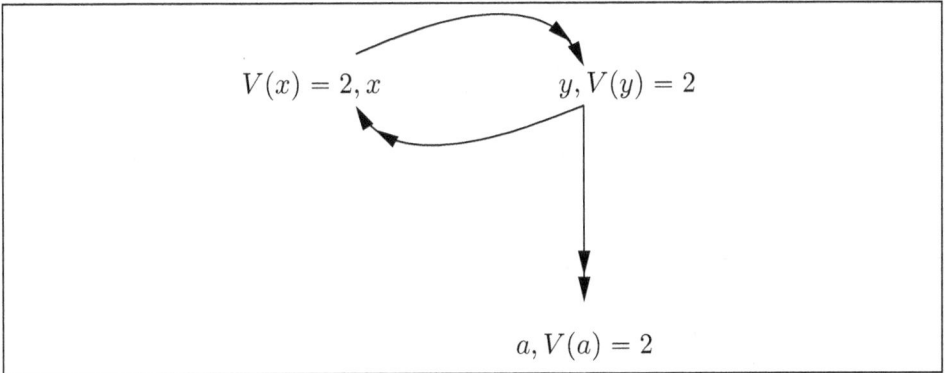

Figure 15

$V^\sharp(x) = V_i'(x)$ *where* x *belongs to the cycle* (S_i, R_i) *and* V_i' *is the function obtained when the cycle was resolved at the appropriate step. Recall that we resolve a higher cycle without affecting any lower cycle and only after it is resolved do we let it act as a set of external attackers on the next lower cycle.*

Example 6.6 (Two ways of resolving a network). *This example illustrates our policy of resolving a higher cycle completely before we pass the attacks to a lower cycle. Consider Figure 15.*

We have two choices for the front for the top cycle $\{x, y\}$*. Choose* $F = \{y\}$*. This will define one possible extension. We cycle through* $\{x, y\}$*, we do not attack* a *yet. We first resolve the top cycle* $\{x, y\}$*.* y *attacks* x *gives* $V_1(x) = 1$*,* x *attacks* y *gives* $V_2(y) = 1$*,* y *attacks* x *gives* $V_3(x) = 0$*. We are stable with* $V'(x) = 0, V'(y) = 1$*. We now let the cycle be an external attacker on* $\{a\}$*. We get* $V'(a) = 1$ *(because* $V'(y) = 1$*), so we get the following extension for the choice of* $F = \{y\}$*:* $V'(a) = V'(y) = 1$ *and* $V'(x) = 0$*. If we choose* $F = \{x\}$*, then by symmetry for the cycle* $\{x, y\}$ *we get the values* $V''(x) = 1, V''(y) = 0$ *and therefore since* x *does not attack* a *we get* $V''(a) = 2$*. So the network* (S, R, V) *of Figure 15, has two complete extensions which are also networks, namely, the networks* (S, R, V') *and* (S, R, V'')*.*

Note that we must mention the valuation as well because for example in V''*,* a *is in with 2 lives.*

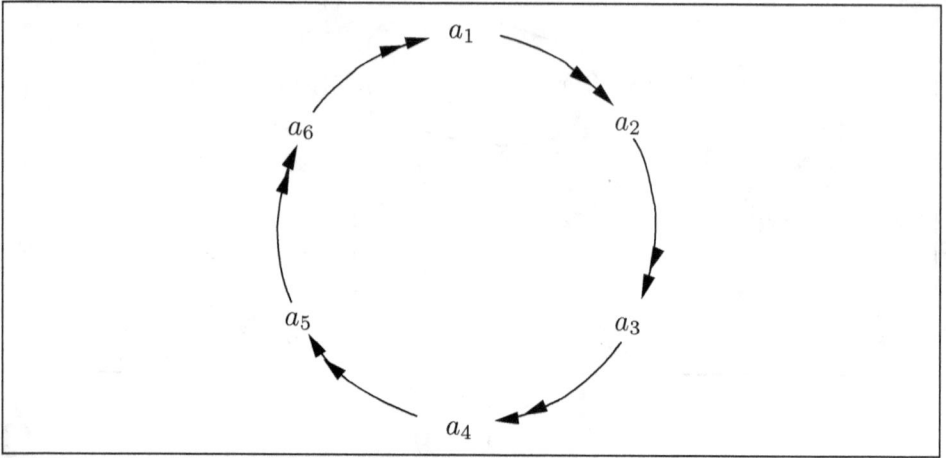

Figure 16

We could follow an alternative policy and just let y attack wherever it can. We get

y attacks $x : V_1(x) = 1$
y attacks $a : V_1(a) = 1$
x attacks $y : V_2(y) = 1$
y attacks $x : V_3(x) = 0$
y attacks $a : V_3(a) = 0.$

The different extension we get for the starting choice $F = \{y\}$ is $V^*(y) = 1$, $V^*(a) = V^*(x) = 0$.

Example 6.7 (Comparison with CF2 semantics). *CF2 semantics also resolves loops by taking maximal conflict free sets. When the valuation $V \equiv 1$, we get the ordinary Dung extensions for acyclic networks. What do we get for cycles?*

Consider Figure 16. This is a 6 cycle.

The CF2 semantics allows for the extension $\{a_1, a_4\}$. Is there a set F of starting points which can get it in VW-semantics? The answer is yes.

$$F = \{a_4, a_5, a_1, a_2\}.$$

What happens in general if we limit F in the top cycles to only one point? I do not know.

Remark 6.2 (Connection with weighted argumentation [41]). *This is an important remark. The material in the machinery of this Subsection has two components:*

1. *The nature and meaning of the numerical valuation V arising from the sex offender area and still to be refined and adjusted. See Footnote 22.*

2. *The step by step propagation protocols of Definitions 6.6, 6.7 and 6.8. These definition apply to any weighted system. All we need to tell the machinery of these definition is how to propagate the values of V from attackers to the target and get a new V' on the target. In other words we need to give a new definition replacing i the item (d) of Definition 6.7 by l formula giving a different mathematical formula for a new $V(z)$ in terms of its own original $V_0(z)$ and the V_1 Values of its attackers (and possibly values of its attack arrows).*

All we have here is a bunch of numbers attacking another number, yielding a new number. See [42] and [43], for example, for many ways of executing such attacks. The method of [41], however, is different from what is discussed in those papers and is different from what we are doing in this paper (the many lives approach). To explain the difference we can use two methods:

1. *We need to devise an example which can be addressed by all views, especially the view of [41] and the view of the many lives of this paper and show how the methods differ in this example.*

2. *Find a more general approach which can contain all candidates for comparison and embed/translate these candidates into this general approach and do the comparison there.*

The next Example 6.8 uses method (1). Method (2) is more complex and could be a subject for a separate paper. We do, however, give you an example in the spirit of method 2, namely Example 6.9.

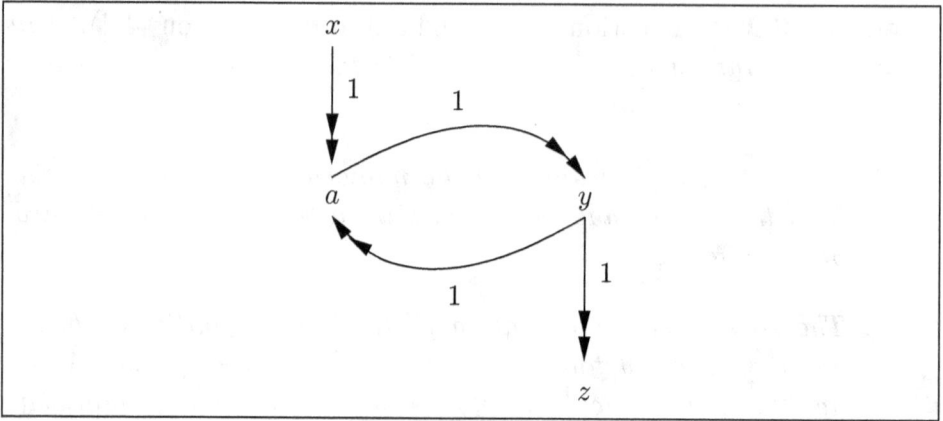

Figure 17

Example 6.8. *We chose an example which can be addressed by both our method of many lives and the weighted approach of [41]. Consider the network of Figure 17:*

1. *First consider this figure as representing a weighted network in the sense of [41]. This has the form $(S, R, \mathbf{w})$, where $S = \{x, a, y, z\}$. $R = \{(x, a), (a, y), (y, a)(y, z)\}$ and $\mathbf{w}$ is a function from R into the positive real numbers $(0, \infty)$, which in this case is giving the identical value 1 to all attacks as shown in the figure 17. This system $(S, R, \mathbf{w})$ conforms with Definition 4 of [41, page 462].*

   Viewed as a many lives network we still have to say what the values 1 annotating the attacks mean. We read them as saying the attack is live. We do not yet say how many lives each node has. We can highlight the first technical difference between our paper and [41]

   (d1): We annotate nodes with number of lives. [41] annotates attacks with strength. In itself this is a small technical difference but the real difference is what is done with the annotation. [41] uses the strength to cancel some weak attacks while we use the annotation to refuse weak attacks. [41] is part of the numerical world view and we are part

of the non-monotonic logic world view. This will become clearer later in this example. The two approaches are orthogonal to each other and can be combined together . The strength of attacks can be aggregated and still be refused by the target.

Let us go on. [41] adds a number β which they call "inconsistency budget", which roughly means that any group attacks of combined strength less than β can be ignored. This is definition 5 in [41]. We can understand this number as our many lives number. Let us choose $\beta = 2$ and understand it in our context as the number of lives for each node. [41] now continues in a unique way. To define extensions for the network of Figure 17, it chooses an arbitrary set of attacks such that the sum of the weights of the attacks in this set is less than β. Since $\beta = 2$ and the attack weight is 1, this means we can arbitrarily choose in our case a single arrow. [41] then proceeds to do the following:

(a) cancel the chosen arrow

(b) proceed with the rest of the network (without the chosen arrow), ignore the weight and compute traditional extensions.

These are definitions 5 and 6 of [41] applied to our Figure 17.

[41] justify their approach in Section 3.2 of their paper

(d2): Note that [41] uses the weights as a licence for attacks to enter the traditional computation of extensions. So for $\beta = 2$, we can choose that the attack $\{y \twoheadrightarrow a\}$ is cancelled and now we compute traditional extensions for the remaining network. We can choose to cancel any single attack in the figure, and after such a choice is made, [41] does not use the weights any more.

Let us now see how we deal with Figure 17 in our paper.

2. If we follow finding extensions in the spirit of Definition 6.4 (even though Figure 17 contains cycles), we get the extension

$$E_b = \{x = 2, y = 2, a = 0, z = 1\}.$$

3. *If we follow the computational approach of this Subsection, we get the following steps:*

Step 1. $x = 2$, *(x not attacked)*

Step 2. *x attacks a, so* $a = 1$

Step 3. *a attacks y, so* $y = 1$

Step 4. *We need to complete the cycle $\{a, y\}$. So y attacks a (we ignore the attack on z), so $a = 0$. This completes the cycle giving $a = 0, y = 1$.*

Step 5. *y attacks z, so* $z = 1$.
The extension we get is

$$E_c = \{x = 2, a = 0, y = 1, z = 1\}.$$

Example 6.9. *This example shows how we can combine the many lives approach with the general weighted numerical approach. Let $(S, R, \mathbf{w})$ be a general finite network with $R \subseteq S \times S$ and $\mathbf{w}$ a function giving weights in the real numbers $[0, \infty)$ to both arguments and attacks, namely,*

$$\mathbf{w} : S \cup R \mapsto [0, \infty).$$

Consider the general configuration for a node a shown in Figure 18. We agree on the following interpretation:

1. *For node $x, \mathbf{w}(x)$ is the strength of node x*

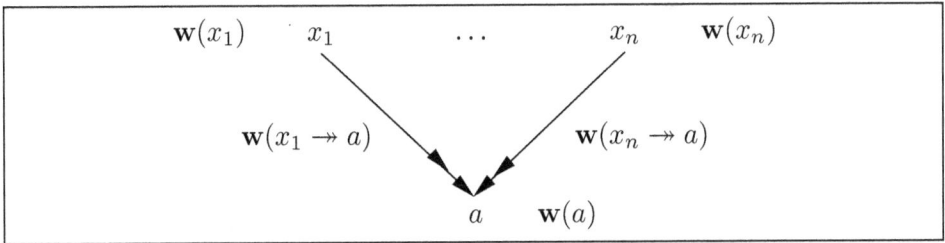

Figure 18

2. For the attack $y \twoheadrightarrow z$, $\mathbf{w}(y \twoheadrightarrow z)$ is the strength of the resistance of the transmission of the attack of y on z. So if $\mathbf{w}(y)$ is greater the attack on z is greater, but if the resistance $\mathbf{w}(y \twoheadrightarrow z)$ is greater, less of the attack passes through and the effect of the attack is smaller.

3. The strength of the end result of the attack from y onto z is given by $\boldsymbol{f}(\mathbf{w}(y), \mathbf{w}(y \twoheadrightarrow z))$, where $\boldsymbol{f}(\alpha, \beta)$ is a continuous function satisfying the following

   (a) $\boldsymbol{f}(0, \beta) = 0$

   (b) $\boldsymbol{f}(\alpha, 0) = \alpha$

   (c) $\boldsymbol{f}(\alpha, \infty) = 0$

   (d) $\boldsymbol{f}(\infty, \beta) = \infty$

   For example we can take $\boldsymbol{f}(\alpha, \beta) = \frac{\alpha}{1+\beta}$.

Let us now look again at Figure 17 and give it the following values as in Figure 19.

The method of propagation remains along the same (using the function $\frac{\alpha}{1+\beta}$) sequence of steps:

Step 1. The value for x is 3, as it is not attacked.

Step 2. The value of the attack of x on a is $\frac{3}{2}$ (using the function $\frac{\alpha}{1+\beta}$, $\alpha = 3, \beta = 1$).

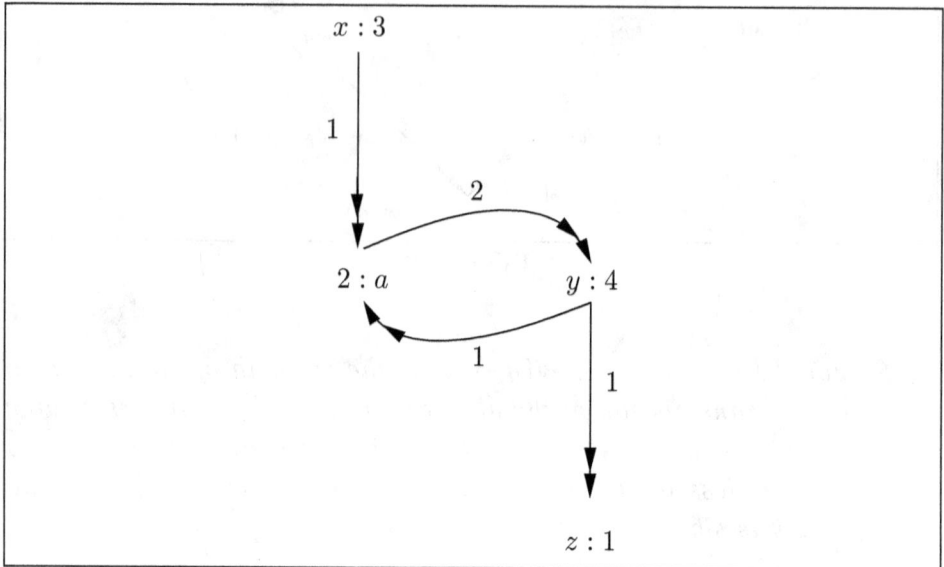

Figure 19

Step 3. *The new value of a is* $2 - \frac{3}{2} = \frac{1}{2}$

Step 4. *a attacks y the value of the attack is* $\frac{0.5}{1+2} = \frac{0.5}{3} = \frac{1}{6}$.

Step 4. *The new value of y is* $4 - \frac{1/6}{3} = 4 - \frac{1}{18} = \frac{71}{18}$.

Step 5. *y attacks a with value* $\frac{71}{18.2} = \frac{71}{36} = 1.9722$.

Step 6. *The new value of a is* $\frac{1}{6} \div 1.9722 = 0$. *Thus the stable loop solution is* $a = 0, y = 1.9722$.

Step 7. *The value of the attack of y on z is* $\frac{1.9722}{2} = 0.98611$.

Step 8. *The new value of z is* $1 - 0.98611 = 0.013888$.
    The final extension is approximately/practically $x = 3, a = 0, y = 2, z = 0.013888$.

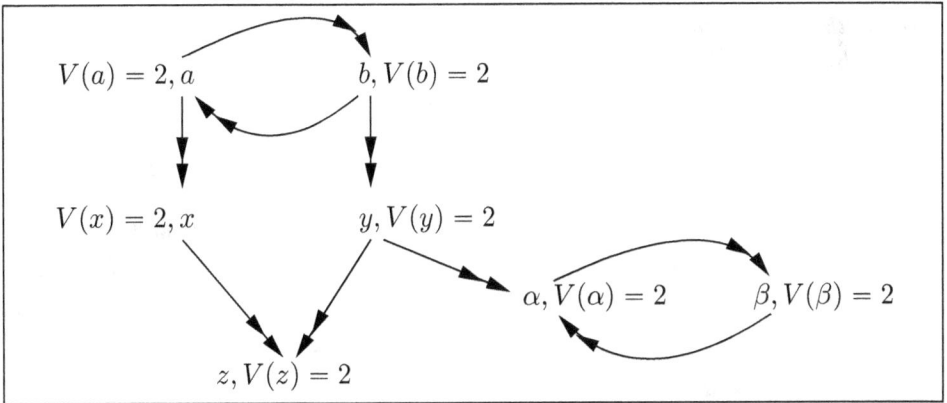

Figure 20

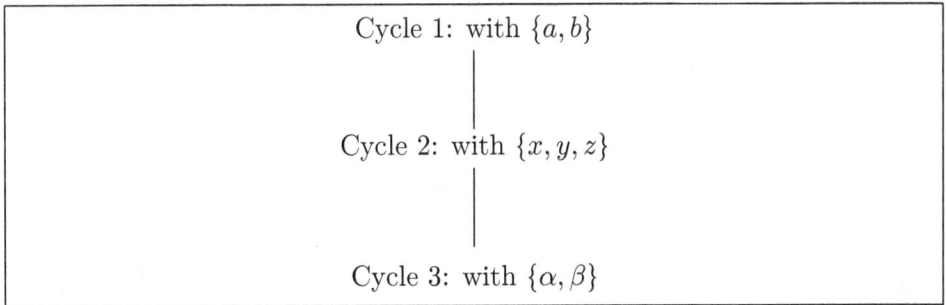

Figure 21

Example 6.10. *Consider the network of Figure 20 (see Definition 6.5).*

The SCC ordering of cycles is Figure 21.

We now calculate the extensions. We start with the top cycle 1 and work our way down the cycles. The steps are intended to define a new valuation V'. Each step modifies V into a new V_i. The index "i" increases as we move forward in steps along the arrows of the cycle.

Step 1. *Choose an element in cycle 1. Different choices would lead to possibly different extensions. Let us choose a. We have that a attacks b. We get that the original $V(b) = 2$ changes into the new $V_1(b) = 1$. Notice that even though a attacks x in cycle 2, we do not attack cycle 2 until cycle 1 is completely resolved by cycling through it*

359

until no more changes to the valuation are introduced. It is then and only then that we use the new stable values we get for the elements of Cycle 1 to attack Cycle 2. This is our policy in defining extensions. There are other policies possible.

b attacks a, we get a new value for a which we call $V_1(a) = 1$. Now we continue to move in Cycle 1 and observe that a attacks b and so the value of b changes again from $V_1(b) = 1$ to the value 0. Since this is another change in the value of b we increase the index of V and call it $V_2(b) = 0$.

We stop, because b is no longer capable of attack. We are stopping with values $V_1(a) = 1$ and $V_2(b) = 0$. We realise we have a minor accounting problem with the indices of V. On the one hand we want to emerge from Cycle 2 with a clearly named new valuation and on the other hand we need to increase the index of V as we run through the cycle. So let us adopt the highest index used, in this case the index is "2" and so we upgrade the index of $V_1(a) = 1$ to be $V_2(a) = 1$. Cycle 1 is resolved with the exit valuation V_2 with $V_2(b) = 0.V_2(a) = 1$.

Step 2. *We now resolve cycle 2. We have a attacking x, so let us give it a new value $V_?(x)$. The question is what index to give V. This problem of indexing V needs to be systematically and formally defined and we do give a definition later in this Section. Meanwhile we just want to go on with our example, just to show the reader how the steps work. So let us use the index 3, since V_2 is expanding from Cycle 1 to the next cycle. So we have that the original $V(x) = 2$ becomes the new $V_3(x) = 1$. x attacks y, so $V_3(y) = 1$, y attacks z, so $V_3(z) = 1$. z attacks x so $V_3(x) = 0$. We get $V_3(y) = 1$ and $V_3(z) = 0$. So we continue and summarise our exit value from Cycle 2. Let us call it V_4. We have $V_4(y) = 1$ and $V_4(z) = 0$ and $V_4(x) = 0$.*

Step 3. *We now approach cycle 3. We have $V_4(y) = 1$. y attacks α, so $V_5(\alpha) = 1$. α attacks β, so $V_5(\beta) = 1$. β attacks α and so $V_6(\alpha) = 0$.*
Stop.

The extension we got is V', with

$$V'(a) = V_2(a) = 1$$
$$V'(b) = V_2(b) = 0$$
$$V'(x) = V_3(x) = 0$$
$$V'(y) = V_3(z) = 1$$
$$V'(z) = V_4(z) = 0$$
$$V'(\alpha) = V_6(\alpha) = 0$$
$$V'(\beta) = V_5(\beta) = 1$$

Note the following:

1. *Jumping indices $V_1, V_2, \ldots$ is just for the purpose of keeping track of how we go through the loops/cycles.*

2. *We first choose a top cycle and an element in the top cycle and cycle through until stable. We do this for all top cycles and only then do we attack the next level. This is our choice of how to find complete extensions. The reader can choose otherwise. The reader will obtain in such a case a different semantics.*

We continue this example, and choose to start with b this time. By symmetry the top cycle 1 will be resolved with $V_2(b) = 1$ and $V_2(a) = 0$. We continue. Since b attacks y, we get $V_3(y) = 1$. y attacks z so we get $V_3(z) = 1$. z attacks x so we get $V_3(x) = 1$. x attacks y so we need to increase the index of V for y since the value changes and we get $V_4(y) = 0$. So we get, since the value of y is now 0, y cannot attack and so that the value of z does not change and remains $V_3(z) = 1$ and so $V_4(x) = 0$. We summarise and get $V_4(x) = 0, V_4(y) = 0, V_4(x) = 1$.

We now have to continue and make a choice with the third cycle 3. Since $V_4(y) = 0$, we can choose to start with α or with β. We can use a rule if we want, to start with the geometrically marked α or ignore the geometry and also allow to start with β.

If we start with α we get $V_5(\alpha) = 1, V_5(\beta) = 0$, and if we start with β we get $V_5(\beta = 1$ and $V_5(\alpha) = 0$.

The full extension will be with

$$V'(x) = V'(b) = 1$$
$$V'(a) = V'(y) = V'(z) = 0$$

*and say if we choose in Cycle 3 to start with α then $V'(\alpha) = 1$,
$V'(\beta) = 0$.*

7 Better modelling of the case study

7.1 Initial discussion

The preliminary discussions of the previous Sections allow us to
present a better model for universal distortions. Let us summarise
what we got:

1. We agree that the arguments we use must be instantiated in
   the form of $\pm aOb, \pm C(a), \ldots$ where $a, b, \ldots$ come from some
   universe of discourse $U = \{a, b, \ldots\}$, and $O, C \ldots$ are predicates.
   This allowed us to use analogy on arguments. Let us accept this
   observation and from now on talk about arguments in general,
   accepting that these arguments are structured/instantiated as
   described above.

2. We accept the general schema of Figure 1. We have two systems
   P_1 and D, and D is connected to P_1 and is able to distort
   it. The question is what form does D take, and how does D
   communicate with P_1.

3. We accept that we should start with P_1 and D being argumentation networks and that there is an argumentation type
   connection from D to P_1. A change in D causes a universal distortion in P_1. So if $P_1 = (S_1, R_1)$ then this connection might be
   through a $D = (S_2, R_D)$, where S_2 is a set of additional points
   and R_D is a subset of $(S_1 \cup S_2)^2$. Another possibility for the
   connection is that P_1 is a network with a valuation, of the form
   $P_1 = (S, R, V)$, as in the previous Subsection 6.4 and D is a new
   different valuation $D = V_D$.

4. D must be compatible with intuition. It cannot be just any
   formal network. It must be intuitive and the connection with

P_1 must be intuitive. Thus D must be general enough to include/model the annihilator type distortion of Section 3.1 as well as the non-monotonic type distortion of Section 3.2 or the valuation type distortion of Subsection 6.4.

We shall see that the concept of the attack as informational input, [11], is a suitable concept for our purpose.

5. We agree that any formal connection from D into P_1 must be compatible with the following intuition:

> (intuition): Let x be an argument in P_1 and let $A_x = \{y|yRx\}$. Then the disturbance coming from D tends to mitigate the force of the attack from A_x to x.

This principle models the tendency of the offender to mitigate certain "in" arguments in order to allow himself to feel better and have a tolerable view of himself.[19]

Example 7.1. *Let us recall Figure 9 and the node z being attacked by $\{y, u\}$. Imagine that our offender would like to keep z "in", and would like to make it as difficult as possible for the attacks from y and u to be effective (z might be the statement that he, the offender, is an exemplary citizen, and y and u are counter-examples to that statements). He might say the following:*

1. *Both y and u must be "in". One attacker "in" is not enough, (i.e. requiring more "lives" for "being good citizen" one occasional failing does not refute it).*

---

[19]Gadi Rozenberg notes from his experience that sex offenders tend to respond to any immediate attack on them by deflection, rather than using logical interference (in the sense of Dynamic Argumentation) with the network in order to weaken the attack. So for example, if we have the attack chain

$$x \twoheadrightarrow y \twoheadrightarrow \text{Victim} \twoheadrightarrow \text{Offender}$$

then the offender (if he is not a lawyer or a logician) will weaken the direct threat ("Victim") rather than be clever about it and weaken argument x or strengthen argument y.

2. *The attack of y is coming from a non-trustworthy source. Maybe a child who has not complained for many years.*

3. *The argument y has a low value $V(y)$.*

The point is that the language must also contain predicates $V_i(y)$, $V_i(u)$, etc. What our offender is not likely to say is that he wants to have exactly one "in" attacker for z to be "in".[20] This is the case of Example 4 of the mathematical paper on Abstract Dialectical Framework (ADF) [27]. See Chapter 1 for further discussion in relation to ADF.

The reader might ask whether models of the form (S, R, V) are adequate, where S is a set of structured/instantiated arguments as discussed in Subsection 6.1 and V being a numerical valuation as discussed in Subsection 6.4. The answer is such models are almost OK, except that the valuations are in many cases not numerical but qualitative (such as "not reliable", "has an interest", "racist", "lying", etc). So we need a model that can take account of such valuations when calculating extensions. So even if we turn "unreliable" into a number, then the valuations $V(x)$ would be two dimensional (i.e. $V = (V_1, V_2)$, one dimension for reliability and one dimension for the number of lives. So if the sex offender says I am a good exemplary citizen and to successfully attack my statement I would require at least two reliable counter examples, then what do we do with three not so reliable such counter examples?[21] Obviously we need to consider n numerical valued functions $(V_1, ..., V_n)$ such that for each argument x we get a vector $\mathbf{V}(x)$ of values $\mathbf{V}(x) = (V_1(x), ..., V_n(x))$. So if x has say k attackers, $y_1, ..., y_k$, then the all values involved can be represented by an $(n \times k)$ matrix $\mathbf{M}(x) = [\mathbf{V}(y_1), ..., \mathbf{V}(y_k)]$. So we need a general function $\mathbb{B}_x$, (which may be dependent/tailored to the node x, that would take a general $(n \times k)$ matrix of numerical values $\mathbf{M}$ (n is fixed but k is arbitrary) and yield a vector of values $\mathbb{B}_x(\mathbf{M})$.

---

[20]The emphasis is on the word "exactly", the offender might want at least 2, or 3 etc but not "exactly 2".

[21] Compare with the weighted approach of [41]. See the discussion in Remark 6.2.

We are not going to offer a detailed model, just the general pattern for the reader to see the direction we are going.[22]

We do adopt the Boolean approach of [27], see Chapter 1, where we attach a Boolean formula $\mathbb{B}_z$ to any z, but we must use a language with valuation predicates V_i as well.

Thus in Figure 9, for the node z, attacked by y and u we look at the equation of the form

$$\mathbf{V}(z) \leftrightarrow \mathbb{B}_z(y, u, (\mathbf{V}(y), \mathbf{V}(u))),$$

and perhaps we allow $\mathbb{B}_z$ to only be monotonic (up or down) in each variable. This view and its generalisations is now illustrated in some more simplified examples.

Example 7.2. *Consider the network of Figure 9. Assume we have further a valuation V on elements of S. Say $V(s)$, for $s \in S$, which gives s a red colour. The Bench-Capon basic approach would be to decide for example that red coloured nodes cannot be attacked by non-red coloured nodes. So for example, if we let*

$$V(z) = V(u) = \top$$
$$V(x) = V(y) = \bot$$

then y cannot attack z.

So we can write for any $s \in S$:

---

[22] There is a clear connection here with what is known as weighted argumentation. Given a network (S, R), associate numerical real numbers value weights $W : S \cup R \mapsto [0, 1]$, and use these numbers in different ways to define new types of complete extensions. This is relevant to us, we can see how to use such systems to model sex offender distortions (they would change W). See [41] for a key paper on weights, with many central argumentation researchers as authors, and look up the references. The connection here will be pursued in a subsequent paper.

See, however, Remark 6.2, discussing the connection, after we give some technical result there.

One of the referees remarked that potentially , one could formalize cases where an argument is perceived as strong when attacked only by weak arguments (in the sense that if only weak, i.e., easily counterable, arguments can be found against an argument, then this argument is perceived as acceptable, or even perceived as stronger than without the weak arguments).

- $In(s)$ *iff* $[\neg V(s) \wedge \bigwedge_{yRs} \neg \ In(y)] \vee [V(s) \wedge \bigwedge_{yRs \wedge V)y)} \neg \ In(y)].$

*The above approach does not allow us to decide point by point whether
we want to allow points y such that $\neg V(y)$ (non-red) points to attack
points x such that $V(x)$ (x red) points.*

We cannot for example ignore V altogether.

*The idea we get from ADF is that we can write a tailored formula
(as far as V is concerned) for each $s \in S$. We have two requirements,
however, which make us different from ADF. The first is that the tailored formula we use must be a formula of predicate logic and not just
a numerical/propositional formula on the "in" and "out" values of the
geometrical attackers of s (namely on $\{x|xRs\}$). The second is that
we, however, do not go as far as ADF and do not wish to change the
basic Dung approach, namely we wish to keep the understanding that s
is "in" iff all of its "tailored" attackers are "out". The technical property we need is that the formula we use be monotonic in the number
of attackers which are "in".*

7.2 Abstract valuation frameworks (AVF), the equational approach

Given a system (S, R, V_i^y), this Section deals with the equational approach for the case where all properties V_i^y, for arguments $y \in S$ are
propositional. The reader is invited to recall Remark 2.1 and Remark
2.2, Chapter 6, for a better understanding of what we are doing here

Definition 7.1.

1. *Let S be a finite non-empty set of elements. We consider a classical propositional language based on the elements of S. The language has the atomic propositions I_n^y, and $V_1^y, \ldots, V_k^y$ for each
$y \in S$. The V_i^y are propositional constants describing properties
of the node y. I_n^y is a constant intended to mean (y is "in").*

2. *We also have a binary predicate xRy, on S. For every $x \in S$,
let $A(x) = \{y|yRx\}$.*

3. Let $\mathbb{B}_{In}^x$ and $\mathbb{B}_V^x$ $x \in S$, be Boolean formulas in the propositional language with $\{I_n^y, V_1^y, \ldots, V_k^y | y \in A(x) \cup \{x\}\}$. We assume monotonicity of these Boolean formulas in the propositions I_n^y.

4. Let Δ be the theory with the axioms (for each $x \in S$)

$$I_n^x \leftrightarrow \mathbb{B}_{In}^x$$

$\mathbb{B}_V^x$ gives a weighted V value for x in terms of I_n and V_i

5. Let **K** be the three valued propositional logic with the truth table of Figure 22. This is Kleene strong logic of indeterminacy [28, 29].

6. A system (S, R, Δ) as defined above, in item 4 of Definition 7.1, is called an AVF with Bench-Capon valuations.

7. Any Kleene model of Δ is called a complete extension.[23]

8. Note that we yet have to address the question of existence of such models for a given choice of Boolean $\mathbb{B}$ as in item 3 above. Existence can be obtained as outlined in item 1 of Remark 7.2.

Remark 7.1. *Consider Definition 7.1, and the system* (S, R, Δ). *Then the following holds:*

1. If the language does not contain any V_i and $\mathbb{B}_{In}^x$ does not contain I_n^x (unless xRx holds) then we get the the simple Boolean fragment of Brewka and Woltran Boolean ADF for the choice of monotonic formulas.

2. If the language contains a single V_1 and the wffs $\mathbb{B}_{In}^x$ and $\mathbb{B}_V^x$ for each $x \in S$ are as below then we get the Bench-Capon valuation system, where

$$\mathbb{B}_{In}^x = [V_1^x \wedge \bigwedge_{yRx \wedge V_1^y} \neg I_n^y] \vee [\neg V_1^x \wedge \bigwedge_{yRx} \neg I_n^y]$$
$$\mathbb{B}_V^x = V_1^x.$$

---

[23]Note that it is not true that the complete extensions of an ADF are all Kleene-models of the given framework. In fact a complete extension needs to be a fixpoint of the given characteristic operator as used in the respective ADF semantics papers. Being a Kleene-model is not sufficient.

A	B	$\neg A$	$A \wedge B$	$A \vee B$	$A \to B$
0	0	1	0	0	1
0	$\frac{1}{2}$	1	0	$\frac{1}{2}$	1
0	1	1	0	1	1
$\frac{1}{2}$	0	$\frac{1}{2}$	0	$\frac{1}{2}$	$\frac{1}{2}$
$\frac{1}{2}$	$\frac{1}{2}$	$\frac{1}{2}$	$\frac{1}{2}$	$\frac{1}{2}$	$\frac{1}{2}$
$\frac{1}{2}$	1	$\frac{1}{2}$	$\frac{1}{2}$	1	1
1	0	0	0	1	0
1	$\frac{1}{2}$	0	$\frac{1}{2}$	1	$\frac{1}{2}$
1	1	0	1	1	1

Figure 22

3. *The Kleene extensions of item 7 of Definition 7.1 are too general and not under control. As solution to equations we do not know what they look like and what they mean. The sex offenders case has more specific properties which can be used to construct better semantics. We shall see this later in the next Subsection.*

Definition 7.2 (Abstract valuation framework (AVF)). *Let (S, R, V_i^x, I_n^x) for $x \in S$ and $i = 1, \ldots, k$ be as in Definition 7.1. Further, let $\mathbb{B}_{In}^x, \mathbb{B}_V^x, x \in S$ be as follows*

1. $\mathbb{B}_{In}^x$ *has the form* $\mathbb{B}_{In}^x = \bigwedge_{yRx \wedge B_V^y} \neg In^y.$

2. $\mathbb{B}_V^x$ *is either monotonic up or monotonic down in each V_i, in any model of Δ, (Δ as defined in item (4) of Definition 7.1) for example $\mathbb{B}_V^x$ might be taken as $\bigwedge_i V_i^x$.*

Example 7.3. *Let us take another look at Figure 9 and try and express using AVF, the restriction that we accept that z is out if at least two in nodes attack z. We do this by having in the V-language a V_s for each $s \in S$ and let $V_s^y = \top$ exactly when $y = s$. So basically we can now talk about the elements of S. All we need to write now for z is*

$$\mathbb{B}_V^z = \bigvee_{E \subseteq A(z) \wedge |E| \geqslant 2} \bigwedge_{y \in E} In^y.$$

We use this B_V^z in B_{In}^z of item 1 of Definition 7.1.

Definition 7.3 (Equational AVF with V constant). *Let (S, R) be an argumentation network and let for each $s \in S$, let $V_i^s, i = 1, \ldots, k, s \in S$ be additional atomic symbols all pairwise disjoint. Consider the atoms of S and V_i^s as variables ranging over $[0, 1]$. We can also view $V_i : s \mapsto V_i^s$ as variables for functions with domain S and range $[0, 1]$.*

Thus we can view the atoms $V_i^s, s \in S, i = 1, \ldots, k$ to syntactically denote the value of V_i at $s \in S$. Let $\mathbf{h}$ be an assignment, giving for each $V_i^s, s \in S, i = 1, \ldots, k$, a particular value $\mathbf{h}(V_i^s)$ and creating a function $\mathbf{h}(V_i)$ from S into $[0, 1]$.

Let $\mathbf{F}_V^s$ for each $s \in S$ be a Boolean function built up from the variables $\{y, V_i^y | yRs \lor y = s\}$ and the functions

$$
\begin{aligned}
\neg x &= 1 - x \\
x \lor y &= \max(x, y) \\
x \land y &= \min(x, y).
\end{aligned}
$$

$\mathbf{F}_V^s$ is a function for a fixed s, with the variables s and y (such that yRs holds) and all the variables $V_i^y, i = 1, \ldots, k$, and all y such that yRs and s.

For a given $\mathbf{h}$, we can substitute the numerical values $\mathbf{h}(V_i^y)$ in F_V^s and get what we denote by $\mathbf{F}_{V,\mathbf{h}}^s$, which is a function without the variables V_i^y, because these have been instantiated with numerical values.

Assume $\mathbf{h}$ is given. Thus all the variables $V_i^s, s \in S$ have numerical values. Consider the set of equations, for each $s \in S$ as follows $\mathrm{Eq}_{\max}(\mathbf{h})$.

$$
s = 1 - \max_{yRs}(\min(y, \mathbf{F}_V^s))
$$

These equations have the elements of S as the unknowns. Let $\mathbf{f} : S \mapsto [0, 1]$ be any solution of the above equations. Then $\mathbf{f}$ is called a complete $\mathrm{Eq}_{\max}$ extension of the network $(S, R, \mathbf{h}(V_1), \ldots, \mathbf{h}(V_k))$ where (S, R) is the geometrical argumentation network and $\mathbf{h}(V_i), i = 1, \ldots, k$ are the fixed $[0, 1]$ valuations.

Remark 7.2.

1. *Note that since all functions of Definition 7.3 are continuous, by Brouwer's fixed point theorem (see Wikipedia [64]), there is always a solution to $\mathrm{Eq_{max}}(\mathbf{h})$. Using this solution we can get a 3 valued model of "in" being value 1, "out" being value 0, and the other values being "undecided".*

2. *Note that if we do not have any V_i, nor any $\mathbf{F}_V$, then the system becomes the ordinary equational approach, equivalent to the traditional Dung argumentation.*

3. *The function $\mathbf{h}(V)$ assigns a function to V from S to $[0,1]$. We normally assign numerical strength valuation as values $1, 2, \ldots$. This is not a problem. There are many continuous functions matching the intervals $[1, \infty]$ with $[0,1]$. For example, $f(x) = 1 - \frac{1}{x}, f(1) = 0$ and $f(\infty) = 1$.*

4. *$\mathbf{h}$ is fixed. So the values V_i^y do not participate in the equations as variables.*

5. *(S, R) does not have initial values $\boldsymbol{f}_0 : S \mapsto [0,1]$.*

Example 7.4. *Consider the network of Figure 5 and assume that*

$$\begin{aligned} V_1(x) &= 0.4 \\ V_1(y) &= V_1(z) = 0.8. \end{aligned}$$

Assume that $\mathbf{F}_V^s = V_1$.

The equations for the figure would be:

$$\begin{aligned} x &= 1 \\ y &= 1 - \min(x, V_1(x)) \\ z &= 1 - \min(y, V_1(y)) \end{aligned}$$

We get

$$\begin{aligned} x &= 1 \\ y &= 1 - \min(1, 0.4) \\ &= 0.6 \\ z &= 1 - \min(0.6, 0.8) \\ &= 0.4 \end{aligned}$$

We get new values for x, y, z but V_1 does not change. We have no equations to get new V_1.

Note that we can introduce distortion by lowering the value $V_1(x)$ of x to $V_1'(x) = 0.1$.

So the value for y would be

$$\begin{aligned} y &= 1 - \min(1, 0.1 \\ &= 0.9 \end{aligned}$$

a much higher value.

Recall that $1 = $ "in", and $0 = $ "out", and otherwise various degrees of undecided.

Example 7.5. *Let us look at the network of Figure 6. First we need to convert the V values into values in $[0, 1]$. We use the formula $V_1(a) = 1 - \frac{1}{V(a)}$. So we get:*

$$\begin{aligned} V_1(y) &= 0 \\ V_1(z) &= 0.8 \\ V_1(x) &= \tfrac{2}{3} \\ V_1(u) &= 0.5. \end{aligned}$$

Using the equational approach, we get:

$$\begin{aligned} y &= 1 \\ z &= 1 \\ x &= 1 - \max(\min(y, V_1(y), z, V_1(z)) \\ &= 1 - \max(0, 0.8) = 0.2 \\ u &= 1 - \max(\min(x, V_1(x)) \\ &= 1 - \max(0.2, \tfrac{2}{3}) \\ &= \tfrac{1}{3} \end{aligned}$$

7.3 AVF, the set theoretic fixed point approach

The equational approach of the previous Subsection 7.2 has two drawbacks. In the general case, the formulas $\mathbb{B}_s$ need to be formulas of predicate logic, involving labels $V_i(y)$ from a possibly different annotation language $\mathcal{A}$. We don't know how to solve predicate equations in Kleene's logic. Even if we did know how to do that, we would not know the meaning of what we are getting as solutions. We need a set-theoretic approach, if possible. So here we go:

We develop our abstract argumentation approach within the framework of labelled deductive systems (LDS), [58]. The basic idea of LDS is to annotate a given system with labels from some algebra of labels and manipulate both units of the system and their labels. Using labels we can unify the treatment of many similar systems as well as get general generic results.

This Subsection generalises the many annotated argumentation systems including numerical and graded systems into one general framework.

In other words, the semantic machinery of this Subsection is much more general than what we need, but we get it at no extra cost (of developing formal approaches)!

Our LDS systems have the form $(S, R, V, \mathbb{B})$ where $S \neq \varnothing, R \subseteq S \times S$, V is an annotation function giving each element of $S \cup R$ (i.e. each argument and each attack arrow) a label value in some given algebra $\mathcal{A}$. To be specific, think of $\mathcal{A}$ as the predicate V_i^y, $\mathbb{B}_s, s \in S$ all reside. $\mathbb{B}_s$ is a formula which updates the V_i^y annotation. For each subset $E \subseteq S \cup R$ we get a new set of annotations $\mathbb{B}_s(E, x)$ to the node x.

Let us leave the meaning and use of $\mathbb{B}$ vague as above, but insist that we are able to define three attack relations between sets $E \subseteq S$ and nodes $x \in S$.

$\alpha_{\mathbf{d}}(E, x), \alpha_{\mathbf{a}}(E, x), \alpha_{\mathbf{p}}(E, x)$.

The index $\mathbf{d}$ stands for administrative attack, $\mathbf{a}$ stands for ordinary killing attack and $\mathbf{p}$ stands for protective attack. Every protective attack is a killing attack and every killing attack is an administrative attack. To illustrate, consider the many lives model. Assume x is a sex offender which is attacked by y. Assume z attacks y to protect x. z attack on y is a protective attack and must be very strong. y attack on x is a killing attack and is not that strong. It may be that both y and x are still alive because they have enough lives to survive. But they have been attacked so they are damaged but not dead. So how

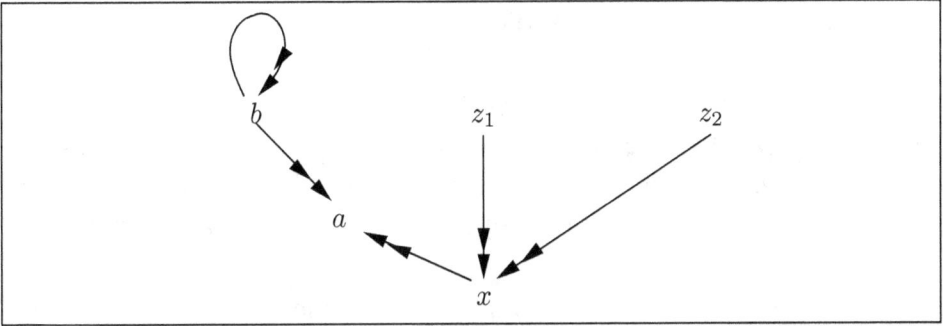

Figure 23

many lives have they lost? The administrative attack will settle that.

We assume $\alpha_{\mathbf{p}} \subseteq \alpha_{\mathbf{a}} \subseteq \alpha_{\mathbf{d}}$.

Assume that for each of these attacks we have:

- E attacks x and $E \subseteq E'$ then E' attacks x.

- E does not attack x and $E' \subseteq E$ then E' does not attack x.

Example 7.6. *Let*

$\alpha_{\mathbf{a}}(E, x)$ *iff* $\exists y \in E, yRx$
$\alpha_{\mathbf{d}}(E, x)$ *iff* $\exists y \in E(yRx \vee xRy)$
$\alpha_{\mathbf{p}}(E, x)$ *iff* $\exists z_1, z_2 \in E$, *such that* $z_1 \neq z_2 \wedge z_1, z_1 Rx \wedge z_2 Rx$.

Consider the network of Figure 7.6

Then in this network we have $E = \{z_1, z_2, b\}$ $\alpha_{\mathbf{p}}$ *attacks* x, *and* $\alpha_{\mathbf{a}}$ *attacks* a *and* $\{a\}$ $\alpha_{\mathbf{d}}$ *attacks* x.

Definition 7.4.

1. *We say that* E *is* at peace *iff for no* Y, a *in* E *do we have* $\alpha_{\mathbf{a}}(Y, a)$ *holds.*

2. *E protects* x *if for every* Y *s.t.* $\alpha_{\mathbf{a}}(Y, x)$ *holds we have* $Y - \{y | y \in Y$ *and* $\alpha_{\mathbf{p}}(E, y)\}$ *does not* $\alpha_{\mathbf{a}}$ *attack* x.

Lemma 7.1. *If* E *is at peace and protects its elements and* E *protects* x *then* $E \cup \{x\{$ *is at peace and protects its elements.*

Proof. Assume not at peace, get a contradiction.

Let $Y \subseteq E \cup \{x\}, z \in E \cup \{x\}$ be such E **a**-attacks x.

Case 1. $x \notin Y, x \neq z$ contradicts E at peace.

Case 2. $x \notin Y, z = x$. We have Y **a**-attacks x. Since E **p**-protects x, E must **p**-attack some elements $y_1, \ldots, y_k$ s.t. $Y = \{y_j\}$ does not attack x. Since Y does attack x, there must be at least one y_1 s.t. E **p**-attacks y_1. But **p**-attack implies **a**-attack, again a contradiction.

Case 3. $Y_o \cup \{x\}$ attacks z and $z \neq x$. Since $z \in E$, E **p**-attacks elements of $Y_o \cup \{x\}$. E cannot attack any elements from Y_0 so E attacks x but this is now case 1, which is impossible.

Case 4. $x \in Y, z = x$. so we have $Y_o \cup \{x\}$ attacks x. Since E protects x, E attacks $Y_0 \cup \{x\}$ but E cannot attack any of its elements. $\quad\square$

Lemma 7.2. *If E admissible and protects x then $E \cup \{x\}$ protects itself because E protects all elements of $E \cup \{x\}$ so $E \cup \{x\}$ does this as well because of the monotonicity condition.*

Lemma 7.3. *There exists an admissible set $E \subseteq S$ s.t. $E = $ all elements it protects.*

Proof. Start with $\varnothing$. It protects its elements and is at peace. Suppose $\varnothing$ protects x then $\{x\}$ protects x and is at peace.

Continue to increase the set using Lemma 7.2, until we reach a maximal st. This is the set E we need. $\quad\square$

Example 7.7. *Start with $\varnothing$. It is at peace and **p**-protects its elements. Suppose the empty set can also **p**-attack. Suppose it protects x. Then x cannot **a**- attack itself. If x **a**-attacks x then $\varnothing$ must **p**-attack x. So $\varnothing$ cannot protect x. Therefore x cannot attack x.*

Definition 7.5. *Let $(S, R, V, \mathbb{B})$ be an LDS system as defined in this Subsection, and assume that we have the notions of **d**, **a**, and **p** attacks respectively to go with it. Using the notions of **a** and **p** attacks we can identify the family of sets E which are admissible and are equal to the set of all the elements E protects. We can now use the notion of **d** attack to update the annotation of each element x in E. Let x be* any element x in E such that E **d** attacks x. Let the new annotation of x be $\mathbb{B}(E, x)$. If x is not **d** attacked by E, leave its annotation unchanged.

Let V_E be the new annotation on E. We refer to the system $(E, R$ restricted to $E, V_E, \mathbb{B}$ restricted to $E)$ together with the **d**, **a**, **p**, respective attacks restricted to E, as an E complete extension of the original system.

Example 7.8. *Consider a network with node b only such that bRb holds . Let us use the **a** attack and **p** attack of Example 7.6. Then node b **a** attacks itself but does not **p** attack itself. Therefore the extension we get is $\{b = 0\}$, not $b =$ undecided.*

If we take the attack notion to be Dung style then b both attacks and protects itself and so must be undecided

We close this Section at this point, in the spirit of Option E3, discussed in Section 1.

8 Comparison with the literature

Given a network (S, R), let us ask ourselves how can we interfere with its traditional complete extensions and semantics. Frankly, since the network is abstract and not instantiated, there is not much we can do. We can ask for some elements in S to be out (resp. in) and vice versae. If we further want to talk about a universal mass interference, (or universal distortion as we call it) we can talk about interfering with a large set of elements.

This is not satisfactory because we cannot give a good qualitative description of the interference, since all the elements of S are atomic. All we have is the Geometry of the relation R. For this reason papers in the literature tend to be technical and deal with local interference (recall Remark 2.2).

Looking at application areas where universal distortions occur, we saw that we need some logical description of the interference and for this reason we resorted to Bench-Capon valuations, or non-monotonic attacks . We can consider major distortions arising from some qualitative considerations in any instantiated approach such as ASPIC or ABA.

As far as we know, we are the first to deal with universal distortions.

Let us now compare with some specific sample of papers in the literature. Fortunately in many cases it is easy just to quote their abstract, to see how different (from our Chapter) and local the interference is.

1. **Papers [30]–[32], of Baumann and Brewka.**

   In these papers the authors study a Dung-style argumentation framework by adding finitely many new arguments which may interact with old ones. They study formally what can happen. These papers are mathematical studying questions like the effort needed to enforce a set of arguments E, measured in terms of the minimal number of modifications needed to turn an argumentation framework (AF) A into a framework A'. such that A'. has an extension containing E. These papers deal with local distortion from our point of view.

2. **Argumentation meets AGM revision.**

   Here we have a series of papers applying revision theory to argumentation. An argumentation network can be revised by adding one more argument, like paper [21] or by adding (integrating with) an entire new argumentation network like in [20]. We stress again that these papers are technical following a generic recipe for mathematical research. We may be able to use them in applications of universal distortion in some areas.

3. **Dynamic Argument Systems.**

   As we have already said, Argumentation Dynamics deals with local distortions, taking a few arguments out, or disconnecting some attacks or adding new elements or new attacks. The distortions we look at are global and follow a meaning . The reader can look up other papers under the dynamic argumentation title in the references.

We do want to mention one paper of Brewka which we particularly like. It is the Brewka's paper [33] which seems promising for applications. It is not directly related to universal distortion but it can serve as an example which can easily be distorted, a sort of realistic semi-theoretical example.

It presents a formal model of argumentation based on situation calculus. It models interacting agents seeking common ground. There are protocols for such interactions and so distortions in our sense can be viewed as changing or distorting the protocols.

9 Conclusion

We discussed universal distortions in reasoning networks. We gave formal definitions but most importantly followed a real life case study of a sex offender. The case study presents a challenge to the argumentation community. We need to expand our research horizons and model the rich world of logical and argumentation therapy for sex offenders. Let us explain why. Start with a simple network with $\{a, b\}$ and aRb and bRa. Assume we have valuation V with values $V(a) > V(b)$. Thus there is only one extension $a =$ in, $b =$ out. Our "offender" suffers from a universal distortion such that his value function V' has $V'(b) > V'(a)$. Hence his extension is $b =$ in, $a =$ out. This is certainly compatible with our case study, where the sex offender tried to minimise the severity of his offences. Our challenge is how to use therapy and correct the situation? We can appeal to the notion of audience of [6], and say to the sex offender everyone but you thinks that $V(a) > V(b)$. This will not work in practice. We need a much richer "envelope space" to operate with, as our patient resists therapy. The professional therapist discovered experimentally that in the case of sex offenders they, the offenders, can see the distortions in other offenders but not in themselves. So they use group therapy. The offenders themselves are used as the envelope space. It is significant that the therapist community actually use argumentation as a major instrument of remedy. We need to define such a space and within its framework define therapsuting steps. In our case it is an attack on V'. We can get ideas from practice in case studies. We thus develop a formal theory of Enveloping Spaces which go beyond audiences. The sex offender case is relatively easy. The area is well established. There are more difficult cases. How about a universal distortion caused by belief in ISIS ideology? What envelope space do we use? Note that ISIS employs positive propaganda for itself. Would our sex offender's therapy work successfully if there were international organisations saying this is a good thing? Think of the time of Plato and relationships between older men and young children. That was the expected norm then. How do you use therapy there?

On the technical side, we developed the following generalisations of argumentation networks:

1. The concept and idea of Universal Distortion in a reasoning system, of which this current Chapter is just a start. The concept exists in human reasoning and behaviour and so can be applied in formal systems which attempt at modelling such reasoning. These include Abstract argumentation but also ASPIC and ABA and indeed many other systems.

2. We generalised Bench-Capon valuation systems and looked at many lives valuations (Subsection 6.4) and generalised to Matrix valuations (Subsection 7.2). More importantly, we can come up with a new notion of proof theoretical semantics where the semantics for a valuation network is another valuation network. Further development of these ideas in a subsequent paper. We also showed that it is connected to Abstract Dialectical Frameworks. The formulas $\mathbb{B}_s$ we need to use for AVF are predicate formulas and need be monotonic in the $I_n(x)$ predicate appearing in the formula. Our way of presentation for our area makes these parts of ADF more intuitive for the sex offender case.

3. We showed a connection with several communities who use logic and argumentation, opening strong application opportunities for the Argumentation community. These are vast exciting future prospects. There is a danger that the argumentation community will become too technical and start feeding upon itself. This is

a dangerous, but natural, development which has already happened to the non-monotonic community and was recognised as such by that community. We hope it will not happen to us if we keep looking at applications.

Acknowledgements

A short version of this Chapter was submitted to COMMA 2016. We thank the three COMMA (Conference on Computational Models of Argument) referees for their careful reading of the text and helpful comments. We also thank the Journal referees for a very thorough and detailed three rounds of outstandingly penetrating reviews.

References

[1] D. Gabbay and G. Rozenberg. Reasoning Schemes, Expert Opinion and Critical Questions: Sex Offenders Case Study. *Ifcolog Journal of Logics and their Applications* (FLAP), 2017.

[2] D. Gabbay. *Meta-logical Investigations in Argumentation Networks*. College Publications, 2013, 770pp.

[3] T. Bench Capon. Value-based argumentation frameworks, 2002. https://arxiv.org/ftp/cs/papers/0207/0207059.pdf

[4] T. Bench-Capon and K. Atkinson. Abstract argumentation and values. In *Argumentation in Artificial Intelligence*, I. Rahwan and G. Simari, eds. pp. 45–64. Springer, Berlin, 2009.

[5] D. Gabbay and A. Garcez. Logical modes of attack in argumentation networks. *Studia Logica*, 93(2–3): 199–230, 2009.

[6] Trevor J.M. Bench-Capon, Sylvie Doutre, and Paul E. Dunne. Audiences in argumentation frameworks, *Artificial Intelligence* 171 (2007) 42–71

[7] D. Gabbay. An equational approach to argumentation networks. *Argument and Computation*, 2012, 3:2–3, pp. 87–142.

[8] S. Modgil and H. Prakken. The ASPIC+ framework for structured argumentation: a tutorial. *Argument and Computation*, 5(1), pp 31-62. February 2014.

[9] F. Toni. A tutorial on assumption-based argumentation. *Argument and Computation*, 5, pp. 89–117, 2014.

[10] D. Gabbay and O. Rodrigues. The numerical approach to argumentation with Bench-Capon values. In preparation (research ongoing, no written draft draft yet).

[11] D. Gabbay and M. Gabbay. The attack as information input. December 2015. Short version appeared in *Proceedings of COMMA (Conference on Computational Models of Argument) 2016*, Pietro Baroni, Thomas F. Gordon, Tatjana Scheffler and Manfred Stede, eds., pages 311-318. Long version draft available.

[12] Jean-Guy Mailly. Dynamic of Argumentation Frameworks. in *Proceedings of the Twenty-Third International Joint Conference on Artificial Intelligence*, IJCAI-13, 2013, pp. 3233–3234

[13] Steven Schockaert and Henri Prade. Solving conflicts in information merging by a flexible interpretation of atomic propositions *Artificial Intelligence* 175 (2011) 1815–1855

[14] Beishui Liao, Li Jina, Robert C. Koons. Dynamics of argumentation systems: A division-based method. *Artificial Intelligence* 175 (2011) 1790–1814

[15] G. Boella, S. Kaci, L. van der Torre. Dynamics in Argumentation with Single Extensions: Abstrac- tion Principles and the Grounded Extension, In: *Proc. of ECSQARU 2009*, LNAI 5590, 2009, pp. 107-118.

[16] Guido Boella, Souhila Kaci, Leendert van der Torre. Dynamics in argumentation with single extensions: attack refinement and the grounded extension, In: *Proc. of AAMAS 2009*, 2009, 1213-1214.

[17] M.Capobianco, C.I.Chesnevar, G.R.Simari. Argumentation and the Dynamics of Warranted Beliefs in Changing Environments, *JAAMAS* 11(2) (2005) 127-151.

[18] D.V.Carbogim. *Dynamics on Formal Argumentation*, Ph.D Thesis, University of Edinburgh, 2000.

[19] C.Cayrol, F.D.de St-Cyr, F.Dupin, M.Lagasquie-Schiex. Change in Abstract Argumentation Frameworks: Adding an Argument, Journal of Artificial Intelligence Research 84(2010) 49-84.

[20] R. Baumann and G. Brewka. AGM Meets Abstract Argumentation: Expansion and Revision for Dung Frameworks. In *Proceedings of the Twenty-Fourth International Joint Conference on Artificial Intelligence (IJCAI 2015)*, pp 2734–2740.

[21] Sylvie Coste-Marquis, Sebastien Konieczny, Jean-Guy Mailly, and Pierre Marquis. On the Revision of Argumentation Systems: Minimal Change of Arguments Status. In *14th International Conference on Prin-*

ciples of Knowledge Representation and Reasoning, KR 2014, Vienna, Austria, 20–24 July 2014. AAAI Press (2014).

[22] M. Caminada and D. Gabbay. A logical account of formal argumentation. *Studia Logica*, 93(2-3): 109-145, 2009.

[23] S. Modgil. Reasoning about preferences in argumentation frameworks. *Artificial Intelligence* 173 (2009) 901–934.

[24] Rakefet Dilmon. Between thinking and speaking—Linguistic tools for detecting a fabrication. *Journal of Pragmatics* 41 (2009) 1152–1170.

[25] S. L. Sporer. The Less Travelled Road to Truth: Verbal Cues in Deception Detection in Accounts of Fabricated and Self-Experienced Events. *Applied Cognitive Psychology*, 11 (1997), 373–397.

[26] R. Dilmon and U. Timor. The Narrative of Men Who Murder Their Partners: How Reliable Is It? *International Journal of Offender Therapy and Comparative Criminology*, XX(X) 1–25, 2013.

[27] Gerhard Brewka and Stefan Woltran. Abstract dialectical frameworks. In *Proc. KR'10*, pages 102Ð–111. AAAI Press, 2010.

[28] Wikipedia article: https://en.wikipedia.org/w/index.php?title=Three-valued_logic&oldid=735045520 (last visited Sept. 22, 2016).

[29] Merrie Bergmann. *An Introduction to Many-Valued and Fuzzy Logic: Semantics, Algebras, and Derivation Systems*. Cambridge University Press, 2008.

[30] R. Baumann and G. Brewka. Expanding argumentation frameworks: Enforcing and monotonicity results, in *COMMA 2010*, pp. 75–86, (2010).

[31] R. Baumann. What does it take to enforce an argument? Minimal change in abstract argumentation. In: *ECAI.* (2012) 127–132

[32] R. Baumann and G. Brewka. Spectra in Abstract Argumentation: An Analysis of Minimal Change. *LPNMR*, pp. 174-186, (2013). http://dx.doi.org/10.1007/978-3-642-40564-8_18

[33] G. Brewka. Dynamic Argument Systems: A Formal Model of Argumentation Processes Based on Situation Calculus, *J. Logic Computat.*, Vol. 11 No. 2, pp. 257–282, 2001.

[34] Reference for Gadi Rozenberg case study. This case is from 2014 in Israel. The details are protected by law.

[35] G. Brewka, S. Ellmauthaler, H. Strass, J. P. Wallner, and S. Woltran. Abstract dialectical frameworks revisited. In *Proceedings of the Twenty-Third international joint conference on Artificial Intelligence* (pp. 803-809). AAAI Press. (2013, August).

[36] S. Polberg. Extension-based semantics of abstract dialectical frameworks. In *STAIRS 2014: Proceedings of the 7th European Starting AI Researcher Symposium* (Vol. 264, p. 240). IOS Press. (2014, August).

[37] H. Strass. The relative expressiveness of abstract argumentation and logic programming. In *Proceedings of the Twenty-Ninth AAAI Conference on Artificial Intelligence (AAAI)* (S. Koenig and B. Bonet, eds.), (Austin, TX, USA), pp. 1625–1631, Jan. 2015.

[38] H. Strass. Expressiveness of two-valued semantics for abstract dialectical frameworks, *Journal of Artificial Intelligence Research*, vol. 54, pp. 193–231, 2015.

[39] H. Strass and J. P. Wallner. Analyzing the computational complexity of abstract dialectical frameworks via approximation fixpoint theory, *Artificial Intelligence*, vol. 226, pp. 34–74, 2015.

[40] Pietro Baroni, Massimiliano Giacomin and Giovanni Guida. SCC recursiveness: A general schema for argumentation semantics, *Artificial Intelligence*, Volume 168, Issues 12, October 2005, Pages 162–210.

[41] Paul E. Dunne, Anthony Hunter, Peter McBurney, Simon Parsons, and Michael Wooldridge. Weighted argument systems: Basic definitions, algorithms, and complexity results, *Artificial Intelligence* 175 (2011) 457–486.

[42] H. Barringer, D. M. Gabbay and J. Woods. Temporal, Numerical and Metalevel Dynamics in Argumentation Networks. *Argument and Computation*, 3(2-3), 143–202, 2012.

[43] H. Barringer, D.M. Gabbay, and J. Woods. Temporal dynamics of support and attack networks: From argumentation to zoology. In *Mechanizing Mathematical Reasoning*, LNCS, vol. 2605, Springer-Verlag, Berlin, Germany, 2005, pp. 59–98.

[44] D. Gabbay, G. Rozenberg and Students of CS Ashkelon. Introducing Abstract Argumentation with Many Lives. Submitted to *Argument and Computation*, Draft January 2017

[45] Vokey, M., Tefft, B. and Tysiaczny, C.: An Analysis of Hyper-Masculinity in Magazine Advertisements, in *Sex Roles* (2013) 68: 562. doi:10.1007/s11199-013-0268-1

[46] D. Grossi and S. Modgil. On the graded acceptability of arguments. In *Proceedings of IJCAI'15*, p. 868–874. volume 2015.

[47] W. L. Marshall and C. Hollin. Historical developments in sex offender treatment. *Journal of Sexual Aggression*, 21(2), 125–135, 2015.

[48] T. Ward, S. M. Hudson, L. Johnston, and W. L. Marshall. Cognitive distortions in sex offenders: An integrative review. *Clinical psychology*

review, 17(5), 479–507, 1997.

[49] F. L'osel and M. Schmucker. The effectiveness of treatment for sexual offenders: A comprehensive meta-analysis. *Journal of Experimental Criminology*, 1(1), 117-146, 2005.

[50] E. Beauregard and B. Leclerc. An application of the rational choice approach to the offending process of sex offenders: A closer look at the decision-making. *Sexual Abuse: A Journal of Research and Treatment*, 19(2), 115-133, 2007.

[51] W. D. Pithers. Relapse prevention with sexual aggressors. In *Handbook of sexual assault*, (pp. 343-361). Springer US, 1990.

[52] T. Ward, K. Louden, S. M. Hudson, and W. L. Marshall. A descriptive model of the offense chain for child molesters. *Journal of Interpersonal Violence*, 10(4), 452-472, 1995.

[53] Roger T. Webb, Jenny Shaw, Hanne Stevens, Preben B. Mortensen, Louis Appleby, and Ping Qin. Suicide Risk Among Violent and Sexual Criminal Offenders. *Journal of Interpersonal Violence*, 27(17) 3405–3424, 2012. DOI: 10.1177/0886260512445387

[54] Gerhard Brewka and Stefan Woltran. GRAPPA: A semantical framework for graph-based argument processing. In *ECAI 2014 - 21st European Conference on Artificial Intelligence*, 18-22 August 2014, Prague, Czech Republic - Including Prestigious Applications of Intelligent Systems (PAIS 2014), pages 153-158, 2014.

[55] D. Gabbay and Co-authors, Abstract Generic Argumentation (AGA), February 2017.

[56] Sylwia Polberg. Understanding the abstract dialectical framework. In *Proceedings of the 15th European Conference on Logics in Artificial Intelligence (JELIA'16)*, 2016.

[57] P. M. Dung. On the acceptability of arguments and its fundamental role in nonmonotonic reasoning, logic programming and n-person games *Artificial Intelligence* Volume 77, Issue 2, September 1995, Pages 321-357.

[58] D. Gabbay, *Labelled Deductive Systems*, OUP, 1986

[59] D. Gabbay, G. Rozenberg and L. Rivlin. A research monograph based on the current paper and on [1], [11, long version], [44] and [55], in preparation for College Publications, 2017

[60] D. Gabbay, G. Rozenberg and Students of CS Ashkelon. Temporal Aspects of Many Lives, draft available May 2017.

[61] http://www.bbc.com/news/uk-england-merseyside-37542868, accessed May 17, 2017, 1100 hours UK time

[62] http://www.ncbi.nlm.nih.gov/pubmed/16040578.

[63] https://en.wikipedia.org/wiki/%C3%89mile_Durkheim, accessed May 17, 2017, 1100 hours UK

[64] https://en.wikipedia.org/wiki/Brouwer_fixed-point_theorem. Accessed May 18, 2017, 1500 hours UK.

Sex Offender Practice

CHAPTER 7
DISTORTION DATA OF SEX OFFENDERS

1 Investigative interviews with child victims of sexual offences

Children have a universal distortion in the sense that when very young children do not have clear boundaries between reality and imagination. Therefore when a sexually abused child is interviewed special care must be taken to make sure the fact are identified.

The following are guidelines for how such interviews are conducted in Israel, as well as experimental facts about how children respond in interviews.

What is the best way to conduct an exploratory interview with the sexually abused child?

Sexual abuse is unique because usually the victim-witness is a child, pitted against an adult perpetrator who has an overpowering interest in covering up the offence.

Sometimes a criminal charge comes about as a result of a false accusation from a spouse — for instance, a wife might try to reinforce her claim for a divorce by making a false accusation which could be of financial benefit to her.

Under Israeli law, the interview has to be recorded, with the place, participants and beginning time all noted aloud at the start. At conclusion, the interviewer has to note the end time. The length of the recording must correspond exactly to this timing.

During the interview, the investigator must specify the non-verbal articulations that occur in the room: for example, noting that the child looks down to the a ground in response to a certain question, or recording instances of the mother making signals to the child, etc.

The general principle of the investigative interview is to create an environment which allows the child freely to communicate past experience and outline specific aspects related to the offence from the general to the specific.

The investigator should be prepared for the interview. That means he needs to read all the background he needs to know, the charges (if any), and any other information. He needs to think ahead about which questions to ask, for which he must make a model. However, he should also be sufficiently flexible to adapt to what is happening in the room. For example the perpetrator may suddenly start crying during the interview.

Stages of the interview

1. *Building a connection with the child*
   At this opening stage the researcher must devote sufficient time to make contact with the child and this will differ between individual children. If the child is agitated, the investigator must calm and reassure him.

   At this stage, the researcher should pay attention to the child's linguistic abilities and what happens to him when he becomes excited or when he is talking about unpleasant or sad things.

   The interviewer should ask the child to introduce himself and talk about a pleasant event he has experienced, or about his kindergarten. In other words, the child should be encouraged to relax by describing things and events unrelated to the offence. This stage should help the investigator in assessing the child's abilities and so in planning the future of the interview.

2. *Checking a child's ability to distinguish between reality and fantasy*
   This is a critical step. The interviewer needs to clarify to the child that he should tell the truth and only the truth. In such cases it is permissible to use the terminology of children's stories suitable to the mental age of the child. For example, it could be

that the interviewer asks the child to remember what happened to Pinocchio when he lied, and that it is not only not permissible to lie but that the interviewer will be able to see a lie for what it is. It is, though, important to make sure that the child knows the difference between true and false. If the child cannot distinguish these two, it is a problem which the investigators should take into consideration.

3. *Presentation: The purpose of the interview*
   The interviewer should explain to the child that he is assisting in an investigation and needs information about what happened, in order to help. The investigator should not make promises he cannot fulfil. It is very common for a child witness to ask about secrecy and it is especially important not to promise the child that such secrets will remain in the interview room. The investigator should also tell the child what will happen later in the process and explain the role of the police −= to the extent that the child can understand. If the investigator considers that this is a problem, he will have to arrange special care for the child.

4. *The child's free expression in describing the event*
   The interviewer should allow the child to speak about the event or events, giving his perceived version and his opinions about what had happened, in his own words and without hindrance. This is a stage where no questions are asked and the researcher should let the child say, without interruption, what he remembers even if it is obvious that there are contradictions. This step should be conducted according to the pace of the child, patiently, calmly, without corrections or questions despite any opinions the investigator might have about inconsistencies and discrepancies

5. *Questioning: direct and indirect*
   The purpose of this step is to expand the knowledge of the investigator with respect to events described during the free narrative. The researcher needs to think about the child's open-

ing narrative and expand the information. Leading questions should be avoided. For example, asking a child what happened on his birthday implies that something did in fact happen on his birthday. It is not permissible to instruct the child with such questions. The investigator should also use the language of the child. For instance, children have their own names for their genitalia. At this point questions should be formed in the nature of requests — "Can you tell me more about this event?", "You can tell me about the man, you said". The investigator should keep in mind that events can be a vague memory if the incident remembered at the age of 13 actually happened when the child was 6. The investigator should pay attention to the child's response to questions such as what happened, when it happened, and who participated.

The child's responses should come without direct questions. Investigators often worry if an answer is incomplete but the question can be revisited at a later stage. If the child does not answer the question the investigator needs to go back possibly putting it in another way, but without pressure, so that the child will not give a response just because he feels stressed.

The investigator must remember that leading questions can reduce the admissibility of the testimony during a trial. There are exceptions for children who cannot express themselves but in such cases anatomical dolls, drawings and other props may be used in the form of a game to help with data collection.

6. *Ending the interview*
   At this stage the interviewer must thank the child for his participation regardless of the outcome (even if the researcher did not achieve his goal). The interviewer must ask the child if he has questions and if the child does have questions the interviewer should answer them if possible. The interviewer must keep open a line of communication in the event that the child remembers something — for example, giving a telephone number.It is recommended to end the session with something neutral (as at the

beginning of the interview) and tell the child that they might meet again. The second phase, following the interview, is done without the child, in which the researcher listens to the recording.

Evaluation of the investigative interview

Assessment of the interview has several aspects:

1. Interview material is important in terms of whether the material gathered contains enough information in order to build a court case.

At this point the interviewer should consult with the police interrogator to assess whether or not there is enough accumulated material to file charges. If necessary, the researcher can summon the child again but the recommendation is that there is as little time as possible between the first and the second interview.

2. The Children's Investigator will determine the extent of the injury to the child.

This will determine the treatment plan as well as the legal position of the child. Physical evidence will be found from medical examinations and can support the case. The test for mental health is more complex. Some idea of the state of the child's mind will be indicated in the first interview and the interviewer has to assess this and the effect it will have on the child's evidence. The investigator will have to evaluate whether it is necessary to visit the scene of the crime, whether the child is capable of identifying a criminal in a line-up, or if the child could be stood as a witness in court. In other words, a balance has to be struck between the operations of the Law and the possibility of further damage to the victim. In most cases researchers refuse to let the child appear in court, which can expose him to the possibility of being a victim a second time. However, occasionally it is deemed a restorative experience, giving the child the chance to experience the reward of showing strength and courage. This will happen only when a child is over the age of 11 years, when the incident is very serious and when the offender is not known to the family.

3. Evaluation of reliability.

The investigator must check whether the child's evidence is reliable. A distinction should be made between the sort of minor mistakes which are made in normal discourse and a full-fledged lie. Eligibility is also a factor. Eligibility is the child's ability to give any sort of a report about a past experience. This is a necessary first requirement but it is not a guarantee of reliability. There is a debate about the competence of children to testify in court, in four main areas. Eligibility is the legal framework which defines the ability to distinguish fantasy from reality and truth from falsehood.

The four main areas are as follows:

- **Imagination** — the conventional wisdom is that small children have difficulty distinguishing between fantasy and reality. This in itself is often used as a reason to doubt their testimony. Various studies have produced no definite answer as to what age children can distinguish between an event that really happened and a construct of the child's imagination. Most studies indicate that children over the age of 6 resemble adults in their ability to identify whether or not the source of the event is a product of inner thoughts or the result of an external factor.

  The consensus is that adults with their more developed cognitive skills and life experiences will always be able to perceive the difference between truth and fantasy. An adult will know, for instance, when some incident is contrary to the laws of nature. Studies show that despite a general doubt as to the ability of children to make such distinctions, clinical experience has demonstrated that children seldom invent significant events that happened in a way that trigger doubt about whether the act described actually happened. This is especially true when allowing the child to describe the act in his own words without help or prodding from instructors or without the use of tools that may implant presuppositions from the questioner. We must distinguish between fabrication and falsity because the fabrication and false report is related to the reliability, not to competency. In case of fabrication and false reporting the lying is done intentionally for a variety of different reasons.

- **Language** — Children have a limited vocabulary which is much less descriptive than adults. A child's vocabulary is usually very dependent on the environment in which he has grown. If the child has been raised in an educationally impoverished environment, the child's vocabulary will be more restricted than if he were raised in an enriched environment, but whatever the environment language at the disposal of the child is more meagre than it would be for adults. It is therefore more likely that his testimony would be interpreted wrongly or that the child will understand the questions and the investigator wrongly. Therefore the researcher should carefully evaluate the linguistic ability of the child and to be aware of nonstandard Hebrew, accepting it rather than trying to change it. It should be remembered that the child may be influenced by the investigator's style of speech so it is important to ask the questions in his own language style.

- **Memory** — the debate on the eligibility of children to serve as witnesses often concentrates on the child's ability to remember and describe their experiences. Research on memory has generally been carried out in laboratories, which means we are talking about experiments which might differ materially from real life experiences or the trauma of having to testify in court. The relationship between memory and age is affected by various factors. There is a connection between memory and stress. Studies have also shown that small children are more likely than older children and adults to have memory distortion. These distortions are larger the further back the event is in the past. When the researcher discusses it he must not ignore the element of trauma. People who go through a traumatic event tend to remember it in more detail than they would recall a commonplace incident. It is the same for small children, there are many events during the day although a child will encounter many experiences during a normal day, if he is injured, he will remember that particular incident more clearly than the others. Certainly he will remember sexual trauma.

As for the contention that stress affects memory, various studies have been conducted examining how memory operates under stress but results are still inconclusive. Some studies seem to show that the memory is even sharper, while others show that the memory completely shuts down. We can therefore say that the results are not conclusive in either direction. However, not all sexual abuse is painful and traumatic. In other words, not all sexual abuse may impair or enhance memory. Even when recalling events which we did not understand or we have suppressed, information which we recover after the passage of time is problematic. For example, Flynn took her research subjects from different ages, showed them an emotional element to an event, which would help them remember the event, and they found that after five months, children reported less information than they initially reported. That is, as time goes on we lose information. It was also found that children six years old reported less information than nine year old children (who could remember relatively more), meaning that younger children are losing more information. Memory can facilitate the interview stage. Memory is a building process in which adults and children are trying to recover the memory.

- **Suggestibility** — regardless of the memory capacity of children, there are doubts about their eligibility as witnesses in light of their exposure to the effects of leading or misleading questions. Suggestibility can also result from impaired memory and also from misinformation from the interviewer. Studies have shown that you can easily affect the memory of adults by using leading questions, particularly when the subjects are asked, using a leading question, to recall events with a personal meaning.

It is well known that the way you frame a question can determine the outcome of, for instance, a referendum. Drafting the questions in a certain way may create biases. Adults questioning children can also create biases. In the case of kindergarten-age children who were exposed to some form of suggestion or interference after an incident, this reportedly influenced the children's accounts of the incident some time afterwards. In other words, they absorbed incorrect information being aired during the interference. Among schoolchildren there is less suggestibility but it is unclear by how much. Most experiments dealing with suggestibility cannot be set up using real trauma and even if it some sort of trauma is involved it will be one experienced in a group rather than alone. However, child sexual trauma is experienced alone. We have no studies which can give an answer about the memory of a child who has experienced a true trauma of this type. But there is consensus on one thing: when a subject is recalling a key traumatic event, leading questions will have less affect in distorting memory than they would if they were directed at a subject remembering comparatively ordinary happenings.

2 Therapy groups of sex offenders

We mentioned in the Chapter that therapy groups of 8-14 sex offenders takes about 20 months. We also said that the objective is to make sex offenders realise, by means of argumentation and logic, that they are doing wrong. This Section explains how this works. It is hoped that researchers in argumentation will get a better view of how the sex offenders therapist community works and hopefully get involved in offering help and interact with this community. This Section introduces two types of dedicated therapy groups for sex offenders and examines the pros and cons of each method of therapy by addressing different aspects of each and examining its logical and practical charactristics.

In order to achieve this goal, we focus on two types of groups: "closed groups" (groups with fixed membership) and railway Groups (open groups with a non-fixed, variable membership).

In jail, the treatment groups usually include about 14 adult sex offenders which is an enormous number and there is definitely a price for dealing with such a large group.

In the opinion of Dr. Rozenberg, the optimal number of participants in groups is no more than 8.

We asked ourselves many times what are the advantages and disadvantages of each type of group and what is the preferred method. To answer this question first of all we explain what a dedicated group is for sex offenders

We are talking about a therapy group with cognitive behavioural orientation (CBT). The patients are treated by this method because other methods were not found to be effective enough. Probably the reasons are:

1. at least to begin with, sex offenders are poorly motivated to accept responsibility for their behaviour

2. sex offenders perceive themselves as victims and they can talk incessantly about their victimization

3. We also believe that in dynamic group therapy we get insight and this is important, although insight is only a part of the therapeutic process, but at this point we can say that the CBT sex offender therapy has a lot of dynamic elements, which we shall discuss later on.

Why group therapy? The simple answer is that statistically it is the most effective treatment. The logic behind it is that sex offenders feel exceptional. They know they are deviants and in group therapy the feeling that they are not alone and that there are other people with similar problems often allows them to become more open and receptive ("openness"). The group helps mitigate the embarrassment and helps the therapist to confront patients. Sex offenders are much more receptive to each other than to other people and so if the group works well, things that patients say to each other are sometimes more important and have more influence than the therapist.

The treatment, based on a model of Bengis (1986) is called "relapse prevention". The goal of the treatment is to prevent future attacks on other victims.

Main goals:

A. raising the level of awareness of the sex offender about the range of behaviour options available to him.

B. developing coping skills and strategies of self-control.

C. creating a sense of control in the sex offender over his life.

Basic assumptions:

A. There is no cure for the disorder, but the subject can learn how to avoid repeating the offending behaviour (therapists are not magicians and do not know how to effect a complete cure for such things as perverted fantasies. Neither can we completely heal mental disorders).

B. When a perverse thought appears, the individual still can choose to avoid violence.

C. model SUD = seemingly unimportant decisions . Sometimes even taking seemingly unimportant decisions may lead to an offence.

For example, a sex offender with a paedophilic disorder who is asked to deliver packages on a regular basis to an office which happens to be next door to a kindergarten. The seemingly unimportant decision to accept the job may lead to abusive behaviour.

D. the offence is planned and not impulsive

E. Not every lapse is a relapse. A violation of one of the risk factors = lapse and relapse= makes a new sex offence. Even if the patient made the wrong decision, and got himself into a dangerous situation (lapse), still he can recognise it withdraw, and thereafter can learn to avoid it, and not say to himself, I stumbled, what can I do?

Let us take this opportunity to express what we believe. We are not sure that all therapists would agree with us. We believe we should talk with simple language to the offenders and not expect drastic changes in their emotional ability in a short time. We have to remember where these people came from and to where they return. For us, it is important that the therapeutic content will survive in everyday life. Also, we do not think we have a mandate to change the culture of the person and have no right to tell the patient whether or not to get engaged, whether or not, to marry several women or one. We believe we are allowed to intervene only if we see a direct correlation between the offender's approaches to future possible offences.

The character of groups: "Closed Groups" mean that you cannot get new patients to join during the group process and the group therapy therefore has a clear beginning, middle and end. Open (railway) groups are groups where patients can join at different stages The generally accepted practice is that the group has two therapists: a man and a woman. It is good for modelling and shows how two can communicate, and can even disagree but can still respect each other without using stereotyped behaviour. The underlying idea of this approach is to reduce extremes of disagreement. It should be pointed out that research on the juvenile probation service found no difference in the success of the group according to the gender of the practitioner

After a brief review of the nature of the groups and the basic assumptions/ let us examine the advantages and disadvantages of open and closed groups.

We will divide this as follows:

1. Nature of the group,

2. Planning

3. Sorting of candidates

4. Therapist

5. Integration of new participants

6. Completion of the therapy.

1. The nature of the group. A closed group is a set with linear characteristics, where the group members are supposed gradually to change, reach enlightenment, to gain knowledge (although this word is not good enough to explain the process because it becomes experiential work, requires emotional and mental involvement, etc.). The patients should identify major patterns within their personalities and learn new ways to conduct themselves. This is circular work. The group works on a particular subject and only after all the members finish working on a specific theme are they allowed to move to the next item.

An open group however, is spiral and the group moves inconsistently. The progress often moves backwards and forwards through the beginning, middle and the end. This method allows us to examine the same issue in different periods and to attack it at different times. The patient is able to do so in a different developmental stage of the group and of himself. While a closed group is relatively regular, the open group is characterized by much lability. In order to understand the movement and development of the open group we must remember the "moods" of Bion theory.

The open group "mood" swings could be dependent, aggressive, avoiding observation, watching, rational, emotional, controlled, uncontrolled, shrinking, and expanding. Sometimes the movement of the group resembles a sort of tango, where the progress is three steps forward and two steps to the side, but it has no distinct development phases. In the open group the therapist must hold all the required information and all the time to must try to combine all the puzzle pieces together and pick up on necessary issues.

2. Planning. Due to the nature of the closed group the therapists in the planning stage can construct a series of gradual and consistently logical contents and to stick to them, adjusting is required only when you need to make changes to the content in on them the working group Before the therapist works in depth on a particular topic it is best to start with less intimidating content, to create links between the participants and gradually to go from easy to hard. For example, in prison we allow patients to work on trust, to describe experiences of their lives. The therapist should express empathy and increase collaboration, in such a way as to reduce defensiveness and allow them to develop direct reactions of empathy towards the victims. In the case of a closed group we do not have a created situation in which a new patient drops in and might be asked to relate to risk factors before working on the assault cycle and before he figures out the trigger, or manipulation and planning of his act. Such a situation certainly can happen in an open group .

3. Sorting of candidates. In a closed group, sorting and choosing the candidates must be very rigorous as if we make a mistake, nobody else can fill the space until the end of the group process and we cannot allow someone else who needs the treatment to take advantage of it. Regarding a decision about the maximum number of participants in a group we think there's room for some flexibility. We have observed over time, mainly because of the need to meet the demand, that we needed to increase the number of participants in our therapeutic groups. We took into account that larger groups meant (mean) mean slower progress and a longer time for the group therapy to reach its end The average treatment lasts for almost two years (in the past, with fewer participants, the process was shorter).

In an open group, we can be less meticulous in choosing the candidates or in cases of doubt we can consider allowing a candidate temporarily to join the group. We can and do, tell him that he goes into a two months trial period, and then later on we make the final decision regarding suitability. Of course, this situation involves difficulties. The patients have problems with frequent changes, which challenges their need for a sense of order and regularity. We should also mention that letting in a person into therapy and then removing him could further undermine his self-image, increase a sense of failure (many patients have a very low self-image) and of course it must be noted that for a patient, the retirement from dedicated treatment (whatever the reason) also increases the sexual risk level. Despite all of that, at least this is something less drastic, less cutting, knowing that it is possible to get a person an opportunity. We feel that a therapist sorting candidates for an open group can make a more relaxed assessment and can take more time for important decisions.

The closed group does not have this advantage, even if we set up a preparatory short group prior to the closed group meetings in order to examine who is ready for the process and who is not.

4. Role of the therapist. When dealing with an open group the therapist must be much more alert and make sure that the complicated group dynamics, with people coming in and out and going around in circles in treatment will ensure the proper treatment of all the

In the case of Open an open group the therapist does not have the luxury of a gradual and systematic progress. A study conducted in 2006 by Dr. Avraham Ofek which explored youth sex-offending groups says: "this railway method (open group) is good for youth and hard for therapists" the difficulties he found being those of regression and repetition. Research indicates that stress on the therapists is much heavier in open groups than in the closed groups. In open groups the therapists have to become acrobats, ensuring that all the patients, although they joined the group at different times, will be treated on all the basic issues of a dedicated sex offenders group. Already at this point we can say that one way to ensure that most of the required content has been addressed is to perform a number of assault cycles for every patient (it is difficult to believe how circles seem different at different times).

In our estimation, closed groups are more appropriate for training new therapists at the beginning of their careers, because with the closed group, the new therapist has the opportunity to observe and learn consistently about intervention, and identifying the logic behind the gradual construction of the contents and development of a group. Probably learning from an open group can be confusing and over-whelming., For that reason, a trainee therapist might consider joining the open group during later stages after viewing closed groups.

5. Integration of new participants. About joining new patients in open groups, there is the fear that the veteran (i.e. senior members) would get bored, having to repeat procedures they have already learned. The therapist must be very creative and try to reach the same goal using various techniques and adjusting the techniques to form the group. Yalom claims that, adding a new member successfully, depends partly on timing: there are better and less good times to add members. During a crisis and struggles it is harder to integrate a new member or if he is admitted, the fear is that the energies are directed to the new member which may disturb the flow of conversation about a burning issue. The most convenient time for admitting a new member, according to Yalom, is when the group is not really moving forward.

A social worker named Tamir Ashman adds that the number of joiners is important. According to Ashman, it is important to induct two members simultaneously, in order to facilitate the process. According to Ashman, when the group is experiencing a period of crisis, it is hard to absorb a new single member but if two or three join, they are able to create leverage and start a fresh viewpoint. Even if the burning issue is forgotten, the process will return to it in another way, with the perspective of the new people. This can create dynamics that allow a more productive viewpoint. Ashman claims that two patients joining at the same time is the optimal number In my experience I have not been able to find a formula definitively to establish what is the optimal number. It depends greatly on the members' personalities and the state of the group. I can also say that if the group is in crisis, with no trust between members and no significant progress, it is a mistake to insert a new person. We have to focus on the obstacles and only then, to add. The fear is that the entrance of a new patient will delay the group work and hurt the trust and intimacy which has been created, but certainly it can be said that the integration of a new person at an appropriate time allows "freezing" of the image of the group and gives an opportunity to review the progress. For example, with the joining of new participants all are asked to describe their offences and we can see a difference in description, with the adding of relevant information, etc.

One of the significant advantages of an open group is that it allows for a change of status and position in the group. In a closed group almost everyone establishes a position in the heirarchy of the group and it is very difficult for an individual to change status thereafter. In open groups every patient has more opportunity for a change of status. A 'back-bench' member of the group might very well become a veteran who can give the newer members the benefit of his experience. The very important thing is that the new patients constitute for the veteran a kind of mirror of where they were to where they are now (I cannot count the number of times I have heard a patient say "In the past, I thought the same as you...") and this is certainly a very new experience for both. One sees where he should strive to get and the veterans have a chance to see the progress they have made, which helps improve their self-image. This lets the veterans look "sideways" on themselves

In the past I had very clear positions on specific situations, for example that new patients are not allowed to join during the process of the departure of an old patient. Today I am really not sure about this. Life is always more powerful than us and all kinds of different situations develop. I thought that somebody's introduction during veteran's separation was unfair to both with neither getting enough attention. But I found that this situation can make a substantial contribution to the treatment of all the patients in the group. Letting a person enjoy telling about the process he just finished and giving him the responsibility of introducing the new one to the group. This actually helps to reduce anxieties and fears, while the new member sees that it is possible to finish the treatment and to change himself. Mostly, new patients enter naturally into an existing commitment and we can see fewer power struggles with the therapists. This is the type of barometer that shows the patient where he wants to go and the others to demonstrate what they have done.

I talk a lot about how the nature of the group affects patients, but I should say that for therapists it is an exhausting task of trying to assess whether or not it is time to add a new patient. What did we miss? How much time must we devote to each subject, etc. Also, due to the nature of the group we have to decide on the recommendations to shorten or extend the duration of the stay. About the finish — I will deal with this after I have described the group.

The Therapy process For closed groups, Treatment focuses on the emotional, cognitive and behavioural aspects and the steps upon which the Group focusses are as follows

1. Familarization. Everyone gives his name and important facts about himself

2. "I and the other" patients are asked to use drawings of themselves and of significant figures in their lives and the work on drawings allows focussing on the life history of the individual. Other methods at this stage are guided visualisation, therapy cards, "Anibi".

3. Empathy or identification of emotions. Work on the different feelings and emotions of the patient and the others. This can be done using pictures with facial expressions, writing a letter to a victim and a letter from a victim, reading the testimonies of the victims and so on.

4. List of sex offender's arguments during therapy. These were listed in Subsection 7.5 of Chapter 6.

5. Addressing emergency pressures. Aggressiveness, assertiveness, passivity. The patients are asked to describe such challenges and how they dealt with them

6. Sexual functioning. It is imperative that we convey to the patient not only what is prohibited, but also what are the alter-

natives.

7. Offence Cycle, see [1]. This can be identified as a summary of the therapy progression.

   (a) trigger: this is what sets of the pattern of behaviour. The trigger need not have any sexual connection.

   (b) feelings and thoughts (cognitive distortions)

   (c) disinhibitors: alcohol, drugs or pornography

   (d) planning

   (e) focus on the offence

   (f) reconstruction of the process.

   Finally we work on how to deal with risk situations. We use films and other means of engaging the senses

8. The closed groups make several "stations" in the process These stations involve stopping and getting feedback which allow the patients an opportunity to test themselves while the therapists can use this to assess the progress and identify points to reinforce and improve.

6. Completion of the therapy. I now come to the last part: finishing closed groups. Closed groups have a set time to finish the process. If treatment can be described as linear, at the end of it we can look at the path we have taken and then disengage. As a group we wind up the process and go our separate ways. This is a very exciting process of termination and hope. We must add that the professional literature talks about how the patients can enhance the effectiveness of the treatment by constantly reviewing the content they worked on in the group. Therefore each client receives the file containing all of his homework, confessions and other things he worked on. We succeeded in organising a reunion for one group of graduates from a prison programme and I can say it was exciting to see the patients after a short period and I am sure it was studied and also significant for us and for them. I would definitely recommend that these class reunions be continued. In open group therapy the situation is different. According to Tamir Ashman, one of the disadvantages of an open group is that it does not deal with the group "here and now" and there is no processing of a separation experience because in contrast to closed groups the patients do not say goodbye to the group, but the group says goodbye to the patient.

In a closed group the therapist can say goodbye to all the patients in a winding-up ritual and thereby has the chance not only to review the progress that the participants have made but to take satisfaction from seeing that the patients have more insight and from the anticipation that they can be better, more well-integrated members of society than before they started the treatment. In railway groups the problem is that there is no such opportunity for a formalised parting nor will any parting take effect with all participants simultaneously. The therapist merely says goodbye to one and then immediately receives a new patient. Sometimes this makes it very difficult for the therapist to gain a sense of achievement and job fulfilment.

Finally, I want to say that I talked about advantages and disadvantages of both methods and tried to propose solutions to reduce the disadvantages of each method, but one thing I have not been able to find a solution to is the experience of the therapist in an open group. In a closed group, the therapist handler finishes with patients, then has a break before taking on a new group. However, in an open group the therapist works through the process, seeing patients grow but instead of finishing the treatment, but then having the therapist then has to start work with other patients in an unending flow with no break, no grand finale, but only a continuous conveyor belt of patients.

3 Child sex offenders, adults and children in the UK, by Steve Spurr, Retired Social Worker▋

3.1 Background

The author is an independent social worker in the UK who had previously worked for children's services departments in London and the South East of the UK for over 30 years dealing with child protection and allegations of abuse against adults working with children. In London the author developed a project that assessed the risks posed by juvenile child sex offenders and also a scheme to identify and protect teenage victims of sexual exploitation by adult offenders. This section details the agencies involved in this work in the UK and the systems and procedures guiding their work from a social work perspective. Common examples of distorted thinking by perpetrators, institutions and employers are listed along with case studies of adult and child offenders. The Chapter concludes with the author's personal view of the reasons why greater success in eradicating child sex abuse has not been achieved and what might be done to overcome this.

3.2 Introduction

The prevalence of sexual offences against children in the UK has received a great deal of attention in recent years due to a growing understanding of the scale of the issue and its consequences for victims. During the past 30 years, the discourse has changed from complicit suppression of the problem to a state of moral panic and confusion over how to deal with the rising number of offenders and the risks they present. The growth of social media on the internet has provided new opportunities for offending and for offenders to normalise their cognitive distortions in relative privacy.

The range of types of sexual abuse against children is wide, extending from abuse within families and close relationships, through abuse by trusted adults in schools, churches and other institutions, to community based exploitation by gangs and organised groups.

The response by government and agencies in the UK has been to encourage professionals to work together to identify children most at risk and to prevent known offenders having access to children. Preventative and educational programmes have also dominated the landscape, but the consensus remains that offences and victims are increasing in number and that sexual abuse is a common experience for many children.

Faced with this situation, the UK government set up the Munro Review of Child Protection[1] and two formal inquiries, firstly the Inquiry into Child Sexual Exploitation in Gangs and Groups[2] which looked at organised networks of adults who had systematically sexually abused hundreds of vulnerable teenage girls in urban cities and secondly the Independent Inquiry into Child Sexual Abuse[3] to work with victims and survivors to determine the scale of the wider problem and to make recommendations for action.

This recent history and the ongoing investigations of institutional employees and the Crown Prosecution's determination to prosecute historic offenders, including high profile celebrities triggered by the Savile[4] affair, has led to widespread anguish that the sexual abuse of children is deeply ingrained in UK society and that a long term strategy is needed to effect change.

3.3 How the UK child sexual abuse system works

The legislative framework for the UK has been created through the criminal law of offences against children, by the Children Acts 1989 and 2004 and the publication of statutory guidance which followed.

---

[1]Professor Eileen Munro's review reports, (2011) resulted in recommendations to free up local authorities from bureaucratic and compliance burdens and to give them more scope for professional judgement, enabling them to create new solutions

[2]Inquiry into Child Sexual Exploitation in Gangs and Groups, Office of the Children's Commissioner 2011 to 2013

[3]An independent statutory inquiry established by the Home Secretary under the 2005 Inquiries Act with the aim of conducting an overarching national review of the extent to which institutions in England and Wales have discharged their duty of care to protect children against sexual abuse.

[4]The celebrated disc jockey and children's BBC television presenter Sir Jimmy Savile was accused.

The most significant interagency guidance is called Working Together to Safeguard Children[5], which along with its supplementary guidance gives detailed advice to a range of agencies on how to proceed with all types of abuse, including child sexual abuse. Other disciplines such as medicine, police, probation and social work have developed guidance to approach specific requirements regarding investigation, treatment and prevention. Universities run research programmes using evidence based theory and are informed by the voices of children and families. The result is a well-developed expert system, operated by many thousands of practitioners in the UK, but which is largely unknown to the general press or public.

3.4   Statutory requirements

Section 11 of the Children Act 2004 places duties on a number of specific organisations which provide services to children requiring them to safeguard and promote the welfare of children.

The agencies are:

- Local authorities and district councils, public health, housing, sport, culture and leisure services, licensing authorities and youth services;

- National Health Service organisations;

- Police

- National Probation Service

- Governors/Directors of Prisons and Young Offender Institutions

- Directors of Secure Training Centres

- Principals of Secure Colleges

- Youth Offending Teams/Services

---

[5]Working together to Safeguard Children — A guide to inter-agency working to safeguard and promote the welfare of children. Most recently revised in March 2015

These agencies are co-ordinated in a local authority area by Section 13 of the Children Act 2004 which requires the local authority to form a Local Safeguarding Children Board to oversee joint working; by providing policies for all aspects of safeguarding children work, to ensure staff are trained, that outcomes are monitored and effective, that services for children are planned and that inquiries into child tragedies may be carried out and lessons learned. Individual agencies retain their own accountabilities and responsibilities for service delivery but Local Safeguarding Children Boards are able to make clear to an organisation where improvements are needed.

In day to day practice the requirements to safeguard children and report abuse or suspected abuse are devolved to every organisation dealing with children and overseen by the governance of the area's Local Safeguard Children Board. It is important to note at this point that the UK does not have a mandatory reporting requirement for child abuse other than that placed on teachers and other children's professionals to report suspected or actual Female Genital Mutilation.[6]

3.5 Schemes to regulate and assess offenders

The investigation of offences and the assessment of offenders is high tariff work with severe consequences for victims, perpetrators, society and professionals and so many scientifically validated tools and scales have been developed in many countries.

Of greater interest here are the organisational models for professionals to work together on the identification, investigation and harm reduction of offenders in the UK.

Such schemes have typically been developed by Police and Probation services, the community of therapeutic professionals working with victims and offenders and the large number of Local Authorities

---

[6]Section 5B of the Female Genital Mutilation Act 2003 places a statutory duty upon teachers along with regulated health and social care professionals in England and Wales, to report to the police where they discover that FGM appears to have been carried out on a girl under 18.

who are tasked with protecting children and vulnerable adults in local areas of the UK.

- The Government's Disclosure and Barring Service which requires professionals and volunteers working in Regulated Activity[7] to undergo regular criminal records checks

- MAPPA or Multi Agency Public Protection Arrangements. These are groups of professionals from agencies involved in safeguarding such as Health, Housing, Education, Probation, and Social Work, led by the police and tasked with making risk assessments and risk management plans for sex offenders and other dangerous adults.

- Local authority multi agency panel meetings concerning young offenders similar to the MAPPA model, convened with a view to early intervention and prevention.

- Local authority procedures to hold Child Protection Conferences[8] and to initiate Court Proceedings where the risk to a child is directly attributable to the actions or inactions of the parents or carers.

3.6   How investigations are managed

Reports of abuse, or referrals as they are commonly known, are made to the local authority or the police where the child lives or is found.

Local authorities have teams of social workers in Children's Services departments (previously known as Social Services) which allocate them to qualified workers for an assessment and to talk to the children.

---

[7]Regulated Activity is work which involves close and unsupervised contact with vulnerable groups including children, and which cannot be undertaken by a person who is on the Disclosure and Barring Service's Barred List.

[8]A meeting of parents and professionals to decide whether a child protection plan is needed to protect a specific child, held under the guidance of Working Together to Safeguard Children 2015.

Police forces have specialist child abuse teams and sexual offence teams which investigate allegations of criminal activity.

The local authority and police always work jointly where the referral concerns sexual abuse of a child where:

- the child and perpetrator are in the same family or household

- where a child sexually assaults another child or adult

- where the adult perpetrator is in a position of trust with children, for example teachers, scout leaders etc. and all professionals and volunteers who work with children

- where a child is being targeted by an organised network for sexual exploitation.

(Note that in the UK a child is anyone under the age of 18 years.)

The police in the UK will investigate reports of sexual assault without the assistance of children's services only where there is no family or trusted adult connection to the offence or where the offence against a child is an historic one. Health professionals are often involved in supporting the work of social workers and police, particularly in cases where medical and psychological evidence may be sought. The following are examples of referrals where police and children's services work together:

A child tells her teacher that her father tickles her private parts whilst he bathes her
A child says that his grandfather rubs his bottom when he places him on his lap
A father of young children downloads indecent images of children

A child tells her mother that a boy at her school has placed his hand up her skirt
A boy forwards on indecent images of his girlfriend to his friends

A boy grabs the breasts of a woman walking home

A teenager complains that her teacher is sending her inappropriate texts
A 17 year old student tells his friend that he has had sex with his college tutor
A child is touched indecently by her music tutor

A parent is worried that her daughter is missing overnight and returns with new jewellery
An older girl encourages younger girls to meet her boyfriend's adult friends
A taxi driver offers teenagers free rides in return for sexual favours

A 14 year old girl tells her friend she has had sex with an adult man she likes
A teenage boy is asked to help run a disco by a man who then tries to have sex with him
A teenage boy who thinks he is gay has consenting sex with an adult man

There are many other variations but these serve to illustrate the wide range of possible case types which are jointly investigated by police and social workers. The examples can be grouped into 5 categories, viz:

1. Abuse within the family, extended family or household

2. Abuse by one child on another

3. Abuse by an adult in a position of trust

4. Abuse by an organised network

5. Abuse by a stranger or non-family member

The procedures and methods of each of these types of investigation are broadly similar but differ in detail as follows.

Type 1) Abuse within the family, extended family or household

These referrals are typified by the risk being from within the child's family unit and where an assessment is required to ensure the child's protection.

Common features encountered during investigation are:

- Family denial in order to keep the child

- Child accused of fabrication

- Older child rejected by family

- Child forced to retract allegation by family

- Older child retracts voluntarily

- Inconclusive outcome with no prosecution possible

Type 2) Abuse by one child on another

It is now known that most adult sex offenders begin offending in their teenage years and thus early intervention and prevention of further offending can be extremely helpful as challenges to distorted thinking is more likely to be successful at this age.

Common features include:

- Distinguishing sexually harmful behaviour (victimising) from sexually problematic behaviour (inappropriate behaviour)

- Distinguishing experimental behaviour from exploitative behaviour

- Distinguishing sexually motivated and offensive behaviour from silly behaviour (e.g. mooning)

- Distinguishing between consensual and non-consensual (or compliance and non-compliance)

- Family denial

- Family rejection

- Exclusion from education

- Precursors include neglect, lack of boundaries, attachment problems,

- Not always a victim of prior sexual abuse.

Type 3) Abuse by an adult in a position of trust

These referrals involve the rapid evaluation of risk to other children and a consideration of the employment position of the alleged perpetrator.

Commonly encountered features include:

- Disbelief by colleagues, parents and employers

- Previous unreported incidents emerge during investigation

- Complex management of confidentiality and publicity

- Containment of public and employer anxiety

- Attempts by organisation to minimise or cover up

- Feelings of guilt by the victim.

Type 4) Abuse by an organised network

These referrals often involve teenage children who have been groomed or controlled and may not be aware they are being exploited.

Commonly encountered features include:

- Victims not trusting authorities

- Victims labelled as out of control rather than at risk

- Poor mental health and self-harm

- Exclusion from education

- Frequently missing from home or care

- Victims controlled by use of public or family shaming, money, alcohol and drugs

- Victims encouraged to recruit further victims.

Type 5) Abuse by a stranger or non-family member

These referrals may be the result of online grooming or from direct contact and involve helping parents and other significant adults to protect children, who as in the previous classification, may not be aware they are being exploited.

Commonly encountered features include:

- Victims' use of social media

- Low self-esteem of victims

- Risk taking behaviour of victims.

3.7 Unconvicted and suspected offenders

People who are suspected of committing sexual offences and those who have been tried for offences and have been found not guilty by a Court pose particular challenges.

For those who have never been charged and where there are grounds to suspect they have a sexual interest in children which may lead to an offence being committed, there are limited options available. Where a Court can be persuaded that an offence is likely, then it has the power to grant a Sexual Risk Order[9] which may also include a ban on foreign travel. Local Multi Agency Public Protection Panels may also put a risk management plan in place to monitor such individuals in order to reduce their risk of offending.

People who have been brought to trial and found not guilty will sometimes proclaim their innocence, particularly those who are public figures such as entertainers, politicians and church leaders. Their protestations of innocence often include statements that they have been "cleared completely" by the Courts, whereas in strictly legal terms, the prosecution has failed to prove the case "beyond reasonable doubt", a high threshold for criminal proceedings which exceeds the "balance of probability" threshold which is applied in civil proceedings and all child protection safeguarding work. Indeed many public figures who have endured lengthy and publicly reported police investigations into suspected historic child sex abuse and who are ultimately not charged will use this declaration of innocence as a means to denigrate the police, the prosecutors and often the complainants themselves. There are cases currently being reported in the UK where former suspects are suing the Police for wrongful investigation but also cases where suspects have been convicted of sexual offences many years after initial charges or investigations had been unsuccessful.

3.8   Offenders' cognitive distortions and defences

Offenders commonly lie and deny their offending behaviour as their first strategy, but when challenged will often rationalise their conduct in a number of ways. They will deny completely any wrongdoing and place the responsibility on the victim and investigators to provide evidence, they will minimise the extent of the offence and its harm,

---

[9]Sexual Risk Order — granted by a magistrate on application by the police where a conviction or caution has not been made but there is a risk of harm. Made under Part 2 of the Sexual Offences Act 2003.

(sometimes in order to distract from a more serious offence) or they will justify offending based on distorted thinking.

All of these defences are important to the offender to enable him to overcome his natural internal constraints in order to begin offending and to continue by justifying his behaviour which runs counter to universal social norms concerning sex with children.

Early intervention will help to reduce the effect of distorted thinking and prevent the offending behaviour becoming entrenched and reinforced through the reward of pleasure. This principle underlies the importance of intervening positively with young offenders who will be more open to change and will enable a lifetime of sexual offending to be curtailed.

- Outright denial:

  - They are lying or fantasising
  - They want to make money out of me
  - They are suffering from a false memory
  - My ex-partner has coached her into making the allegation
  - They've got together and ganged up against me because of a grudge

- Partial denial:

  - I had no idea they would take it that way
  - I was asleep/drunk at the time and didn't know what I was doing
  - I was only trying to teach him how to hold the musical instrument/cricket bat
  - I must have been very naïve and didn't mean any harm
  - I won't be so trusting again because it leads to trouble

- Justification:

  - The child wanted to do it and enjoyed it

- – I taught her about sex in a kind way

- – She took advantage of me

- – He had a crush on me and I didn't want to upset him

- Entitlement:

  - – I deserve sex as a reward as my job is so worthy and demanding

  - – God has granted me this behaviour as I am so dutiful to Him

  - – I was abused as a child and so I am only doing what was done to me

Such feelings of entitlement occur elsewhere, for example a motorist who cannot drive through on a green light because a driver in front has failed to notice the lights have changed, is tempted to drive through the red light because he feels disadvantaged by the actions of the other motorist. The motorist does not consider he has committed a traffic offence in driving through the red light in these circumstances.

3.9   Case studies

Adult offenders, sexual exploitation, family incest, child on child

The following examples have been provided to demonstrate the range and variety of case types that occur, rather than to represent "typical" cases. They are all from practice experience in one London borough and have not been embroidered or altered in any way so as to give a plain description.

Adult offenders

1. Serial offender, child-like wife, sought ways to re-engage with children The offender N was convicted of indecent assault on 2 children who subsequently gave evidence as adults. N was working as an IT teacher in a London primary school where he

had access to over 300 children aged between 5 and 11 years. He was seen at school by a teacher colleague to be viewing an image of a naked child on his computer monitor and this was reported to the head teacher who contacted police and children's services. A joint visit by police and a senior manager from children's services confirmed the presence of a small number of indecent images of young children on his school computer. The school's hard drives were removed for copying and further investigation and his home was searched and computers seized.

The searches revealed thousands of indecent images as well as discs containing a history of direct sexual abuse on children going back 30 years. Another disc contained stories of abusing children which were thought to be "fantasy diaries".

N admitted the offences of making indecent images of children but denied any direct contact abuse. Police investigators made extensive enquiries in the UK and other countries where N had worked in order to identify some of the victims. Five adult males were identified and acknowledged they had been abused by N when he and his wife acted as babysitters for their families. Only two of the victims were willing to make statements and give evidence against N, the others said they did not want the memories of the abuse re-awakened.

N was convicted and sentenced to four and half year's imprisonment, serving half of this. He attended sex offender treatment programmes whilst in custody and on release sought employment and activity which would have offered him further access to children.

N was required to sign the Sex Offenders Register[10] and be monitored for a period of 10 years by the local Multi Agency Public Protection Panel, (a group of agencies working alongside police specialist officers tasked with continuously assessing risk to the public from sex offenders). This panel subsequently au-

---

[10] A requirement under UK legislation which enables Police and Probation Services to monitor convicted offenders through knowledge of the home address

thorised disclosure of his offending history to a Buddhist temple, a teaching staff agency and a tourist guide agency in order to allow these organisations to be able to carry out their own risk assessments. He was subsequently expelled from these groups as a result of the disclosures.

N applied for permission to travel abroad with his wife, herself a diminutive and child-like woman. This was refused and a sex offenders travel ban order issued by a court as he was considered to be a high risk of offending in other countries.

Children's services were concerned that N would continue to seek contact with children and so made an unannounced visit to his home where children's toys and games were seen, which was a concern as the couple were childless. A follow up visit by police with a software tool for determining whether a computer had been used to download indecent material resulted in *N* confessing he had been accessing indecent images of children on the internet. *N* was immediately recalled to prison to serve out the remainder of his sentence and was later convicted of further downloading offences resulting in an additional sentence.

N continues to be monitored in the area where he lives.

2. Secondary school teacher, sex with school cleaner and a student, student afraid to report, information passed to next school.

   D was a secondary school science teacher in a mixed school for pupils aged 11 to 18 years. He was an attractive, charismatic and confident man who was popular with staff and students. His teaching subject was in great demand because of a national teacher shortage for the sciences.

   Another member of teaching staff had discovered *D* and a 15 year old student after school hours in a classroom resources cupboard which had been locked by a key from the inside. *D*'s explanation was that he and the pupil had been searching for equipment and denied the door had been locked. The pupil was interviewed by senior school staff and a social worker but made no complaint or allegation against D. The pupil later disclosed that D had been giving her extra tuition with regard to a forthcoming critical examination and that she had gone into the cupboard voluntarily when they were interrupted by the other staff member. The pupil refused to make a statement or be interviewed by police about the matter and it was considered in her best interests to let the matter rest until after her examinations when she would be asked again.

D was suspended by the school pending an investigation and it was discovered that he had been flirtatious with many of the school female staff including the school cleaner who said that she was having occasional sex with him after borrowing money from him for a pressing debt.

The pupil would not confirm that anything improper had occurred and D resigned from the school without any action being possible against him.

D obtained another teaching appointment in another secondary school in a different district and was due to start the next term. He had not given honest details about his previous employment and so the new school did not seek a reference from his last school and were therefore unaware of the cupboard incident.

A link made by the police that D had applied for the new teaching post was made through the school applying for a criminal conviction check. The check was clear as he had never been convicted of an offence but it triggered knowledge of the incident which was disclosed to the new school and led to their offer of employment being withdrawn.

3. School caretaker, conspiracy, community alarm.

    R was a male school caretaker aged 58 who had worked in several schools in the district over the past 35 years. He was well known in the area and via the local press for decorating the garden of his school based bungalow with festive figures at Christmas time.

    A woman living in Australia emailed the headteacher of her old secondary school in the UK to say that she had been sexually abused by R when he was the caretaker of that school some 30 years previously. The email was passed to the Police and Children's services who advised the headteacher to seek the woman's permission to pass the email to the authorities to respond to. The headteacher sent a supportive email to the woman who consented to the proposed action and who subsequently made a lengthy disclosure to Police in Australia who passed on her statement to the UK Police.

The UK Police sought out class contemporaries of the woman complainant who were then interviewed and three of them made similar allegations of abuse. It was determined that none of these women had retained contact with each other since leaving the school.

The caretaker was suspended and told to leave his school based bungalow accommodation. His wife and son mounted a vociferous campaign against the school and local education authority resulting in anxiety and panic amongst the school staff and local community.

Special support was given to the staff of the junior school where the caretaker had last worked and measures were taken to determine whether any of the children had been abused by R. The risk or likelihood of them having been abused was considered to be low as the children were all under the age of 11 years and were therefore always accompanied by teaching staff. The nature of the allegations against R was that he had groomed several 15 year old girls who had come to see the ponies he kept on the secondary school grounds. It was therefore thought unlikely he could have had such unsupervised access to younger children in his last school, although the possibility was never completely excluded.

R protested his innocence saying that the girls had borne him a grudge for not allowing certain privileges and that they had conspired together. Evidence of letters written between the first complainant and R produced at the trial conclusively convinced the jury of his guilt and he was sentenced to six years imprisonment. R attempted to appeal his conviction but the appeal was disallowed.

R's family never accepted his guilt and continued to hector and harass the local education authority and some school staff. Legal action to protect individuals and to evict the family from the school accommodation was necessary.

4. Solo GP, work experience victims, airplane complainant.

P was a General Practitioner (GP) who practised alone in a small surgery in the local area. His wife was the practice manager and nurse. He had been practising for many years and also undertook visits to the US to take part in clinical research.

Two 14 year old girls were given short work experience placements with P at his surgery following his offer to their local secondary school. The girls were given administrative tasks in the surgery and complained to their parents that they had each been touched sexually by P.

The Police evaluated the girls' statements and recommended to the General Medical Council that P be suspended from his general practice. P's wife began a public campaign outside of the surgery to protest her husband's innocence.

The Police looked into P's background and discovered that several years previously he had been arrested at London airport on his return from the US as a result of a female passenger alleging that he had touched her sexually during the flight as he sat next to her. The woman complainant was due to take a connecting flight out of London and was reluctant to miss her connecting flight in order to make a formal police statement. The complaint against P was therefore not investigated nor pursued.

This information could not be used in P's trial for the assault against the girls as it was not legally admissible and the woman complainant did not wish to resurrect the matter. However P

was convicted of indecent assault and sentenced to two years imprisonment and a fine of £50 to be paid to each girl. *P* appealed his conviction and his barring from the medical profession unsuccessfully.

5. Dentist

K was a local dentist with a thriving practice. He was detected by Police to have downloaded indecent images of children on his home computer.

K was arrested and his professional association suspended him from practice.

The images that *K* had downloaded were of the most serious kind as they depicted the sexual torture of young children.

K did not have any children of his own but there was considerable concern that he had access to children in his work and that they would at times be unconscious through anaesthetic.

K was convicted and received a nominal suspended sentence. The judge made an order banning him from treating children but remarked that he hoped this would not affect his successful business. *K*'s professional association attempted unsuccessfully to erase his registration.

6. Teaching Assistant, previous concerns, church work, university application.

F was a 23 year old teaching assistant working at a local secondary school. He had previously volunteered at a church youth club and his mother was a special needs teacher at another school. A friend of a 14 year old girl at his school reported that she had been told that her friend had been in a sexual relationship with *F* and had felt that it was not right. The 14 year old was interviewed and confirmed that she had been groomed by *F* over a period and that she had recently had sexual intercourse with him.

F was arrested but not before he had been alerted through social media and he had been able to delete material on his home computer.

The investigation discovered that there had been several occasions when the church youth club and the school could have reported *F* for inappropriate behaviour which would have alerted them to his sexual interest in children. The church and school managers became defensive and feared a legal action against them by the victim and her family.

F was sentenced to four years imprisonment and on release sought to volunteer in another youth club and also applied to a youth work university course. Disclosures were made by Police to the church and university and these options were refused to him.

A sexual exploitation case

M was 15 years of age when she came to attention for going missing frequently from her family home for several days at a time. Her father was abusive to her mother and had left the home 3 years previously. *M* associated with older men in the local area and there were serious concerns that she was abusing alcohol and drugs and receiving these and money from these men. A risk management plan was put in place but this only served to increase professional anxiety about her risky behaviour. A Child Abduction Notice[11] was sought which named several adult males and had the effect of warning them not to associate with *M*. *M* continued to behave riskily and was taken into the care of Children's Services after being found in a drug user's household and alleging rape.

M was placed in a children's home under a Secure Accommodations Order[12] in a children's home in Wales, many miles distant

---

[11]Granted by a court under section 2 of the Child Abduction Act 1984 which bars an adult from associating with an under 16 year old child, usually used in cases of kidnapping but relevant here to protect children from undesirable adults.

[12]Made under the Secure Accommodation Regulations 1991 and which satisfy

from her home area. She resumed her education and her health and emotional well-being improved. She applied to be released from the Secure Order and was returned home eight months later whereupon she immediately resumed her risk taking behaviour with her former associates. The Child Abduction Notice was now no longer effective as M had reached the age of 16 years which is the upper limit under the Act — an anomaly with the Children Act definition of a child being up to the age of 18 years.

M continued to lead a chaotic and troubled life and her suspected existence through prostitution and pornography was confirmed. She became pregnant at 17 and decided to see the pregnancy through. She was placed in a highly nurturing mother and baby unit some distance from her home where she gave birth and was successfully re-housed with her child.

A family sex abuse case

T was a 13 year old girl who lived with her twin brother, her older brother and mother and father. The father appeared to be quiet and unremarkable and was employed; the mother was a chronic user of alcohol and she later died. T disclosed that her father had been sexually abusing her to the extent of full sexual intercourse for several years. He had also forced T and her twin brother to have sex together. The twins were removed from their home and placed in separate children's homes and the father was charged with several serious sexual offences. The father was required to live separately from his wife and older son and it transpired that he had been sadistically sexually abusing his wife throughout their marriage.

T's fury at her father led her to exaggerate and fabricate some of the evidence she gave against him, which led to him being found not guilty of all of the charges.

T grew up despondent and angry seeking the affection of strangers through sexual contact. T openly prostituted herself and became

---

section 25 of the Children Act 1989 where a child can be deprived of their liberty if they are at serious risk of harm to themselves or others.

addicted to alcohol and drugs. She subsequently had several babies all of which were removed by Children's Services owing to the inability of T to safely care for them.

T's twin brother managed to complete his education and became a youth worker.

A child on child case

J was an able 15 year old boy who had come to his school's attention over the previous 2 years for a number of incidents which were; touching a girl's breasts, making simulated sex noises in class, "humping" or simulating dry sex with a girl, and being found in the girls' toilets.

A was a 14 year old Eastern European girl with low self-esteem at the same school who had an interest in J and who had previously given him oral sex in a local park.

J persuaded A by text to meet him and 3 other boys in the school toilets where she was encouraged to provide all 4 boys with oral sex in turn.

The incident resulted in A being bullied by other girls and she left the school and her family moved to a different area.

The 4 boys were known to the police for being gang members involved in criminal activities. Their parents refused to co-operate with the school or children's services. The police could not investigate the incident as A said that she had acted willingly in an effort to please J.

There were concerns that A might have been particularly vulnerable as it had been suspected that she had been sexually abused by adults whilst in homeless person's accommodation upon arrival to the UK.

The girl's behaviour was thought to be sexually problematic behaviour (inappropriate) in relation to her willingness and the boys' was considered to be sexually harmful behaviour (victimising) due to the coercive and exploitative nature.

3.10   Issues for practitioners

Practitioners are advised to be aware of many factors which although not exclusive, are strongly connected to sex offenders.

- The scale and extent of offending is not accurately known but is likely to be vast

- The details of the offence must be obtained from official sources and not from the offender or their family

- Offenders commonly admit lower order offences to conceal more serious ones

- Offenders groom the protective adults first

- Half of all offenders don't have a gender preference and many will offend against both sexes and all ages

- Non-contact internet offences doesn't imply contact offences haven't already happened

- Women are also sex offenders, but often because of a chaotic lifestyle, are ill, are manipulated by a male or because they have been victimised

- Offenders often have a distorted sexual development so change is unlikely to be easy

- Most offenders aim to please their victims as well as control them

- Offenders lie, so best to approach with "healthy scepticism" and "respectful uncertainty"[13]

- A systematic assessment of risk is far better than gut instinct.

---

[13]Lord Laming in his report The Victoria Climbie Inquiry, 2003

3.11   How failures can occur

Lack of experience and confidence amongst professionals:

> Almost all cases of adult child sex offenders have features
> where previous concerning behaviour had been noted by
> colleagues or family members and where these concerns
> were not acted upon. With child offenders the situation is
> more complex as many behaviours might be better judged
> as inappropriate, problematic or just silly. Offenders know
> that they are more likely to go undetected if they groom
> protective adults before attempting to groom and abuse
> a child. Behaviour such as gradually pushing boundaries
> in relation to contact with children to suggest that they
> are trustworthy might be the offender testing what he is
> able to get away with. Reporting a colleague for worrying
> behaviour is a big step for many staff who may be junior
> to the suspect and who may not have had any training.
> Reporting a child and risking labelling them a sex offender
> is a difficult decision.

Professionals and employers relying on offence information from offenders:

> Offenders are often highly skilled at presenting the details
> of their offence in the most positive and minimal light.
> Crucial details such as the true age of the victim, the use
> of coercion or substances, and the premeditated nature of
> the offence are rarely volunteered and professionals and
> employers can be persuaded to give the offender a second
> chance.

Misplaced sympathy, rule of optimism:

> Most people including trained professionals are naturally
> sympathetic and wish to see the best in others to the ex-
> tent that their judgement can be affected when dealing

with cases. Nobody would want a family unit to be destroyed through an unfounded allegation of abuse. Offenders are skilled at portraying themselves as the victim and attracting a sympathetic and therefore an inadequate response from the employer, investigator or assessor. This is true even for the Courts in the UK as in the case of the dentist above where the judge expressed the hope that he could continue to practice, a comment that may have helped him successfully appeal against erasure from his professional register.

Arrogance of employers, church, institutions, BBC:

Large organisations have traditionally thought themselves capable of dealing with internal problems due to their inflated view of their own abilities and because of the reputational harm that would ensue if the matter was passed to external authorities. The history of sexual abuse of children by members of the clergy does not need any repetition here and the recent UK experience of the BBC in the UK which failed to control Jimmy Savile over many decades is an example of where even the most reputable of organisations can fail absolutely.

Poor employment practices:

Although good employment practice should prevent individuals from resigning during investigations into misconduct, it still commonly occurs, even in organisations such as the Police where the practice is allowed in "exceptional cases". Similarly, employers are often tempted to avoid an investigation altogether on the basis that a "compromise agreement" is reached with the employee suspected of abuse. This serves the employer well in avoiding bad publicity and potential costs, but is at the expense of future victims. The current practice of not providing meaningful references to future employers other than to confirm dates of employment can also serve to protect abusers from exposure. There is no mandatory duty to report suspected abuse as already discussed and this remains a disputed issue amongst safeguarding experts in the UK.

3.12 Recommendations for future management

There have been many recent examples of sound action to improve the performance of professional staff such as better training and tighter regulatory frameworks. However this has inadvertently served to increase the burden and workload on these groups which has resulted in referral thresholds rising unacceptably and delays and oversights occurring.

Several high profile investigations in the UK have also led to the whole system being discredited by the press in the eyes of the public through very lengthy police investigations which at their conclusion led to no prosecution. The Independent Inquiry into Child Sexual Abuse is currently struggling to continue its work having suffered the loss of its first three chairpersons and the withdrawal of victims' groups and a series of senior legal resignations.

Commentators have said that making incremental changes and additions to an already over complex system is unlikely to work and that a total reform and redesign process is required. This is most unlikely to occur soon in the UK where the current preoccupation is on the withdrawal from the EU and the associated economic issues.

In a similar way, calls to make reporting of suspected abuse mandatory and for court procedures to be improved will be ineffective if the agencies and courts are not well enough resourced to implement the changes fully.

Alternative and practical suggestions to improve the effectiveness of the overall system should include leaders in central and local government and the professions giving firm direction to the need to protect children more actively. The novelty of the realisation of the extent of child sexual abuse has begun to fade in recent years with the result that an assumption has been made that everything is under control; this needs to be urgently re-addressed.

The government looked thoroughly into the need for change in the Munro Report which amongst many recommendations promoted the importance of early intervention through early help to families under pressure. This is an important principle and highly relevant to child sexual abuse prevention but one where double funding streams would be required to ensure the present arrangements can be maintained whilst the changeover to a preventative rather that reactive system is implemented.

Austerity measures, currently in place in the UK, have been shown to have an undue impact on the most vulnerable members of society which largely features women and children. It is hoped that a possible future change of policy will result in greater investment in quality staff and services which can operate a well-designed if not highly complex system. Investing in and promoting the status of the services that operate the system will result in better performance and morale.

3.13 Acknowledgements

Much of the author's knowledge has been based on the work of the Lucy Faithful Foundation, a UK charity dedicated to the eradication of child sexual abuse, providing training to professionals and to support victims of abuse.

The author is also indebted to many professional colleagues in his former authority who worked together setting up projects to make the area safer for children and young people.

Finally, sincere thanks are offered to the children, teenagers and family members who contributed to the growth of professional knowledge through their courageous commitment to the truth.

References

[1] T. Ward, K. Louden, S. M. Hudson, and W. L. Marshall. A descriptive model of the offense chain for child molesters. *Journal of Interpersonal Violence*, 10(4), 452-472, 1995.

CHAPTER 8
REASONING SCHEMES, EXPERT OPINION AND CRITICAL QUESTIONS. SEX OFFENDERS CASE STUDY

1   Background and Orientation

This Chapter examines in detail the argumentation features in the domain of sex offender with some applications to the scheme of "Argument from Expert Opinion". We build a model for reasoning schemes, critical questions and expert opinion on the question of "the degree of risk of a sex offender".

We discover that in order to properly model expert practice in this area we need to use numerical argumentation as well as the new notion of "Attack as Information Input". The model is generic and we believe is not restricted to the sex offence area of expertise.

Our Chapter also offers a more detailed example for Walton's argumentation scheme of Expert Opinion as well as a bridge between the argumentation community and the community dealing with sex offenders. We offer an introduction to the student on the subject of determining the degree of risk of sex offenders. We also look at standard international tools for determining the risk of sex offenders and see how the argumentation community can integrate these tools.

This Chapter is the first of a series of papers (see also our paper [33] on Universal Distortion, also Chapter 6 in this book) dealing with Argumentation, Logic and Sex Offenders (ALSO), a topic which lies in the borderline of three vibrant communities:

1. The community of therapists, psychologists and experts involved in dealing with sex offenders. This is an important community

involved both in the medical, social and legal aspects of sex offenders. You have such people in every large community in every country and on their expertise hangs the freedom/health/suffering of many people. This community uses logic and argumentation not only in their interaction with the courts of law but also in their therapy methods (although they are not fully aware of the formal aspects of what they are doing).

2. The old established community of debate/dialogue/arguments/ fallacies. This community is today, let us say, associated with say the Springer journal entitled: Argumentation – An International Journal on Reasoning and the Amsterdam ISSA-conferences, the latest being `http://cf.hum.uva.nl/issa/conference_2014.html`.

3. The computational argumentation community, associated with the COMMA conferences `http://www.ling.uni-potsdam.de/comma2016/` and the Journal *Argument and Computation*.

There is some communication between the communities 2. and 3., but no awareness between communities 1. and {2,3}. Communities 2. and 3. connect with the general logic community. Community 3. is a younger community and do not seem to fully appreciate, (in my opinion, D.G.) the enormous contribution of community 2.

We see the value of this Chapter as connecting between the communities 1. and {2,3}, and between communities 2. and 3. internally. The benefits we envisage for the communities involves are we hope as follows:

1. The sex offender community use logic and argumentation extensively without being aware of it as such. By coming together and observing the practice of the sex offender community, the argumentation communities can get ideas for new models of argumentation, rooted in actual practice. This can lead to further development of logic and argumentation. In our humble opinion, the situation is analogous to observing the flow of water in pipes and developing new turbulance differential equations as a result.

2. The numerical tools and models used by the sex offenders community, when put in formal forms and in the abstract language of the argumentation communities, look very simple and the argumentation community can develop better tools for the sex offender community.

The structure of our program, of which this Chapter is a modest beginning, is very simple, using a repeated iteration of Steps 1–8 below:

Step 1. Start with the first iteration of our approach with an initial package of logical tools, which in this initial Chapter is comprised of reasoning schemes, critical questions and Walton argumentation scheme for Expert testimony. This package is familiar to community 2. and is also being recently studied in community 3.

Step 2. See how this is practiced in community 1. Get new ideas how to improve the package.

Step 3. Export the initial results back to community 2.

Step 4. Try to express what you see in community 1, using methods of community 3.

Step 5. Discover that community 3 needs to sharpen and extend their formal logic and argumentation tools.

Step 6. Export the new tools to discuss issues in community 2.

Step 7. Try to interest community 1 and offer community 1 better tools.

Step 8. Go to Step 1 and start the process again, with the new package of tools you accumulated along the way in the previous iteration.

We put emphasis in informing communities 2. and 3. on how community 1. works. So we give a lot of details about the processes of community 1. This is a natural flow of information. The findings about how the therapists work are surprising. They actually use logic without explicitly being aware of it. See [33]. So the benefit to argumentation, communities 2 and 3 is further development of ideas and tools. How to inform and interest community 1. in argumentation is another problem. We need to show them the benefit of formal modelling, but for that we need to develop tools. Remember community 1 are therapist, they want to see results! It is a challenge to com-

munities 2. and 3., are they just theoreticians or can they generate
some benefits? Fortunately there is a challenging way forward. The
the approach in community 1 is stylised which therefore lends itself to
automation. There are argumentation tools available by communities
2 and 3 and so one can provide community 1 with tools to use. Of
course community 1 will test these tools over a period of time much
like the testing of a new medicine.

1.1 Introduction

The aim of the Chapter is fourfold:

1. We consider the applied area of dealing with sex offenders where
   the expert testimony plays a central role. The assessment of the
   risk is an important tool for the courts for making decisions
   and judgments about sex offenders. The courts consult senior
   experts specialising in assessing risk of sex offenders according
   established international standards. The courts in Israel have a
   stylised process in consulting the experts. Both in establishing
   the qualifications of the exports and in examining his testimony.
   The aim of this Chapter is to model in logic and argumentation
   this stylised process. We believe that this process actually applies to any expert testimony in Israeli courts and not just for
   expert testimony of the risk of sex offenders.

2. The second aim is to export to logic and argumentation ideas
   and formal tools which may emerge during our attempt to build
   the model in the first aim. Such export need not necessarily
   be connected with Argument schemes. In fact in our case the
   export is the idea of the attack as information input [6], and the
   theory of argumentation under universal distortion [33].

3. Our third aim is to provide the experts working with sex offenders with unified view of how to approach finalising their testimony about the risk of sex offenders. This will become clear
   as we progress in the Chapter. There are various internation-

ally developed tools available (see Appendix C), but there is no unifying approach (Super View/Super tool).

4. The fourth aim is to refine the Walton scheme of argument from expert testimony and to set an example for refining other Walton schemes.

We therefore in this Chapter analyse and model in detail real case studies of the use of expert opinion (obtained by G. Rozenberg, who is an expert) on the question of the degree to which a sex offender is dangerous to society (expert opinion on risk evaluation of a sex offender). The modelling is based on theoretical foundations of [2]. The reader can consult an Appendix to this Chapter giving a formal background from argumentation), and on 11 year's experience of the second author.

The practical situation can be described schematically as follows:

1. The court is considering a a sex offender and it needs to determine his/her degree of risk.

2. The court asks expert $e1$ for a report. The defence lawyers the sex offender also ask their own expert $e2$ for another report

3. The experts produce their respective reports, respectively arguing for a judgement about the degree of risk of the sex offender. For an expert e, let us denote his report by R_e. We thus have two reports on the table; R_{e1} of expert $e1$ reporting for the court and R_{e2} of expert $e2$ reporting for the defence. The reports have the form of a written document.

4. We have a court scene in which there is the Judge, the experts, a prosecutor, and a defence attorney. The procedures involve attempts at establishing two types arguments:

   (a) Is the respective expert indeed an expert?

   (b) Does the expert respective report indeed establish the claimed degree of risk?

Item 4(a) involves two argumentation networks Q_{e1} and Q_{e2}, each representing the respective discussion/debate about whether the respective expert is indeed an expert witness. Item 4(b) represents two argumentation networks S_{e1} and S_{e2}, each representing the discussion/debate about whether the respective expert report does indeed establish its respective conclusion. (The arguments in R_e appear/are absorbed in S_e respectively).

5. We thus get two pairs of argumentation networks, one for each expert, these are

$$M_1 = (S_{e1}, Q_{e1}) \text{ and } M_2 = (S_{e2}, Q_{e2}).$$

We can write $M = (M_1, M_2)$ as one master network.

6. Note that the argumentation networks S_e and Q_e may not be of the same type. In fact we have not described yet what they respectively look like, this will come later in the Chapter, in Subsection 1.2 and Section 2.

Thus our model uses a two level argumentation network. Figure 1 describes the model. The network M shows two experts $e1$ and $e2$. The argumentation networks Q_{e1} and Q_{e2} show the debate about qualification of the experts. The networks S_{e1} and S_{e2} show the arguments involving the degree of risk of the single accused sex offender.

The system M_1 is the package related to expert 1 and M_2 is the package relating to expert 2. The Judge in court sees both M_1 and M_2. We call the overall system seen by the Judge M. This is the initial position. Expert 1 is the main expert hired by the court to give testimony. Expert 2 is hired by the defence. Expert 1 presents an argumentation network S_{e1} and an extension $E1$ (see Appendix A for the concept of extension) which includes the key argument/assertion

$A_1 = $ "The sex offender is dangerous to society and to what degree".

Expert 2 introduces S_{e2} and extension $E2$ which excludes the statement A_1 but includes his own statement A_2.

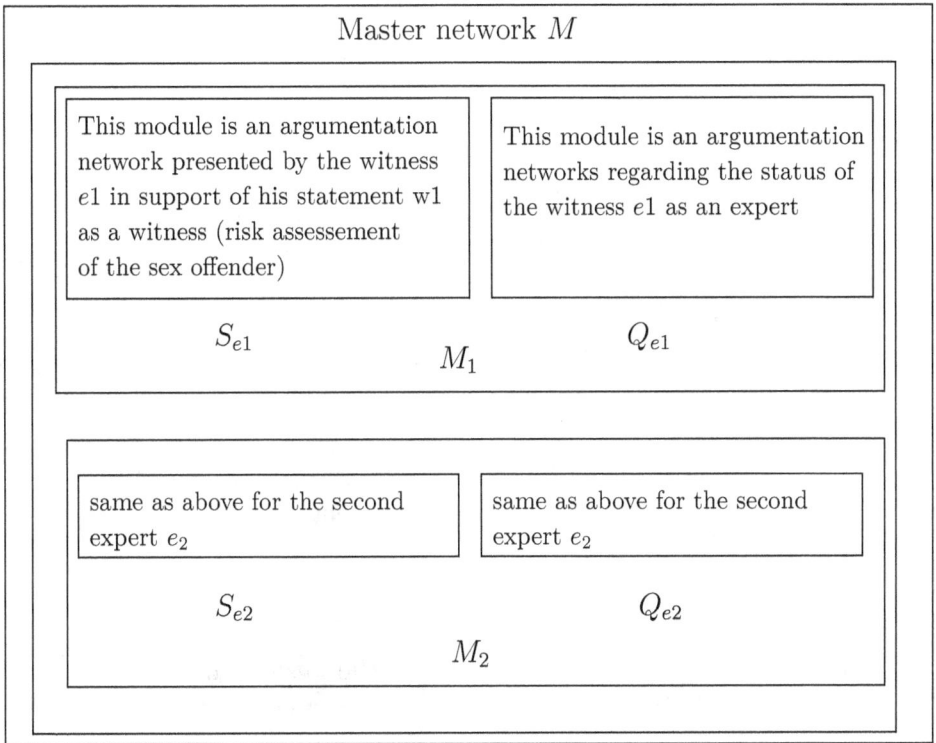

Figure 1: Two level argumentation network

The defence, prosecution and experts argue. The arguments can be attacked on Q_{e1} and Q_{e2} which establish their expertise or attacked from S_{e1}, S_{e2}.

Our modelling will be from real cases of how the system works.

The Judge sees M (he sees the interaction/debate which ends up in M) and in his own mind transforms M to a final $M*$ which the Judge uses to make a decision on. (We cannot model M^*, it is in the Judge's mind), we need to do field work to interview many judges.

We invite the reader to consult the Appendix about the formal machinery of argumentation networks, however, in order to keep the flow of this section going let us say that that formally an argumentation network has the form (S, R), where S is a finite set of arguments and $R \subseteq S \times S$ is the attack relation. An extension $E \subseteq S$ is a set of conflict free arguments which satisfy certain conditions.

d = sex offender is dangerous to society
to degree x

↑

a = sex offender is a new born religious
believer and is repentant, the last offence
taking place after he repented

↑

b = sex offender has three previous
convictions of rape

↑

c= sex offender has undergone
extensive special treatment

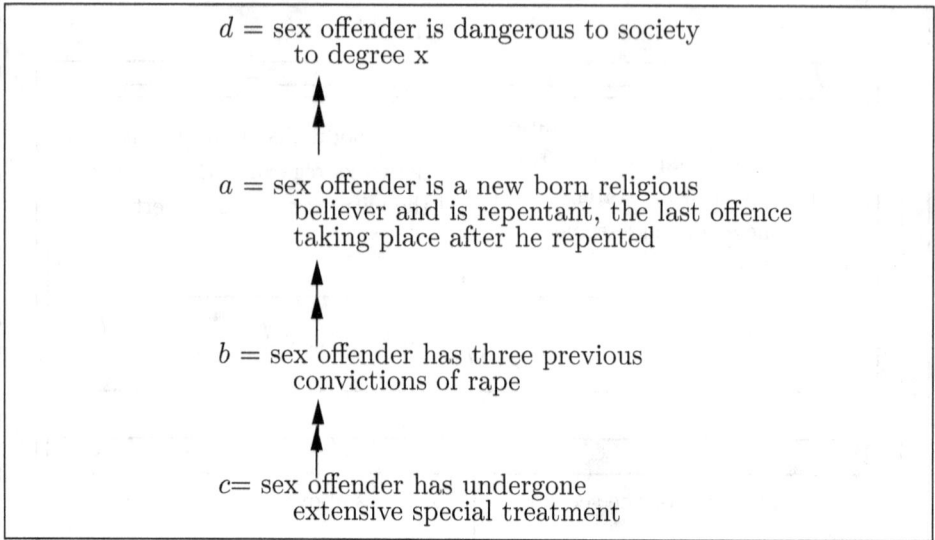

Figure 2: Sample argument chain

These concepts will all be defined in the later formal background section. Figure 2 gives an example of the defence claims.

$$\text{Extension } E2 = \{c = \text{in}, a = \text{in}, b = \text{out}, d = \text{out}\}$$

The expert $e1$ claims that after interviewing the sex offender he believes $d =$ "the sex offender becoming a newborn religious believer" is just an act and should be ignored. Therefore his S_{e1} is Figure 3.

$$\text{Extension } E1 = \{\alpha = \text{in}, c = \text{in}, b = \text{out}, a = \text{out}, d = \text{in}\}$$

Remark 1.1. *We need to place this Chapter within the framework of existing research of logic, argumentation and law. The reader can consult [40].*

Let $\mathbb{CC}$ *be a transcript of a court case proceeding and* $\mathbb{M}$ *be a logic/argumentation model of whatever sort (see [40]) which can address/model some or all aspects of such a court case* $\mathbb{CC}$*. We consider in this set up a single clear statement of fact E which requires an expert testimony to establish it. E plays a role in* $\mathbb{CC}$ *and its logical*

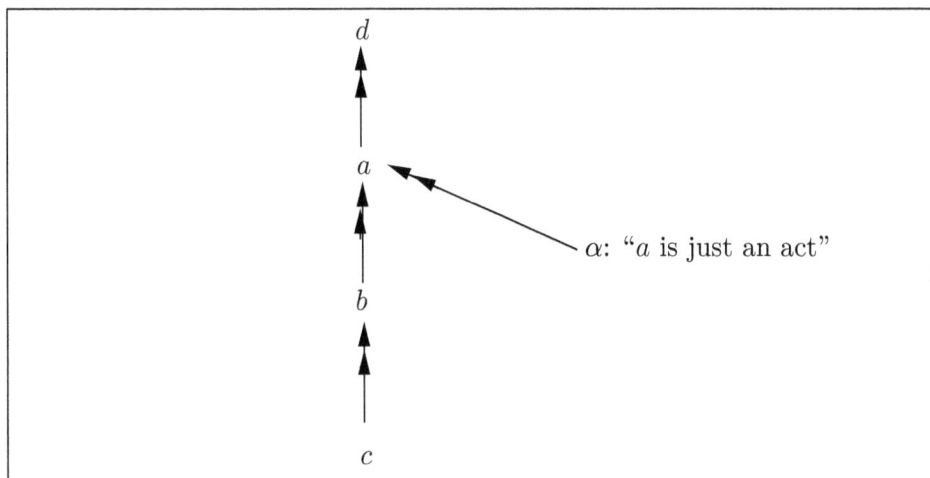

Figure 3: Sample attack on the chain of Figure 2

modelling M. *But the procedure for establishing E through the expert testimony may be different and is disjoint from that followed in* CC. *The following are examples of this concept.*

1. *In a court case involving taxation it is important to establish when a deal was concluded. The "expert" testimony in this case is a bank record of the payment transaction.*

2. *In a rape case of a minor it is important to establish the age of the victim. The "expert" is a birth certificate.*

3. *A convicted prisoner seeks an early release from prison on account of argument E, where*
   E = prisoner has terminal cancer and has at most 30 days to live.
   Here we need an expert testimony which may involve medical data and a separate cross examination of data and expert.

4. *A mother has three babies who die in their sleep, and seems to be a natural unexpected death. E might be a statement that the probability of coincidence is almost nil. The expert might be a Bayesian maths Professor and the considerations mathematical.*

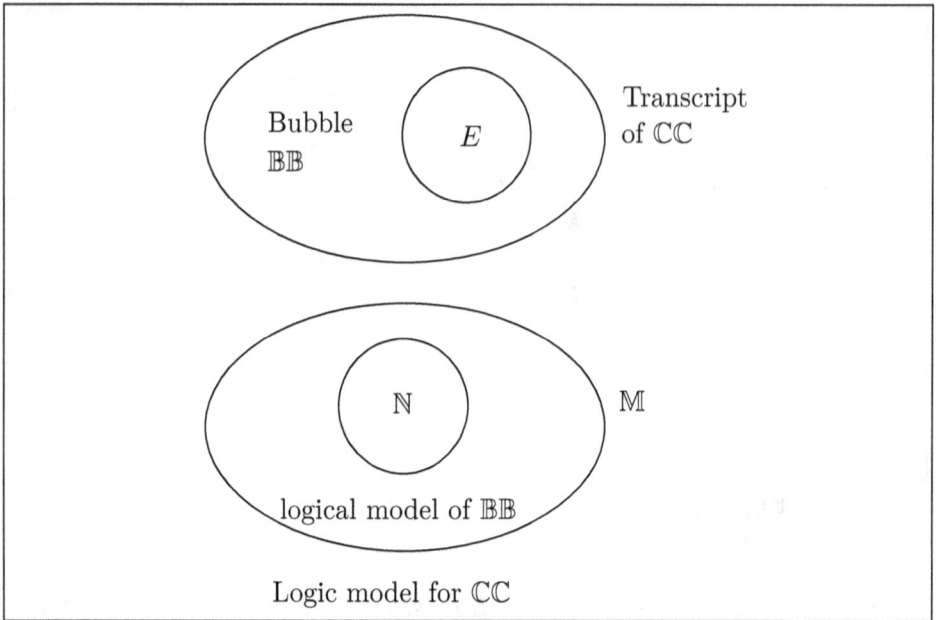

Figure 4

5. *The statement E might be a component in a trial of a sex offender where E is*
   E = the risk assessment for the offender to offend again within 6 months of his release is very high.
   Again we need an expert to assert E.

Figure 4 explains schematically the role of the expert in the overall trial $\mathbb{CC}$. *The expert's participation is a bubble* $\mathbb{BB}$ *in* $\mathbb{CC}$. *It is a mini-procedure not affected nor related to* $\mathbb{CC}$. *It can be modelled by* N *which may be completely independent of the modelling* M.

Our Chapter deals with the procedures followed in Israel for bubbles $\mathbb{BB}$ *for establishing via expert opinion the statement E of the degree of risk of a sex offender. By examining the data and procedures of such cases we offer ideas towards modelling* N *for* $\mathbb{BB}$. *If we want to compare our models with other works in the literature (see survey [40]), then we need to compare with discussions relating to models* N *and not to models* M *(see Figure 4).*

We shall discuss this further in the conclusion section.

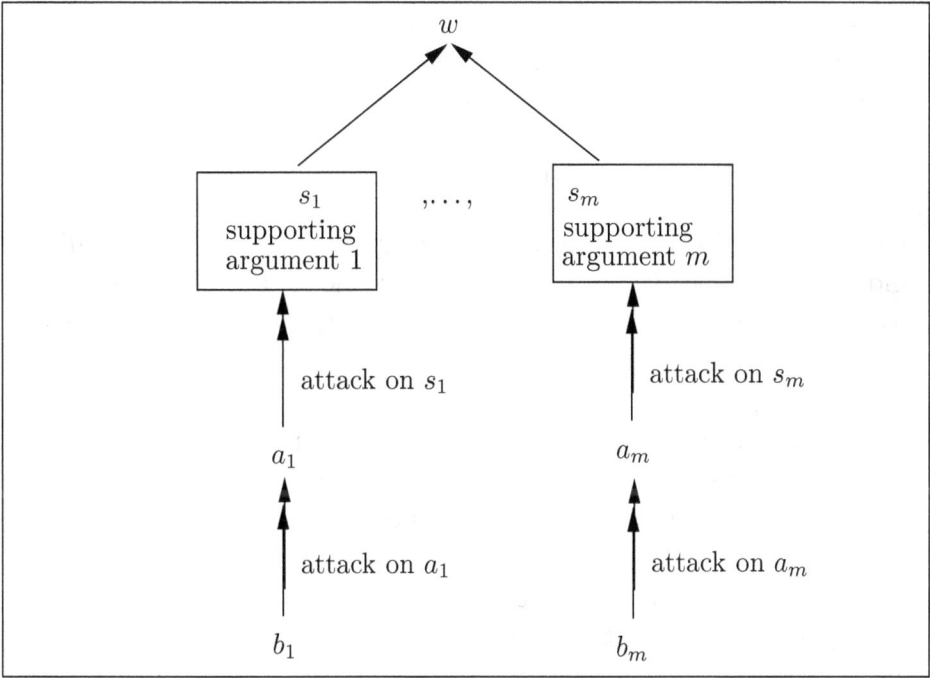

Figure 5: Expert witness e: The structure of the main argument Module S_e

1.2 The expert witness networks

We now discuss and compare overall view of the role of expert witness in argumentation networks, as discovered and presented in our Chapter.

An expert witness network (EW-network) has a very specific form, as in Figure 5. We use "$\twoheadrightarrow$" for attack and "$\rightarrow$" for support. Note that this is the internal structure of the networks S_{e1} and S_{e2} appearing as black boxes in Figure 1. In technical terms (consult Appendix), we have the form of argumentation network with both attack and support which forms a tree directed at a single top argument w. The tree depth is 4. The main argument is w. The arguments $\{s_i\}$ all support w. Each s_i is attacked by a respective a_i and is in turn defended by a respective b_i.

The actual attacks and support come with strength (either expressed as numerical strength or as qualitative strength such as weak, very weak, medium, etc.). The final decision (extension) of this entire network is a labelled output of the form

$$(\text{strength}; w).$$

The strength of w increases with more supporting arguments available and also increases if the supporting arguments are stronger. So to attack w and lower its strength we can attack and eliminate some of its supporting arguments or attack and lower the strength of such arguments. The most effective strategy is to attack the expert and thus lower the strength or even eliminate him and all his arguments.[1]

The above format applies to the expert testimony for the case of expert for assessing the risk of sex offenders. An actual court case may involve other experts as well, for example medical experts. See Example 1.2. When several experts are involved the Judge will see more pairs of argumentation networks involving all the experts. For each expert the respective network goes into an overall network of the Judge. See Figure 6.

Example 1.2. *To see how the network of Figure 6 works, consider the following simplified story involving two types of experts, one a risk assessment expert and one a medical expert (so actually there could be 4 experts involved, a risk assessment expert and a medical expert invited by the court and another pair invited by the defence):*

An offender attempts to rape a young woman who happens to be 3 months pregnant. The offender attempts to attack her and she pleads with him to leave her alone because she is pregnant. In a rare display of sympathy, the offender goes away. Two weeks later the victim loses the child. The Judge has 3 statements[2] to consider (let us assume for simplicity of the example that the defence does not bring its

---

[1] In practice there are factors such as therapy which actually lower the risk value. We can view them as an attack with a negative strength. The way it is done is discussed in Section 4, presenting the formal model.

[2] The reader should note that we describe the process schematically. The items W1, W2 and J are not arguments as they do not contain details. We are just describing the process the Judge has to go through in his mind.

own respective experts, but is satisfied in cross examining the court experts):

1. *How dangerous is the offender, from risk assessment expert witness 1, who supports statement W1*

2. *Did the attempted attack on the woman cause the loss of the child? This is statement W2 from medical expert witness 2*

3. *Length of prison sentence to give the offender. This is statement J, which is determined by the Judge.*

W1 and W2 are considered by the Judge in reaching his decision about J.

W1 and W2 come with their own networks as in Figures 1. So we have two networks challenging the expertise of the witnesses and in addition we have two more networks, one describing the debate with the risk assessment expert witness which has the form of Figure 5, and another network describing the debate with the medical expert witness. We do not know what structure this network has, because we have not researched and interviewed court cases with medical experts. It could be that medical expert witness networks also have the form of 5.

Remark 1.3. *We note the following important comments:*

1. *The network of Figure 5 is a stylised (in Israeli courts) accepted matrix for expert witness testimony. This means that*

   (a) *There are stylised legal and social factors to decide how "expert"the expert is (in Israel one gets official recognition by the system as being an expert on issue w) and attacks on the expert address these factors.*

   (b) *The argument factors $s_1, \ldots, s_m$ are fixed factors for evaluating w, which different experts must address. This is a fixed matrix $S_{(e,w)}$ for each statement w which requires expertise. So for w = risk of sex offender, we have its set of factors. For F = risk of fire hazard (of a business building), we have a different set of standard factors, etc).*

```
┌─────────────────────────────────────────────────────────┐
│                   Judge's network M                       │
│   ┌───────────────────────────────────────────────────┐  │
│   │  ┌──────────────────┐   ┌──────────────────────┐   │  │
│   │  │       W₁         │   │         W₂           │   │  │
│   │  │ Figure 4 for w=w₁│   │  Figure 4 for w=w₂   │   │  │
│   │  └──────────────────┘   └──────────────────────┘   │  │
│   │     Module S_{e1}           Module S_{e2}          │  │
│   │                                                    │  │
│   │        J₁,         ...,           J_r              │  │
│   │      Other arguments in the Judge's network        │  │
│   └───────────────────────────────────────────────────┘  │
└─────────────────────────────────────────────────────────┘
```

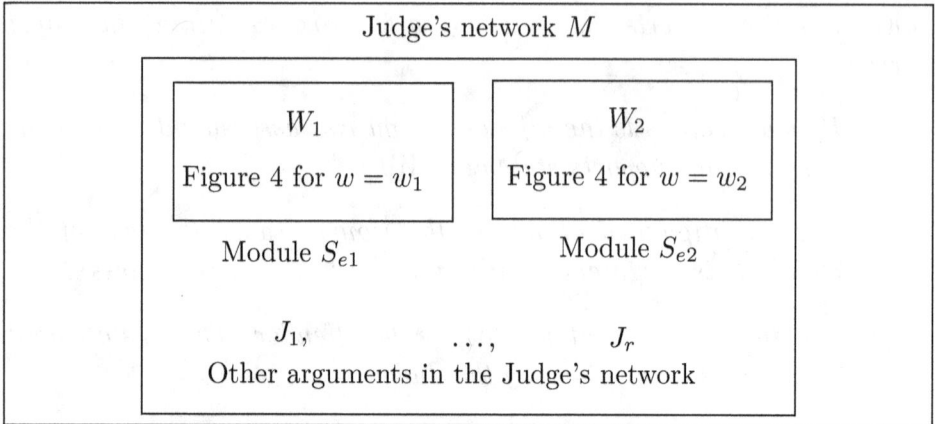

Figure 6: compare with Figure 1

Again such factors are officially recognised and need to be addressed by any expert witness on w. (These factors are listed in Section 3. They are not a sample list chosen by the authors. They are the officially recognised list coming from the international community.)

The attacks a_i, b_i are also standard. The expert witness expresses his use of s_i. The opposition attacks with a_i and the expert answers with b_i.

This is the end of the discussion. There is no more response/attack on b_i. The Judge sees the matrix of Figure 5 and emerges with the single outcome (Strength label: w) and continues with his other items of the Judge's network.

2. *Expert witness testimony is a matrix and so is different from ordinary witness testimony. The latter has no matrix form. An ordinary witness simply describes in his own words what he witnessed.*

2 The attack on the expert

In their book [7], Walton, Reed and Macagno put forward a list of argumentation schemes. On page 310 they have the scheme of argument from expert opinion. We quote:

> *Major premise:* Source E is fan expert in subject domain S containing proposition A.
> *Minor premise:* E asserts that proposition A is true (false).
> *Conclusion:* A is true (false).

Critical Questions

CQ1: *Expertise question:* How credible is E as an expert source?

CQ2: *Field question:* Is E an expert in the field that A is in?

CQ3: *Opinion question:* What did E assert that implies A?

CQ4: *Trustworthiness question:* Is E personally reliable as a source?

CQ5: *Consistency question:* Is A consistent with what other experts assert?

CQ6: *Backup evidence question:* Is E's assertion based on evidence?

The Walton analysis of Arguments from expert opinion is just an initial proposal of 6 critical questions. We agree with the Walton program of argumentation schemes but would like to examine in more detail the scheme of argument from expert opinion and refine this scheme (this is done in Appendix B). Practical experience with expert opinion indicates that a much wider range of critical questions and queries is employed. There also remains the formal question of what is the structure and role of the expert opinion module in the overall master argumentation network (the context network in which the expert opinion is but one argument of many) and how such a module can integrate into this overall master argumentation network.[3]

The Walton model described above, proposes only 6 queries which can be put forward to the expert. Four of these questions relate/attack the expert and his qualifications and the remaining two relate to the substance of his testimony. The sex offender area presents a detailed example for the Walton Scheme. We have seen in the previous section the structure of the network representing the debate of the substance of the expert witness testimony in the case of an expert on risk assessment of sex offenders. We now address the debate about the expert witness qualifications as an expert. In this section we list many more queries and their proposed answers by the expert which have been used in practice by various defence attorneys over the course of 10 years.

The following should be noted.

1. From argumentation point of view, some of the questions involved in the cross examination/attack on the expert (by the defence) are requests for information, preparing the ground for/implying the attack the qualification of the expert. The experts reply by providing further information which deflects such attempts. In fact observing this sort of interplay inspired the idea of the attack as information input, [6].

2. The questions and answers involved in the attack on the expert of the sex offender case are typical in their generic structure and can actually apply, after modification, to attack experts in areas other then sex offenders. See Appendix B. For example, we can apply the generic form to forensic psychiatry experts who specialise in determining the mental capacity of an accused person and even to experts assessing damages for insurance case.

---

[3]The Walton schema did not arise in the Walton, Reed and Macagno book, but first appeared in Woods and Walton, reprinted as chapter two of [37]. Also, this sort of scheme is shown to be (anyhow asserted to be) of very little help in legal contexts. The supporting argument to that effect can be found in [38] in the chapter on "Neutrality and Expertise".

3. Note that the attack and defence in this court case bellow are mostly not direct. In formal argumentation in order to attack the statement "I am an expert", for example, one directly asserts the statement "you studied at a third rate university and such institutions do not train experts". In a question and answer court case interaction the above attack can be implied by the question "which university did you go to?"and it can be defended by the answer "I went to a government recognised and government funded university", which implies that the university I went to is not third rate.

The following is an actual court case in which the second author participated. It lists the questions he was asked and his answers. The text presented is a translation of a transcription taken by the official court clerk of the Military Court in Jaffa, Israel, in a court case of a certain sex offender. The expert witness is Dr Rozenberg. The text is the exact detail of the cross examination questions presented to Dr Rozenberg and his answers. The original is in Hebrew and it was faithfully and accurately translated by the authors. Although our translation is not a legally accepted notarised "authorised translation", it is sufficient for the purpose of this Chapter. Some slight modifications were made to avoid the possibility of identifying the offender.

It is a case study of an attack on the qualifications of the expert. It also has a stylised structure. It is comprised from questions about how the expert acquired/ studied for his qualifications, as well as how he went about (methodology used in) composing his report. There are also questions about specific items in his report, not with a view of attacking the item but with a view to see if the expert understands their significance. Other questions relate to whether the expert understands the limitation and margins of error of his report.

There are also traditional Fallacies, trick questions, tempting the expert to answer in such a way that he appears to be racist, over-confident and full of his own importance, unable to take criticism or timid, hesitant and unsure of himself. We must remember that these questions are asked on the witness stand in front of a Judge and are intended to discredit the expert.[4]

We need to explain to the perceptive reader what is our purpose in quoting the case study below. We just want to give the reader a flavour of what is going on in establishing the qualification of an expert in the case of sex offenders testimony. We are not analysing and examining the text according to some critical methodology but only noting its structure. Something we could do in the future, for example, is to look through many such transcripts and see how the defence lawyers try in a subtle way to personally discredit the expert. We just note in this Chapter that such attempts take place but we do not study the allowable limits (in Israeli practice) and subtlety of such actions.

We encourage the reader to think of a different Expert, say a fire fighter expert testifying that the fire in a factory was caused by faulty electrical system not installed correctly. In this case the "accused -sex offender" corresponds to "the electrical system installation"' and the "degree of dangerousness/risk assessment" corresponds to "the degree of negligence and faulty standards in the electrical system installation". Further,"the sexual offender medical treatment"corresponds to "maintenance of the electrical system" and "abuse of the accused" corresponds to "overloading and misuse of the electrical system". We leave further analogies to the reader.

The next section will give case study examples of attacks on the actual testimony of the expert.

Court case dated 9.4.2013

Question 1.   What is your qualification?

Answer 1.   8 year's experience. I also train young clinical criminologists. I also treat both young and older offenders in jail and outside the jail. I give lectures to various experts in many forums. I have an official licence from the government.

---

[4]Doug Walton and John Woods jointly and separately have written many wonderful books on the Fallacies. The reader should consult the internet.

Question 2. Is there a structured training course that experts in your area have to take?

Answer 2. There is no such course. You can take an MA degree in clinical criminology and then there is a regulated further study where you first assist an experienced expert in his cases and then continue on your own, and in time have your own assistant trainees.

Question 3. What are the steps taken in your evaluation of how dangerous the accused is?

Answer 3. I collect every possible document on the person. I even check his facebook account, get previous convictions, get previous expert reports, military service reports, hospital treatment reports and social services reports. I then conduct a clinical interview with the person and if possible interview and talk to all his relatives.

Question 4. Are you doing all of this alone? It is possible that someone else, had he/she been with you would have reached a different conclusion?

Answer 4. I do it alone. However, there is a legal expert advisor, who is also a clinical criminologist who reads my report. If I have doubts, I consult colleagues. Such colleagues of similar qualification as myself who I expect will reach a similar conclusion.

Question 5. What method do you use?

Answer 5. I use *structured clinical evaluation*.

Question 6. What exactly do you check? How do you put all your findings together and get a result?

Answer 6. I construct a picture and address both static and dynamic clinical factors. You (the lawyer asking Question 6) are asking me to distill to a few sentences, many years of experience, the reading of vast professional literature and clinical expertise. This is not possible.

Question 7. How important is his childhood? Maybe it was traumatic? Maybe he was sexually abused?

Answer 7. His childhood is important. One can later identify if he developed repeating patterns of behaviour. Having traumatic experience does not diminish him being dangerous to society. The trauma can even make him more dangerous if it has not been dealt with and the person can be influenced by it even today.

Question 8. Why for two people who grew up in the same neighbourhood does one become a sex offender and one not? What makes one become a sex offender?

Answer 8. This question has many psychological explanations from the area of social psychology, but the simple answer is that one chose to be an offender and the other did not.

Question 9. Surely you will agree with me that the evaluation of dangerousness is speculative?

Answer 9. I disagree. It is a science which we continually try to improve and make more accurate.

Question 10. What is the percentage of mistake?

Answer 10. We are human and we can make mistakes. In the middle evaluation of risk the error is greater than at the extremities. There is less error in high risk or low risk evaluations.

Question 11. The person is in jail and is angry and frustrated. It influences his behaviour when you interview him.

Answer 11. I work a lot with different people from different backgrounds and I have learnt over the years to isolate this factor and give the factor its proper perspective. I can construct a good picture independent of the local state of mind of the subject. I also try very hard to make the subject comfortable during the interview.

Question 12. What is the effect of the person's confession?

Answer 12. A confession does not influence the risk assessment. It does influence his prospects of successful rehabilitation.

Question 13. If a person does not confess then he cannot be rehabilitated?

Answer 13. Not true. There is also treatment for those who deny wrong doing, and world literature shows it diminishes the risk and dangerousness of the subject.

Question 14. Surely you understand that if a person was drunk or under the influence of drugs and his judgement was temporarily diminished and he offended, then certainly he does not deserve a heavy punishment?

Answer 14. Punishment is not my area of expertise, and I will not say anything about it. I concede that if the person is on drugs then his behaviour is context-dependent, but we need to ask what will happen if he goes on drugs again.

Question 15. Surely you accept that if a person is drunk at a party and touches a girl's breasts, then if he were not drunk he would not have done it?

Answer 15. What if the next day he gets drunk at another party and does it again?

Question 16. For a person to rape a woman she needs to be there, if she were not around he would not have raped her?

Answer 16. Surely you understand that your question is problematic. I hope you are not suggesting that women would stop walking the street so as not to provoke rape?

Question 17. You conducted one interview, very short, half an hour. This is not enough!

Answer 17. I did not record the time, but I am sure it was not just half an hour! I estimate at least an hour and a half to two hours. Had I needed more time, I would have taken more time.

Question 18. My client felt like he was in an interrogation chamber!

Answer 18. I tried in every possible way to make him comfortable. I supported him and encouraged him. I insisted his gaolers take off his handcuffs. I did not agree to interview him handcuffed!

Question 19. You pushed him into a corner.

Answer 19. Part of the interview is to confront him with inconsistencies in his own version of the story. If he says two contradictory things, it is important to find the truth and also to see how he responds when caught lying.

Question 20. Give me some examples of factors which increase the assessment of risk and dangerousness and explain why these factors are relevant and contribute.

Answer 20. Young age of the offender, offending a male victim, offending victim in age group 13–15, offending in a public place, recidivist sex offender, attacking foreign victim (tourist).

Question 21. Is there different risk assessment, say between Russians vs. Ethiopians?[5]

Answer 21. There is no research on this. You need to be familiar with their respective mentality. I have many cases of Ethiopians, so I know their customs.

Question 22. I want you to give me a computer-like table, factor x_1 gives risk increase y_1, factor x_2 gives y_2, etc.

Answer 22. A person has a complex personality, is not a computer. You cannot make a table like that. A table would ignore many important factors unique to each individual's person.

Question 23. Why not use existing statistical analysis following data about different cases?

Answer 23. Please look at this *stat99-tool*. A child can do the table. The table does not include, for example, if the subject had treatment or made an effort to change. Is he hyper-sexual? Does he have ego problems? Is he in a supportive environment? The tool is just numerical statistics.

Question 24. Your answer 23, is it knowledge or conjecture?

Answer 24. I know the literature and I included in my report the unique characteristics of the subject. I try to be up to date in the professional literature.

---

[5]Note that this is a trick question trying to hint at racism.

Question 25. There are research results which contradict what you are saying.

Answer 25. I'll be happy to see them. It is important to understand that there are some researches around that are esoteric and some were conducted on small sample populations.

Question 26. What percents do you give to clinical factors and what to actuarial factors?

Answer 26. I cannot give percentage division. I weight all factors.

Question 27. Don't you think it is arrogant to claim that in a 1.5–2 hour interview you can assess a man?

Answer 27. I make every effort to check everything available about the person and check recurrent behaviour patterns. I do the maximum I can.

Question 28. Perhaps your maximum is objectively really minimum.[6]

Answer 28. I collect every possible relevant information.

Question 29. He was treated badly in prison. Why did you ignore this?

Answer 29. No such thing was reported to me.

Question 30. Why did you not use the test report about the person?

Answer 30. There was no such report.

---

[6]This is another trick question.

Question 31. Why did you not address a treatment the person received?

Answer 31. I addressed it on page xxx of my report. I wrote the subject was not emotionally open to it.

Question 32. Today he understands he was wrong and wants to be treated. Doesn't this make him less dangerous?

Answer 32. The mere declaration of desire for treatment does not reduce risk. If indeed he takes treatment, then we can check if he became less dangerous. But if he starts treatment and then stops before the end of the process, this could be an indication that he is more dangerous.

Question 33. Do you really think that if he gets a prison sentence this would help his condition?

Answer 33. This is not my speciality. This is just a legal judgement question.

Question 34. You asked the subject some very personal questions, like about masturbation, if you were to ask me such questions, do you expect I would answer you?[7]

Answer 34. I ask the question in a specific context. I explain why I am asking and the subject agrees to answer.

Question 35. Did you explain to the subject how important it is that he cooperates?

Answer 35. I repeatedly explained this to him many times. I tried to encourage him.

---

[7] This is another trick question, maybe hinting the expert himself is a bit weirdÉ

Question 36. I assume that there is no chance for someone who does not cooperate to get low risk assessment?[8]

Answer 36. Not true. There were cases of non-cooperating subjects to whom I gave low risk assessments.

3 The attack on the expert testimony

3.1 Background

There is consensus in the international community (ATSA — Association for the Treatment of Sexual Abusers) on the factors which contribute to the assessment of risk of sexual offenders. There are also several actuarial tools to help the expert in assessing the risk of a given patient. The tool asks the expert to evaluate/answer questions about the individual patient and then gives a risk assessment a final grade which is a number x, $k < x < m$, where x, m, n are integers (depending on the tool). The expert can use several tools, as well as some additional clinical factors (determined by the experience of the individual expert) and the expert integrates all these results (in his own mind, as there is no Super Integrating Tool) into a final determination.

There is no super-tool which can integrate/reconcile the results of several existing tools. The expert has to decide which tools to use and how to integrate them. The ATSA list of factors are recognised by the Israeli courts, and the expert witness is expected in court to address these factors and be challenged by the defence attorney of the sex offender. The main tools are listed in the Appendix and the typical questions and answers in court are presented in this section. The list in this section does not represent any particular court case but is based on 11 years practice and thousands of expert opinions put forward by the second author. The courts follow Israeli law. The defence lawyer may invite his own expert to present a possibly different report and different conclusion. In this case the second expert will also appear in

---

[8]This may be another trick question hinting at ego problems of the expertÉ

court and be subjected to the same procedures as the first expert with the prosecutor performing the attacks on the second expert factors.

This section deals with the attacks on the expert testimony. The structure of the testimony is as outlined in Figure 5. Each numbered item below represents an attack sequence on a factor s. Each item comprises of three sub-items;

- what the expert says concerning factor i, denoted by s_i. (The expert can either introduce the factor in his considerations or not mention it at all. The factor may support the increasing of risk assessment of the offender or support the decreasing of the risk assess of the offender. There are international packages which assess the contribution of such factors s_i.)

- the attack on what the expert says, denoted by a_i. (This attack is mounted by the defence. So if the expert does not mention a factor which decreases the risk assessment the defence can ask why? If the factor increases the assessment of risk the defence might add information which makes the increase smaller. If the expert gets his facts wrong, then his entire testimony is at risk and the expert loses credibility. So this does not happen in practice. Note further that the node a_i denotes all of what the defence says which can be comprised of several attacks in the formal sense, or a joint attack or a higher level attack, etc.)

- the experts answer to the attack denoted by b_i. (Many of the answers of the expert are explanations or more information, which motivated the authors to write the theoretical paper [6].)

The factors come with labelling of strengths: low, moderate and strong. We shall see in the Appendix, which surveys Tools which assess the strength of these factors, that numerical strength are assigned to them both positive and negative numbers, the qualitative strengths can be derived from these numbers. Note that the factors s_i can be factors which increase risk or sometimes factors which decrease risk (such as participation in therapy). We still view them as "supports" with negative input, which turns them as "attacks". The examples below show that the "counter attacks" a_i on s_i can either question the strength and the significance suggested by s_i or they can question the validity of s_i in applying or not applying to the sex offender in question or can be factual attacks on the factual part of the factor s. Some attacks a_i are logical fallacies. The replies b_i to the items a_i are more in the nature of explanations, rather than "counter-counter-attacks" on a_i. We shall see in Section 4, when we present the formal model, that this type of sequence, namely:

$$b_i \twoheadrightarrow a_i \twoheadrightarrow s_i \rightarrow w$$

where the nature of the double arrow "$\twoheadrightarrow$" changes in the sequence, requires special attention and formal modelling.

We need to be more explicit here. Let us assume that the expert puts forward factor s_i. Factor s_i has two parts, the factual part and the assessment part arguing its contribution to how dangerous the offender is. For example The lawyer of the defence attacks s_i with counter argument a_i. If the counter argument is successful against the factual part, then the credibility of the expert is shattered, and all his support arguments s_j for all j are destroyed. Take s_{15} for example. The factual part is that the offence was in a public place. The attack $a_{15}(b)$ simply says that the attack was at night at an isolated part of the public place and so the factor should not be used. This is not a factual attack. But if the defence proves $a_{15}(a)$, that the attack was at home, then this is a factual attack and all the support arguments s_j for all j are destroyed. On the other hand if the lawyer's attack $a_{15}(a)$ is factually destroyed by b_{15}, then his other arguments can still be used. The lawyer is not an expert, he is not committed to the same credibility criteria, and he is expected to try all kinds of arguments. a_{15} may be destroyed but his other arguments may survive. The defence lawyer may invite his own expert to present a possibly different report and different conclusion. In this case the second expert will also appear in court and be subjected to the same procedures as the first expert with the prosecutor performing the attacks on the second expert factors.

We finally would like to put the contents of this section (namely the attack on the expert testimony) into a general perspective from the point of view of argumentation: An influential classification of dialogue types is that of Walton and Krabbe [22]. We recall their distinction between persuasion and deliberation dialogue. The goal of a deliberation dialogue is to solve a problem while the goal of a persuasion dialogue is to test whether a claim is acceptable The material of this section falls under the category of persuasion dialogues. In such dialogues, two or more participants try to resolve a difference of opinion by arguing about the tenability of a claim, (in our case the degree of risk of a given sex offender), each trying to persuade the other participants (in our case mainly the Judge) to adopt their point of view. General dialogue systems regulate such things as the preconditions and effects of speech acts, including their effects on the commitments of the participants, as well as criteria for terminating the dialogue and determining its outcome. Good dialogue systems regulate all this in such a way that conflicting viewpoints can be resolved in a way that is both fair and effective [23]. In our case the procedure as we described is a highly stylised tree of depth 4, and the final arbitor is the Judge.

Furthermore the particular arguments used are informational and numerical, as we shall see in later sections.

The reader would also benefit greatly from looking at the important paper of Gordon, Prakken and Walton, [21] and the survey [24].

Let us begin.

Full Matrix/List of Relevant parameters/ factors to assess sexual risk

Note that the attacks on these factors are taken from protocols of actual cases involving Dr Rozenberg and his actual replies. They are not from a single court case but a representative compilation. But each sequence was actually asked and answered in court. The wording describing the node s_i is the authors wording simply saying the factor was or was not introduced in the experts report. We could have written "+" and "−". The entries for a_i and b_i are from transcripts of actual court cases.

3.2 This factor is the age

Sex offender's age taken into account when making the risk assessment. Below is the official table of the age groups and the risk strength assigned to them

Risk factor	Age group
1	18–34.9
0	35–39.9
-1	40–59.9
-2	60 or older

A significant factor with at least moderate importance

- s_1 The expert gives a contribution due to this age factor

- a_1 The attack says that the offender is older so according to the table the risk factor strength should be less.

- b_1 The expert reply: Recent literature shows the relationship between the age of the offender to a level of sexual risk is not so dichotomous, for example, we learn that the dangerousness decline in child molesters is milder and occurs in older ages than among rapists. Also, the person who committed the offence in an advanced age, his age should not be taken that seriously as a risk reducing factor.

3.3 Division/classification of sex offenders by the official definition of the nature of their offence

child molester- victim under age 13
rapist- victim above 13 years old. A significant factor with at moderate importance

s_2 — The expert gives a contribution of risk due to this factor.

a_2 — The attack says that the expert should have taken into account that risk of rapists against the passage of time declines at a faster rate than that of in child molester.

b_2 — Expert reply: Recent literature shows the relationship between the age of the offender to a level of sexual risk is not so dichotomous, for example, we learn that a dangerousness decline in child molesters is milder and occurs in older ages than among rapists. Also, the person who committed the offence in an advanced age, his age as a factor that reduces dangerousness should be taken with a grain of salt.

3.4 Family status

The official classification is as follows:

Bachelor — a person who has not lived with an Intimate Partner nor had a joint household with a partner for a period of at least 2 Years. If bachelor then this factor raises the dangerousness.

This factor is of Low importance.

s_3 — The expert put forward this factor

a_3 — The attack: You can see that the accused person is acquainted with a woman, maybe even married her, and managed a relationship for almost 2 years. Technically he is considered a bachelor but arguably it teaches us about his capabilities and reduces risk.

b_3 — Expert reply: The literature indicates that the fact a person contacted and possibly married is insufficient. Only if he would be able to manage relationships with common household for two years it will show the ability to keep significant relationship.

3.5 Index non-sexual violence (NSV) – any convictions

If the offender's criminal record shows a separate conviction for a non-sexual violent offence at the same time they were convicted of their Index Offence, this factor raise the dangerousness.

A significant factor with at least moderate importance

s_4 — Expert mentions use of violence.

a_4 — The attack: If the offender's criminal record does not show a separate conviction for a non-sexual violent offence at the same time they were convicted of their Index Offence, this factor should be ignored.

b_4 — Expert Reply: Do not ignore the fact that almost all sex offences include aspects of coercion and violence and the choice to convict a person of a crime of violence is a legal issue rather than sex offence issue.

3.6 Prior non-sexual violence – any convictions

Having a history of violence is a predictive factor for future violence. A significant factor with moderate importance

s_5 — Expert did not address this factor, (meaning that in the court case this factor was not mentioned in the expert's report. Since this is a mitigating factor the defence asks why was it not mentioned).

a_5 – The attack: If not convicted, so arguably he usually keeps the law and it is one-time lapse and the current conviction probably discourages him.

b_5 — Expert reply: Sometimes the person tells us himself that once he used violence against family members or others and the absence of conviction of violence does not necessarily indicate that he never used violence.

3.7 Prior sex offences

The best predictor of future behaviour, is past behaviour. A meta-analytic review of the literature indicates that having prior sex offences is a predictive factor for sexual recidivism.

A significant factor with high importance

s_6 — expert mentioned that the person had previous offences which increase the risk.

a_6 — Attack : This was a long time ago. Since then for many years there were no conviction. So previous conviction probably discouraged him.

b_6 — Expert Reply. Criminal that have several conviction at any time in the past is still to be considered dangerous. The existence of a conviction for sex offence often indicates quality of functioning of law enforcement officials and victims readiness and motivation, (if such were indeed), to complain. Also, in law, sometimes for a similar offence the offender can be convicted on different offences, for example, reveals himself in public might be convicted of committing a public indecent assault, but charges may be ether wild behaviour in a public place.

3.8 Prior sentencing dates

This item relate to criminal history and the measurement of persistence of criminal activity. The Basic Rule: If the offender's criminal record indicates four or more separate sentencing dates prior to the Index Offence, the offender is more dangerous. Count the number of distinct occasions on which the offender was sentenced for criminal offences. The number of charges/convictions does not matter, only the number of sentencing dates.

A significant factor is law importance

s_7 — Expert used this factor, even though the past convictions were not sex related.

a_7— Attack: If not convicted before, so arguably he usually keeps the law and it is one-time lapse and the current conviction probably discourages him. We can claim that if the subject made prior offences that teach about his criminal lifestyle, the risk sex assessment should evaluate only sexually dangerous and nothing else and the index offence is one-time lapse. People with criminal life style mostly feel disgusted by sex offences and shy away of it and their self-esteem injured therefore current conviction probably discourages him

b_7 — Expert Reply: A person who has a background of criminal offences shows difficulty to maintain limits and respect the boundaries of correct behaviour and one of the main concerns is that reluctance not to respect the laws and other limits may result in repeated sex offences, too.

3.9 Any convictions for non-contact sex offences

Offenders with paraphilic interests are at increased risk for sexual recidivism. Offenders who engage in these types of behaviours are more likely to have problems conforming their sexual behaviour to conventional standards than offenders who have no interest in para-philic activities. If the offender's criminal record indicates a separate conviction for a non-contact sexual offence, the offender is more dan-gerous.

A significant factor with high or very high importance

s_8 — Expert did use this factor

a_8 — Attack: You can argue that sex is contactless low threshold of severity of injury and despite the offence with high recidivism, even if a person carries the offence again, the damage it can cause to the potential victim not so strong a man performing very offensive offence with contact and entering offences. Typically, offenders who commit-ted Non-contact Sex Offences contact offences are less likely to make contact sex offences.

b_8 — Reply: The person that makes risk sex assessment is not a judge, and it is not his job to determine severity of harm, but to indicate to which group the subject belongs and what are the chances that he will make again sex offences, regardless of the severity of the offence.

3.10 Unrelated victims (victim known to the offender, but not family)

The items concerning victim characteristics. Sex offence on Unrelated Victims related to higher risk assessment. Research indicates that offenders who offend only against family members recidivate at a lower rate compared to those who have victims outside of their immediate family.

A significant factor with high importance

s_9 — Expert used this factor

a_9 — Attack: Offender who harm the victims in his family is less dangerous because he is often perceived as a "lazy" who probably will not look for victims outside the family.

b_9 — Reply: Despite the fact that the person who harm victims within the family hurts somebody outside the family is relatively low, but it still exists. In addition, the offence to be possible because of problematic family climate expressed within weak limits and if the family circumstances do not change, significant treatment, then the individual may return to the same environment that allowed the violation in the past and may again exploit his authority and hurt.

3.11 Any stranger victims?

The Basic Principle: Research shows that having a stranger victim is related to sexual recidivism. If the offender has victims of sexual offences who were strangers at the time of the offence (stranger is defined as a person known to offender for less than 24 hours prior to the offence), is related to higher sexual recidivism.

A significant factor with high importance

s_{10} — Expert says the victim was a stranger.

a_{10} — Attack: A strong connection formed between the offender and the victim, even though they met less than 24 hours (they had intimate conversation before the offence).

b_{10} — Reply: But he hurt the victim, who is not a relative and possibly in future is pushing a minimal introduction to compromise.

3.12 Any male victims?

The Basic Principle: Research shows that offenders who have offended against male children or male adult recidivate at a higher rate compared to those who do not have male victims.

A significant factor with high importance

s_{11} — The expert used this factor

a_{11} — attack- you say that a sex offender attacking male victims is more dangerous than offender who attacks female victims. This is clearly a prejudiced judgement between males and females. You see a man attacking another man as sick and therefore you make him more dangerous.

b_{11} — Reply There is no prejudice here, the observation is based on statistical data.

3.13 Alcohol consumption is clearly associated with violence

This is a strong factor in assessing risk. s_{12} — The expert increased the risk owing to the offender's high alcohol consumption

a_{12} — Attack 1: the offender has rehabilitated, he is no longer drinking.

$a_{12}(b)$ — Attack 2: the man has been alcoholic for a long time without offending, so there is no real connection.

b_{12} — Reply: The expert assertion about use of alcohol is based on the offender report of his use of alcohol, and it is well known that such reports can be unreliable. The offender report of alcoholism could be a cover for some more serious pathological causes.

$b_{12}(b)$ — Furthermore the use of alcohol can cause offence while drunk. This is a worrying factor because he might drink and be inhibited in the future and offend again.

3.14 The use of hard drugs

The connection between being a drug addict and sexual offence is not strong enough. Research identifies two types of drugs (excluding alcohol) contribute to hyper sexuality, namely Cocaine and Meta-amphetamines. To the extent that we get confirming scientific reports about the connection, we will consider drug abuse as a risk factor. At any rate this is a weak factor

s_{13} — The expert mentions this as a factor.

a_{13} - Attack 1 —The offender has rehabilitated, he is no longer drug addict.

$a_{13}(b)$ — Attack 2 The man has been addict for a long time without offending so there is no real connection.

b_{13} — Reply. The expert assertion about use of drugs is based on the offenders report of his use of drugs, and it is well known that such reports can be unreliable. The offender report of drug addiction could be a cover for some more serious pathological causes. Furthermore the use of drugs can cause inhibited behaviour and to lead to offence while under the influence. This is a worrying factor because he might use drugs in the future and offend again. Note that meta-amphetamines do increase /flood the sex drives and therefore might push the man to further offence.

3.15 Sexual offence while the offender was under court order

This could be, for example, a legal trial, conditional sentence, legal restrictions, etc. This is a strong factor

s_{14} — Expert used this factor.

a_{14} — Attack - the offender has been punished and will behave. Furthermore he did not understand at the time the full meaning of legal restrictions but now he does understand.

b_{14} — Reply: Maybe the offender just says he will now behave but this does not ensure that he will not offend again.

Furthermore the effects of the present trial and punishment will wear off as time goes by.

3.16 Sexual offence in a public place

This is a medium strength factor

s_{15} — The expert used this factor.

$a_{15}(b)$ — Attack- The offender made his offence at night at insulated place and the chance that somebody would see him is low.

b_{15} — It is still a public place and even at insulated places people can pass. It is known that offending in a public place indicates a deep difficulty to restrain oneself and control one's drives.[9]

---

[9] One of the referees made the following comment about this case (factor s_{15}), I quote:

> "Why might someone not attack on the basis that it was raining, so there was a lower chance of being interrupted? Or in a place that was not visible to passers-by? Why is the attack a conjunction of night time and isolation — surely isolation could be enough to form an attack? What I would expect is that the typical attacks would be evidenced through reference to the court record — and then I would expect to see many different attacks that might be levelled arranged into groups, classes, or hierarchies perhaps. Similarly with defences

3.17 The use of force while offending

This includes using firearms or the threat of using firearms, or use of physical force, or threat of physical damage or kidnapping.

This is a medium strength factor.

s_{16} — Expert uses this factor.

a_{16} — Threat is not really use of force.

b_{16} — professional literature shows it is it. Threat is definitely count as a use of force. Many times it is enough to compel person to make things that he didn't. Conviction of violence in addition to conviction of sexual offence indicates the offender not only cannot control his sexual drives but also cannot control his aggression.

3.18 The offender subjected the victim to a variety of sexual violations

These include: Penis penetration to vagina, finger into vagina, foreign object into vagina, groping the victim, masturbating over the victim, forcing the victim to grope the offender, forcing victim to masturbate, Forcing victim to give offender oral sex, offender giving victim oral sex, offender exposes himself (excluding exposing for the purpose

---

against those attacks."

We note that s_{15} is a transcript of a case in court. The reader might ask whether we have collected an exhaustive list of transcripts and analysed them and examined them? Maybe the above suggested referee questions were asked in other cases? The answer is we did not assemble a larger set of transcript but a representative one. There is sufficient data and we learnt a lot from these examples already, namely the idea of the attack as information input, see [6].

Let us examine the transcript of s_{15} itself, to show the reader what we mean by representative. The attack a_{15} adds factual information, and tries to say, given this information, then the place was not really public. The response b_{15} is actually saying that the factor's contribution to the risk assessment of the sex offender was determined statistically based on the formal definition of public place (as opposed to the concept of not containing people) and the extra information is not relevant to the statistics. Again b_{15} is an attack by adding information.

In fact b_{15} is also a valid counter-attack to the referees suggestions above ("it was raining", or "it was in a place that was not visible to passers-by", etc...), again because such cases did not go into the statistics!. Compare with b_{20}.

of executing the offence), forcing victim to make sex with a third party/object, penetration of penis to anus, penetration of finger to anus, penetration of object to anus, kiss, forcing the victim to masturbate the offender.

This is a weak factor

s_{17} — Expert lists the offences done by the offender

a_{17} — attack. These should be considered a single offence and not a list of multiple offences. Moreover, almost any rape or other sex offence including a variety of sexual violations. For example, it is almost impossible to rape without groping the victim.

b_{17} — Reply: yes legally it is a single offence, but statistics shows that multiple components increase risk of re-offending in the future. The offender needs multiple stimulations to satisfy his drive. The offender might even commit some unusual acts in the future, and if the indictment detail the violation, than probably is was a different offence and not a basis to perform another offence.

3.19 Sex offender with victims from different age groups

In such a case the offender is considered more dangerous because the offender has a larger group of potential victims.

The age groups are:

$$0–6.99; 7–12.99; 13–15.99; 16 \text{ and above}$$

s_{18} — Expert mentions this factor

a_{18} — Victims may not look their ages so it only an illusion that the offender is not focused on a single age group.

b_{18} —- Reply. As an expert I have a choice and judgement on whether I work like a simple mathematical machine or try to decide on the correct evaluation and scenario. I try to understand the triggers motivating the offence and using that evaluate how dangerous the offender is and to what age groups. I especially examine the significance of cases where the victim's age is near the boundaries.

3.20 Age of victim is 13–15 years

An offender attacking this age group is more dangerous if the offender is 5 years older or more than the victim.

This is a medium factor

s_{19} — Expert mentions this factor

a_{19} — Attack. The age division into group is arbitrary and further teenagers.

Vary in how old they look, and many times 13–15 years old looks like elder.

b_{19} — Reply: the expert exercises judgement. The problem here is that the offender seeks an intermediate age group between children and grownups. There is the danger of a shift into the neighboring age groups. It is offenders responsibility to know the exact age of teenager. And mostly the confusion is a result of cognitive distortion of the offender.

3.21 Offender has not been able to maintain continuous employment up to the offence

This is a medium factor

s_{20} — Expert quotes this factor

a_{20} — There is an objective market difficulty in maintaining continuous employment. Many employers sack people in order not to give them tenure.

b_{20} — This is a statistical observation. The statistics show increase in risk. The statistics does not consider the reasons behind the lack of past continuous employment.

3.22 Offender violated some restrictions imposed by court orders, not necessarily sexually connected

This is a medium factor.

s_{21} — Expert mentions this factor

a_{21} — The past offences are not sexual, why are you mentioning them?

b_{21} — The offender cannot keep to proper boundaries, and his "internal policeman" is weak. If within the boundaries of court orders the offender could not police himself, he might reoffend if we release him now.

3.23   Empathy towards the victim

Weak factor

s_{22} — Expert mentions this factor

a_{22} — The literature shows there is no significant connection of this factor to risk.

b_{22} — If there is no empathy to the victim the offender will not appreciate the damage he is doing, and will not be interested or respond well to remedial treatment.

3.24   Disrespect to authority and institutions

s_{22} — expert mentions this aspect

a_{22} — The literature shows there is no significant connection of this factor to risk

b_{22} — If offender does not respect authority, then the offender if released with disrespect the officer supervising him/her and will try to out-manoeuver the officer and offend again

3.25   Medical treatment to lower the sexual drive

This is an important factor, medium strength, as long as the patient participates

s_{24} — expert mentions this.

a_{24} — The Offender agrees to a chemical castration without being forced to do it. He is risking his body and might have to face side effects. This is a proof of how much he appreciates his wrong doing in the past and shows commitment to be risk free in the future. This must be considered a significant factor.

b_{24} — This treatment affects the offender capabilities, not his personality and tendencies. Therefore without a genuine internal change there is still the risk of further offence, especially if the treatment is discontinued.

Furthermore the offender agreeing to the treatment may be just manipulative and not genuine, and we can be sure only if he continues with it for a considerable period of time. This is why this factor doesn't change the risk assessment in the long term..

3.26 No community or family support for the offender

Low factor.

s_{25} — Expert mentions this factor

a_{25} — Offender can take care of himself

$a_{25}(b)$ — It is bad enough that everyone abandoned the offender, you have also to punish him for it?!

b_{25} — This is not a punishment but the unfortunate fact that the offender will have no support to help him not offend again.

3.27 Offender is mentally retarded

This increases risk, medium factor.

s_{26} — Expert mentioned this factor

a_{26} — This is God's doing, what can the offender do?

b_{26} — Mental retardation leads to dis-inhibition. The offender cannot learn from experience or appreciate vague situations with unclear boundaries.

3.28 Mental illness

Medium factor for increase in risk

s_{27} — Expert mentions this factor

a_{27} — What can he do, it is not his fault.

b_{27} — Mental illness leads to dis-inhibition. The offender has difficulties to learn from experience or appreciate vague situations with unclear boundaries.

We are not supposed to be politically correct but we deal in science and it is proven that mental illness increases risk of re-offending.

3.29 Offender does not accept responsibility for his actions nor expresses regret

Factor of low importance

s_{28} — Expert mentions this factor.

a_{28} — The literature does not consider this significant

b_{28} — If the offender does not accept responsibility of regret he will not be interested in any change. Accordingly, his chance to integrate on treatment and to derive the usefulness from it is low.

3.30 Did the offender plead guilty?

This is low factor.

s_{29} — Expert mentioned this factor

a_{29} — A literature do not attach much importance to this factor with the possible exception of a small group of offenders.

b_{29} — For offences within the family unit this is an important factor.

Furthermore, it is less likely the offender will accept treatment nor benefit from it

3.31 The offender has a distorted way of thinking

Low importance.

s_{30} —- Expert mentions this factor

a_{30} — This factor is not identified in the literature. Besides, everyone has distorted ways of thinking one way or another.

b_{30} — Sex offenders have their own characteristic distortions, that form the basis to rationalise and justify his offences. We know there is a connection between thinking positions and behaviour.

3.32 Offender has low opinion of himself

Medium importance for increasing risk.

s_{31} — Expert presents this factor.

a_{31} — Person with low opinion of self the offender will not dare offend.

b_{31} — On the contrary offender will not dare approach normal relationship and will find someone weak to offend and attack.

3.33 Offender is physically or mentally impotent or is ashamed of his sexual organs

Factor of medium to high importance

s_{32} — Expert mentions this factor as increasing risk

a_{32} — On the contrary, there is no risk, he cannot do it he will not do it.

b_{32} — Not at all, we are dealing with frustration as a basis for action. To prove him-self the offender might prey on the weak such as children.

3.34 Impulsiveness, low tolerance to stimuli

Factor of medium importance.

s_{33} — Expert presents this factor

a_{33} — Usually his impulsiveness is not connected with sex

b_{33} — Impulsive people are unpredictable, you cannot be sure what the offender will do.

3.35 Strong sex drive

Factor of high importance for risk.

s_{34}- - Expert presents this factor

a_{34} – So what, the offender will just be busy masturbate more often and is less likely to offend.

b_{34} – Research shows that on the contrary, increase masturbation enhances existing sex drives and not diminishes them. The offender is more likely to seek real contact.

3.36 Sexual deviation

Such as pedophilia, exhibitionism, proterism, etc.

Factor with high risk.

s_{35} — Expert uses this factor.

a_{35} — The man is sick, he needs hospital, not punishment.

b_{35} — I am not a Judge, the fact is that people with sexual deviation are high risk offenders.

3.37 Offender completed medical treatment

This is medium factor in reducing risk.

s_{36} — Expert did not include this factor

a_{36} — The offender did conclude a treatment why did you not include it as a high risk reducing factor?

b_{36} — The treatment is not effective on some people. They emerge from it with some success but these fade in time. The real test is if the offender continues the program suggested by the treatment.

3.38 Sex offender treatment was interrupted and never completed

High risk factor.

s_{37} — expert uses this factor.

a_{37} — The interruption was due to objective factors such as the offender was sent to prison and was not allowed to complete the treatment.

b_{37} — Even if it is not the offender's fault the fact is that half a treatment is risky and makes the situation worse in confusing the patient.

3.39 Does the offender understand/ know the risk/ trigger situations? Can the offender use adaptive preventive measures?

Medium factor

s_{38} — The Expert said the offender did not know.

a_{38} — The offender did know but when you talked to him he was under stress and could not list them. Anyway there is not enough research about this factor

s_{38} — It is important to know the risk/ trigger situation for offence and learn to avoid them. It is important for the offender to know that even simple, seemingly unimportant decisions can put him at a risk of a trigger situation.

3.40 Personality disorder

Factor of low importance.

s_{39} — Expert mentions this factor.

a_{39} — There is not enough research on this factor.

b_{39} — Sometimes this can be the reason for the offence. For example a narcissist might think the victim actually wants sex and the offender is actually being helpful.

Personality disorders are very difficult to treat.

3.41 The offender has had a long prison sentence

Factor with low strength.

s_{40} — Expert mentions this factor.

a_{40} — Offender did not offend in prison and suffered long enough. Why don't you let go instead of continuing to support punishing him?

b_{40} — I don't deal with punishment. I deal only with risk assessment. Today there is literature that indicates that having served a long prison sentence does not reduce risk but might even increase risk.

4 The formal model

The argumentation model we need is different from the traditional one, see Appendix A, and so we need to begin with an orientation discussion. We have already mentioned, at the end of Subsection 3.1, that the attack on the expert's testimony is an instance of Walton and Krabbe's [22] persuasion dialogue. The nature of the attacks in such dialogues is brilliantly studied in [21]. In the case of our expert testimony, the attacks have a numerical and informational aspects and these need to be modelled. We stress to the reader that the sex offender's risk assessment models express the risk directly with

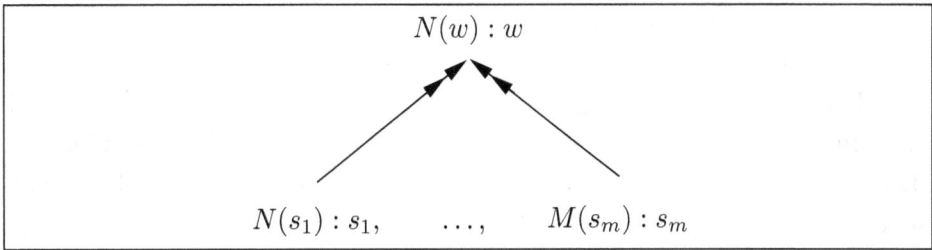

$$N(w) : w$$

$$N(s_1) : s_1, \quad \ldots, \quad M(s_m) : s_m$$

Figure 7

numbers. This is what the community does and this is what their tools do. It is not the case that we have arguments that get weaker by virtue of attacks and we are using numbers in our meta-language modelling to reflect that . The numbers in the sex offender case are in the object language.[10]

4.1 Orientation and discussion

The basic situation we face is that of Figure 7 (compare with Figure 5):

The node w has a numerical strength $N(w)$ and the nodes $s_1, \ldots, s_m$ also have numerical strength $N(s_1), \ldots, N(s_m)$. The Appendix gives some discussion of $N(s_i)$ and $N(w)$ and how to calculate them.

These are the numbers attached to the factors $s_1, \ldots, s_m$, and to w. The numbers $N(s_i)$ aggregate to give the number $N(w)$.

This is the first round in the expert's persuasion testimony, as represented by Figure 7. The issue is the node $N(w) : w$, and the support for it are the factors $N(s_i) : s_i$, as given by the expert. In

---

[10]We mention in passing that Henry Prakken [28] has shown a way to simulate this weakening process, what he calls argument "accrual", using ASPIC+ and similar systems, but his approach causes an exponential blow-up in the number of arguments. The formal argumentation community, especially those members identified with the COMMA conference, are going through what we can call the "combination phase". Combine argumentation with probability, with Bayesian networks, with modal logic, with automata (not done yet),etc etc. This is reminiscent of the Fuzzy Logic community, when they went through this phase, making anything fuzzy. It is a necessary and healthy evolutionary stage which sooner or later it will be over. Our use of number is not a "combination phase"use, we simply model what the sex offender's community does.

our formal model we need to say how $N(s_i), i = 1, \ldots, m$ are joined together to yield $N(w)$.

The numbers $N(x)$ can be positive, negative or zero, and fall within a range $N^- < N(x) < N^+$ where $N^- < 0 < N^+$. We will see in the Appendix, for example that N^- can be -13 and N^+ can be 5 (not necessarily $N^+ = N^-$). We obtain $N(w)$ from $\{N(s_i)|i = 1, \ldots, m\}$, via some mathematical formula. All the tools in the Appendix use the same formula: they simply add the numbers up. $N(w)$ is the strength of the risk w.

If it is a positive natural number $0 < N(w) \leq N^+$ it indicates the degree of the risk of the sex offender. When it is zero then there is no risk and when it is negative there is strongly no risk. The numbers $N(s_i)$ for the factors s_i indicate an increase in risk when positive, neutral when zero and a decrease in risk when negative. All the tools of the Appendix use the mathematical formula

$$N(w) = \sum_{i=1}^{*} N(s_i)$$

where $\sum^*$ is a sum function which truncates the sum at N^+ when positive and at N^- when negative.

So, for example, if we have $N^- = -13$ and $N^+ = 5$, then we have

- $-10 +^* 5 = -5$

- $-10 -^* 7 = -13$

- $5 +^* 3 = 5$

Before we continue any further we must issue a word of caution. The perceptive reader, especially a member of the COMMA community who is well versed with numerical argumentation, will no doubt wonder at the simplicity and naivete of this model and its numerical aggregation method. Well, this is not our (the author's) model. It is a description of what the risk assessment community do in practice, formulated/translated into argumentation language. We do this to show that there is an opportunity for the argumentation community to offer a better model. So the perceptive reader might continue and ask why don't we, the authors, offer a better model? Our answer is that the risk assessment community think like the medical community. Any new offer needs to be tested and monitored over some years. We need to talk to them extensively before we make any new model for them to use. For example the numerical ranges of the various factors are sometimes different. We need to ask and understand why this is so. At the moment we do not know why.

See Appendix C for a more detailed discussion.

Let us add here another word of caution, while we are in this mode. Some readers from community 2 might wonder whether the Abstract Dialectical Framework [30], can be of service here. We accept that the Abstract Dialectical Framework is a powerful tool gaining momentum in the community. However, the answer is no, it is not suitable. It is essentially classical propositional logic reformulated and presented as argumentation. Community 2 people are familiar with classical propositional logic and can see that it is not suitable.

Let us alert the perceptive reader to another benefit coming from the risk assessment community to the argumentation community. There is another point to be aware of. The "attacks" on the factors $s_1, \ldots, s_m$, as practiced in court and reported by the second author, are not numerical but informational. The attack a_i on s_i sometimes asks for clarification and sometimes gives more information. The attack b_i on a_i is again just a clarifying reply. So we have here a network where the attacks are diverse, sometimes they are numerical and sometimes they are non-numerical. This gives us the new idea of the attack as information input (see [6]). It means that the traditional concepts of extension, conflict freeness and attack and defence, may not apply in our case, and we need new formal concepts to deal with such networks. We also need to explain how a node a attacks a node b by sending it information. Notice that if the information is explanatory, then this is support and we get a network with both attack and support (also called Bipolar Networks in the community (see [9] and the references there, be aware that [9] follows a long chain of previous papers by the community of argumentation). See Appendix A for formal argumentation definitions.

We note that a numerical attack can also be regarded as informational, in the sense that it gives new numbers as new information.

So we have three problems here, the first two are under our immediate control and the third requires further consultation with the Sex Offenders community:

1. Develop networks where the meaning of the attack relation may vary in different parts of the network.

2. Explain and define the notion of informational attack, where node a attacks node b by sending it some information which may attack b or support b or may even be consistent with b, but thus changes b. We must compare with [21, 31-33]. This, however, requires a full subsequent theoretical paper.

3. Develop a better numerical model for aggregating risk for the sex offender community. See Appendix C for the challenges involved.

4.2 Formal presentation of the model

We are now ready for the formal machinery. We first define the traditional argumentation networks for finite acyclic graphs. These have only the traditional grounded extension. This will allow us to appreciate the next definition, that of informational argumentation network for finite acyclic graphs.[11]

---

[11]Note that our models are trees of depth 4 anyway, so acyclic graphs are OK for us.

However, there seems to be an issue about the use of graphs with cycles. This issue is already raised and discussed in [21]. See [25, 26]. We quote from [21]:

> "Although argument graphs are not restricted to trees, they are not completely general; we do not allow cycles. This restriction is intended to assure the decidability of the acceptability property of statements. At first sight, the condition that argument graphs be acyclic would seem to be a severe limitation. However, things are not that serious. Firstly, in systems using Dung's approach most cycles in realistic examples are two-cycles between arguments with incompatible conclusions. In Carneades, these can be represented

Definition 4.1.

1. *Let S be a non-empty set and let $R \subseteq S \times S$ be a binary relation on S. We say R is acyclic iff there does not exist a sequence $(x_1, \ldots, x_n)$ in $S, n \geq 1$ such that*

$$x_1 R x_2, x_2 R x_3, \ldots, x_n R x_1.$$

2. *x is said to be a source point if there is no y s.t. yRx. x is an endpoint if there is no y such that xRy.*

Proposition 4.2. *Let (S, R) be a finite acyclic graph. Then for some $x \in S$ we have that x is a source point.*

Proof. Assume that there are no source points. Let $x_1 \in S$, then for some $x_2, x_2 R x_1$. Similarly for some $x_3, x_3 R x_2$. We carry on and get a sequence $x_1, x_2, x_3, \ldots, x_n, \ldots$ such that $x_{n+1} R x_n$. Since S is finite, for some $m, n, m \neq n, m < n$ we have $x_m = x_n$. Thus gives us a cycle $(x_m, \ldots, x_n)$. $\qquad\square$

Definition 4.3. *Let (S, R) be finite acyclic. We define a Caminada $\{0, 1\}$ labelling λ on S as follows.*

Step 1. *Let all source points x be labelled $\lambda(x) = 1$.*

Step 2. *Let y be any point such that for some x, xRy and $\lambda(x)$ is defined in Step 1 and $\lambda(x) = 1$. Let $\lambda(y) = 0$.*

$\vdots$

---

as a pair of arguments pro and con the same statement, which does not introduce cycles into the argument graph. Next, cycles caused by indirectly using a statement in support of itself are also excluded in many other systems. ....in Dung's [8] abstract framework defeat graphs with odd cycles may have no stable extensions. Also, intuitions differ on the proper treatment of cycles [26]. Prior 'relational' approaches to this problem, such as Dung's preferred and grounded semantics as alternatives for stable semantics, are not directly transferable to Carneades, with its dialogical and procedural elements. We therefore leave an extension to graphs that allow for cycles through exceptions for future work".

Step $2n+1$. *Let x be any point such that $\lambda(x)$ is not yet define but for all z, such that zRx, we have that $\lambda(z)$ is defined and $\lambda(z) = 0$. Let $\lambda(x) = 1$.*

Step $2n+2$. *Let y be such that $\lambda(y)$ is not yet defined but for some $z, \lambda(z)$ is defined and $\lambda(z) = 1$. Let $\lambda(y) = 0$.*
  Let S_λ be all points x such that $\lambda(x)$ is defined at any n.

Proposition 4.4.

1. *For the S_λ of Definition 4.3 we have that $S_\lambda = S$.*

2. *The function λ thus defined is unique.*

Proof.   1. Let $x_1 \in S_\lambda - S$. Consider the set of all z such that zRx.

 (a) If the set is empty then $\lambda(x) = 1$ by Step 1 of Definition 4.3.

 (b) If the set is not empty and all its members are in S_λ, then if for some z, zRx_1 and $\lambda(z) = 1$ then at some stage $\lambda(x_1)$ is defined and $\lambda(x) = 0$. On the other hand, if for all z such that zRx_1 we have $\lambda(z) = 0$, then at some stage $\lambda(x_1)$ is defined and $\lambda(x_1) = 1$.

 (c) Therefore there exists an element z such that zRx_1 and $\lambda(z)$ is not defined. Call this element x_2.

 Assume by induction that we have $(x_m, x_{m-1}, \ldots, x_1)$ such that

 $$x_m R x_{m-1}, \ldots, x_{m-1} R_{m-2}, \ldots, x_2 R x_1$$

 and $\lambda(x_j)$ are all undefined. Repeat our reasoning for x_m and get x_{m+1} for which $\lambda(x_{m+1})$ is undefined.

 Since S is finite, we have that for some $m_1 < m_2$ we have $x_{m_1} = x_{m_2}$. This gives us a cycle. A contradiction. So $S_\lambda = S$.

2. It is clear from the construction that λ is unique.

$\square$

Example 4.5. *We use the example of a 61 year old sex offender (age factor s_{15}) and let us present his risk sex assessment. We use one of the tools listed in the Appendix. His age scores as (-3, being the score value for s_1). The offender attacked in a public place (factor s_{15}, value +1) and used force (factor s_{15}, value +1) to subdue a male (factor s_{11}, value +1) unrelated (factor s_{19}, value +1) and unfamiliar (factor s_{10}, value +1) victim (factor s_{18}, value +1) who was 7 years old. It was his first conviction on sexual offence (factor s_6, value 0) and in fact, his first conviction for any offence at all (factors s_7, s_8, value 0). Official documents show that he had lived with an Intimate Partner and had a joint household with a partner for a period of at 26 Years (factor s_3, value 0) and worked for at least 15 years at the same place of work (factor s_{20}, value 0).*

The person successfully completed group sex offender therapy (factor s_{36}, value -2) and today he admitted to the assessor that he knows he has deviant sexual urges (factor s_{35}, value +2) but he learned adaptive coping techniques and knows how to avoid dangerous situations (factor s_{38}, value -1).

The figure for this example is Figure 8. The score values for the factors are given above and this is the information being sent to the node w. The score values sum up to +3. Section 3 lists the factors s_i and also lists a_i and b_i. the reader can see that a_i and b_i add more information.

5 Conclusion and discussion

We begin with an incredible observation we made while writing this Chapter. The sex offender international community is also a community doing therapy. They try to help the convicted sex offenders and in fact to have one's prison sentence reduced one needs to join a therapy group. The second author of our Chapter also has responsibility for conducting and dealing with therapy groups. The most amazing discovery we made is that the therapy uses argumentation and logic. The accepted wisdom in the sex offender community is that besides physical and all the other problems which the offenders have, they also

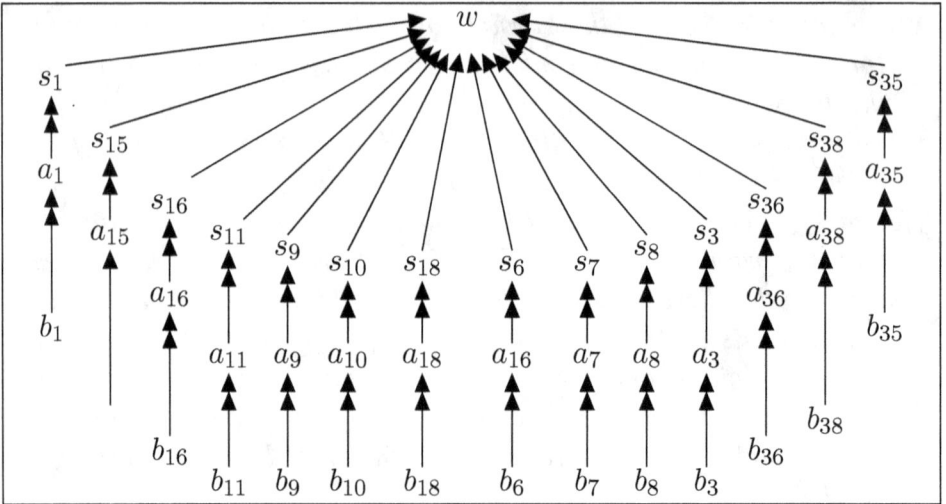

Figure 8: Figure for example 4.5

suffer from reasoning distortions, and in order to lessen the offenders risk to society his reasoning distortion needs to be corrected.[12] To do this they use argumentation. We have also written paper [33] on the the theory of Universal Distortions in reasoning and argumentation. This type of distortion is not restricted to sex offenders, you see it in religiously motivated terrorism, fundamentalism and more. One needs to model it and find ways to counter it.

We now turn to the current Chapter.

Our starting point in this Chapter is current research in the argumentation community on reasoning schemes, critical questions and expert opinion (see [40]), focussing on one of the 60 argumentation schemes in the important book [7] of D. Walton, C. Reed and F. Macagro, namely the scheme:

Scheme 2: Argument from Expert Opinion

Our approach was to look at one expert – the second author Dr G Rozenberg, an expert on assessing the risk factor of a sex offender –

---

[12]It is amazing that in group therapy the sex offenders can identify the distortions in others but not (the same distortion) in themselves. It can take 20 months to make them see the problem.

and see what argumentation experience he has accumulated over the years in court, and try and model it, in the spirit of [7].

We quote the hope mentioned in [7]:

> "It would be very helpful for users of the schemes to have a more refined system of classification, so that the user could search through to find a scheme applicable to her needs..."

We hope that our research in this Chapter is indeed helpful.

The following are the main points of the process and some comments on what we found. We present it from the point of view of the argumentation community, since this community is the one to take the next step in communicating with the sex offenders community. We look at the process of expert witness for risk assessment and we see the following steps:

1. The court appoints the expert to check a specific statement witness. This is standard practice.

2. The expert writes a report and delivers to the court and the defence. Again, it is expected standard practice for the expert to produce a report, but in the sex offenders case in Israeli courts the expert must address an internationally recognised list of factors.

3. The expert is attacked by the defence on two fronts (this is recognised by Walton in a general way but has a specific form in the Israeli courts for the case of sex offenders).

    (a) The expert is attacked personally as an expert, (see Section 2). We listed the type of questions/attacked used against the expert during his 11 years experience. Some of these we incorporated in refining the Walton scheme in Appendix B.

    (b) The report items are attacked, (see Section 3). The form of attack and defence is stylised in Israeli court in the form of a tree of depth 4. Owing to this stylised form there is the possibility of computerising/automating the process, thus exporting a tool to the sex offender community.

4. We discovered that there is a stylised generic form of the personal attack on the expert. Some questions are about his knowledge and methodology and some are trick questions to discredit the expert personally, hoping to show for example that he is a racist, has hidden agenda, etc. We commented on some of such questions in Section 2, and added another question scheme CQ7, to Appendix B.

5. There is also a stylised form of attacks on the report. The report must address certain questions s_i which seems to be instantiations of a list of generic questions applied to the area of expertise. The expert opinion is attacked by the defence on each item s_i by the question a_i and the expert replies to the attacks by reply b_i, and that is all of it. This observation is not trivial observation: The defence is not allowed to further attack the response of the experts and more importantly the defence is expected to ask only about the ATSA recognised factors. It can bring its own expert.

To the extent that the defence bring their own expert, the defence expert follows the same steps as the court expert.

6. The two experts do not interact, that is they do not answer, attack or speak to each other. The whole parallel process is observed by the court. We need to think whether this gives us new ideas about dialogue/debate procedures.

7. When modelling what is happening, we realised that we need to use Informational attack systems, where the attack is information input [5, 6]. The information input can be attack, can be support, or can aggregate to any of the above. This is a new game for the argumentation community.

8. What is more surprising, is that the sex offender therapists community uses logic to treat sex offenders. The community is not

explicitly aware of this. They regard the sex offender as suffering from reasoning distortions and proceed to actually use argumentation to try and correct such distortion and reduce the temptation to offend. Once we, the authors, realised this, we were motivated to study reasoning distortions in general, see paper [33].

What are our lessons from the above?

- The way expert opinion is handled. It is different from mere witness testimony which is not stylised and regimented. We can ask whether this stylised expert witness procedure is an example where we can have a Computerised Judge responding to a stylised computerised form filled in by the expert witness.

- We need to do similar investigation of real life/court practice for each of the 60 argument schemes of [7].

- In practice attacks are informational and so this area of argumentation needs to be further developed, as well as a serious comparison with [21, 29, 31].

- We need to examine the use of fallacies in respectable argumentation systems, especially when numerical or other aggregation is involved. We saw in some of the attacks on the expert, that some gentle hints of Ad Hominem were employed, (Questions 21, 34 and 36) even though in the Israeli stylised system, the expert was recognised. See also the trustworthiness questions in Appendix B.

- Focus more on the works of D. Walton, J. Woods and collaborators and the research of the Informal Logic/ Fallacies communities. Follow papers [21, 22, 23, 24, 27, 31, 33] and the references there.

We conclude this section and the Chapter by comparing with related literature.

As we remarked in Section 1, Remark 1.1, we need to compare our Chapter only with other papers which analyse in detail specific expert witness protocols and modellings. There are not too many such models and detailed procedures tend to be too special anyway. One aspect however, can be compared and is of great importance, and that is the use of statistical reasoning by experts. Such use is problematic, see [41, 42, 43]). Our risk assessment factors in Section 3, rely heavily on statistical packages. This has been criticised by Canadian courts

> "On September 18, 2015 a Canadian court (Ewert v. Canada, 2015) strongly cautioned the continued use of five risk instruments (Hare Psychopathy Checklist Revised [PCL-R], Violence Risk Appraisal Guide [V-RAG], Sex Offender Risk Appraisal Guide [SORAG], Static 99, Violence Risk Scale — Sex Offender [VRSSO]) by the Correctional Service Canada (CSC)1 with Aboriginal inmates. The instruments evaluate risk for future violence (VRAG), risk for sexual violence/offending (SORAG, Static 99, VRS-SO), and the presence of psychopathic traits (PCL-R)".

See [46].

There are three methods for assessing risk of sex offenders

1. The subjective one, by the expert/therapist. This has been criticised as being too subjective

2. The use of international statistical packages, (see Appendix C, and for example [45]).

3. A new method, *Structural clinical judgement*, can be used which benefits from the good aspects of both 1. and 2. See [47]

Acknowledgements

We are grateful to the Journal referees for very helpful and thorough comments and penetrating criticism. A 12 page short version of this paper was submitted to COMMA and we got very constructive and valuable comments from the three COMMA referees.

References

[1] D. Godden and D. Walton. Argument from expert opinion as legal evidence. Critical questions and admissibility criteria of expert testimony in American legal system. In *Ratio Juris*, 19(3), September 2006, pp. 261–286.

[2] D. Gabbay. *Meta-logical Investigations in Argumentation Networks*, College Publications 2013, 770pp.

[3] D. Gabbay, B. Liao and L. van der Torre. Two dimensional argumentation networks: towards modelling of Walton argumentation schemes. Draft, 2015.

[4] New York State Division of Probation and Correctional Alternatives Research Bulletin: Clinical and Structured Assessment of Sex Offenders. http://www.criminaljustice.ny.gov/opca/pdfs/somgmtbulletinaugust2007.pdf

[5] D. Gabbay and A.Garcez. Logical modes of attack in argumentation networks. *Studia Logica*, 93(2–3): 199–230, 2009.

[6] D. Gabbay and M. Gabbay. The attack as information input. December 2015. Short version appeared in pages 311-318 of Pietro Baroni, Thomas F. Gordon, Tatjana Scheffler and Manfred Stede, *Proceedings of COMMA 2016*, IOS press, long version draft available.

[7] D. Walton, C. Reed and F. Macagro. *Argumentation Schemes*, CUP, 2008, 452pp.

[8] Greg Restall. *An Introduction to Substructural Logics*, 400 pp, Routledge 2000.

[9] D. Gabbay. Logical foundations for bipolar argumentation networks, *J Logic Computation* (2016) 26(1): 247-292. doi: 10.1093/logcom/ext027 First published online: July 22, 2013.

[10] M. Caminada and D. Gabbay. A logical account of formal argumentation. *Studia Logica*, 93(2-3): 109–145, 2009.

[11] D. Anderson and R. K. Hanson. Static-99: An actuarial tool to assess risk of sexual and violent recidivism among sexual offenders. In R. K. Otto and K. Douglas, eds. *Handbook ofViolence Risk Assessment Tools*. Milton Park, UK: Routledge, 1020.

[12] H. E. Barbaree, C. M. Langton, R. Blanchard, and D.P. Boer. Predicting recidivism in sex offenders using the SVR-20: the contribution of age-at-release. *International Journal of Forensic Mental Health*, 7, 47–64, 2009.

[13] K. S. Douglas, J. R. Ogloff, T. L. Nicholls and I. Grant. Assessing risk for violence among psychiatric patients: the HCR-20 violence risk assessment scheme and the Psychopathy Checklist: Screening Version. *Journal of Cosnulting and Clinical Psychology*, 67, 917–30, 1999.

[14] D. L. Epperson and J. D. Kau. Minnesota Sex OffenderScreening Tool, Revised (MnSOST-R): Development, Performance, and Recommended Risk Level Cut Scores, 1999.

[15] R. K. Hanson. What do we know about sex offender risk assessment? *Psychology, public Policy, and Law*, 4, 40–72, 1998.

[16] R. K. Hanson and K. E. Morton-Bourgon. The accuracy of recidivism risk assessments for sexual offenders: A meta-analyiss. *Psychological Assessment*, 21, 121, 2009.

[17] A. Harris, A. Phenix, D. Thronton, and R. . Hanson. Static 99: Coding rules revised 2003. Ottawa, ON: Solicitor General Canada, 2003.

[18] R. K. Hanson and A. J. R. Harris. STABLE-2007 master coding guide. Public Safety and Emergency Preparedness, 2007.

[19] M. Rettenberger and R. Eher. Predicting reoffence in seal offender subtypes: A prospective validation study of the German version of the sexual offender risk appraisal guide (SORAG). *Sesual Offender Treatment*, 2, 490–521, 2010.

[20] S. Wong, M. Olver, T. Nicholaichuk and A. Gordon. Violence Risk Scale Sexual Offender Version, 2009.

[21] T. F. Gordon, H. Prakken, D. Walton. The Carneades model of argument and burden of proof, *Artificial Intelligence* 171 (2007) 875–896.

[22] D. Walton and E.C.W. Krabbe. Commitment in Dialogue: Basic Concepts of Interpersonal Reasoning, *SUNY Series in Logic and Language*, State University of New York Press, Albany, 1995.

[23] R.P. Loui. Process and policy: Resource-bounded non-demonstrative reasoning, *Computational Intelligence* 14 (1998) 1–38.

[24] C. I. Chesñevar, A. G. Maguitman and R. P. Loui. Logical models of argument. *ACM Computing Surveys (CSUR)* Surveys Homepage archive Volume 32 Issue 4, Dec. 2000 Pages 337-383.

[25] P. Baroni, M. Giacomin, and G. Guida. SCC-recursiveness: A general schema for argumentation semantics, *Artificial Intelligence* 168 (2005) 162–210.

[26] D. Gabbay The handling of loops in argumentation networks, in JLC special issue on Loops. *J Logic Computation*, first published online February 20, 2014 doi:10.1093/logcom/exu007 (83 pages)

[27] H. Prakkend and G. Vreeswijk. Logics for defeasible argumentation *Handbook of philosophical logic* Vol 4, D Gabbay and F Guenthner Editors, Springer, 2002, 219-318.

[28] H. Prakken. A Study of Accrual of Arguments, with Applications to Evidential Reasoning, *Proceedings of the 10th international conference on Artificial intelligence and law* 85-94 ICAIL'05 June 6-11,2005 Bologna, Italy.

[29] G. Brewka and T. F. Gordon. Carneades and abstract dialectical frameworks: A reconstruction. In *Proc. COMMA'10*, pages 3–12. IOS Press, 2010.

[30] M. Grabmair, T. F. Gordon and D. Walton. Probabilistic Semantics for the Carneades Argument Model Using Bayesian Belief Networks. Computational Models of Argument. *Proceedings of COMMA 2010*, IOS Press (2010), 255–266.

[31] G. Brewka and S. Woltran. Abstract dialectical frameworks. In *Proc. KR'10*, pages 102–111. AAAI Press, 2010.

[32] H. Prakken and G. Vreeswijk. Logics for Defeasible Argumentation, in *Handbook of philosophical logic* vol 4, D Gabbay and F Guenthner, editors, Springer 2002, Pages 219–318

[33] D. Gabbay, G. Rozenberg and L. Rivlin. Argumentation and reasoning under the influence of Universal distortions. *Ifcolog Journal of Logics and Their Applications*, 2017, pp 1769-1900. short version of this paper was submitted to COMMA 2016.

[34] F. van Eemeren and B. Garssen. *Handbook of Argumentation Theory*, 16 Jul 2014

[35] I. Rahwan and G. R. Simari, eds. *Argumentation in Artificial Intelligence*, Springer, Aug 2009.

[36] John Woods and Doug Walton. Argumentum ad Verecundiam, *Philosophy and Rhetoric*, 3 (1974), 135-153;

[37] John Woods and Doug Walton. *Fallacies: Selected Papers 1972-1982*, 2nd edition with a Foreword by Dale Jacquette, pages 11-28, London: College Publications, 2007; pp. 17-24. First published in 1989 by Foris.

[38] John Woods. *Is Legal Reasoning Irrational? An Introduction to the Epistemology of Law*. College Publications, 2015

[39] Douglas Walton. *Witness Testimony Evidence Argumentation, Artificial Intelligence, and Law*, CUP 2008.

[40] Henry Prakken and Giovanni Sartor. Law and logic: A review from an argumentation perspective *Artificial Intelligence*, Volume 227, October

2015, Pages 214–245.

[41] N.E. Fenton. Science and law: Improve statistics in court. *Nature*, 497:36- 37, 2011.

[42] N.E. Fenton and M. Neil. Avoiding legal fallacies in practice using Bayesian networks. *Australian Journal of Legal Philosophy*, 36:114–151, 2011

[43] N.E. Fenton Improving statistical estimates used in the courtroom. http://bayes-knowledge.com/attachments/article/30/(Fenton% 202011)%20Science%20and%20Law_%20Improve%20Statistics%20in% 20Court.pdf, (accessed on December 18, 2016, 0625 hours UK).

[44] D. Grubin. Inferring predictors of risk: sex offenders, *International review of psychiatry* 9, 1997.

[45] A. J. Harris, Phenix, Hanson and Thornton, Static-99 coding rules Revised, 2003.

[46] S. M. Shepherd and R. Lewis-Fernandez. Forensic risk assessment and cultural diversity: Contemporary challenges and future directions. *Psychology, Public Policy, and Law*, 22(4), 427, 2016.

[47] R. K. Hanson, A. J. Harris, T. L. Scott and L. Helmus. Assessing the risk of sexual offenders on community supervision: The Dynamic Supervision Project (Vol. 5, No. 6). Ottawa, ON: Public Safety Canada, 2007.

Appendices

A Refining the Walton argument schemes

We begin with an orientation, clarifying where this Appendix stands in relation to the work of Doug Walton. In their brilliant 2006 paper [7], D. Godden and D.Walton. Argument from expert opinion as legal evidence, Godden and Walton say in their abstract, we quote:

"Abstract. While courts depend on expert opinions in reaching sound judgments, the role of the expert witness in legal proceedings is associated with a litany of problems. Perhaps most prevalent is the question of under what circumstances should testimony be admitted as expert opinion. We review the changing policies adopted by

American courts in an attempt to ensure the reliability and usefulness of the scientific and technical information admitted as evidence. We argue that these admissibility criteria are best seen in a dialectical context as a set of critical questions of the kind commonly used in models of argumentation."

The paper gives criteria for determining what is expert opinion. By comparison, in the 2008 book [6], only 6 questions are represented, (see below, namely CQ1-CQ6). Many more questions could have been included in [6], in view of [7]. We do not know the reason for the exclusion. Perhaps the authors of [6] wanted to include only questions which apply to any kind of expert testimony, and not just to the case of a legal experts. Adopting this explanation/point of view, what we do below is refine the set CQ1-CQ6, in view of the questions in Section 2, choosing those questions which we think apply to any expert. At the end of this appendix we quote more from paper [7].[13]

The Walton critical questions for an expert claiming A, can be refined as follows each question is presented with the refinement sub-questions listed below:

CQ1: Expertise question: How credible is E as an expert source?

* What is your qualification?

---

[13]The sex offender case offers a specific detailed example of handling expert opinion. We have no doubt that Walton could address this case by defining new argumentation schemes which meet the requirements of this scenario. This Appendix informally refines Walton's scheme for expert witness testimony, by listing sub-questions to the critical questions. We note that there is no notion of sub-questions in Walton's conception of argumentation schemes. We do feel, however, that this is the way to go at least in this case. This is not a criticism of Walton Scheme for expert opinion, this is only a minor issue. Computational models of Walton's conception of argumentation schemes have been developed which are capable of generating or inferring arguments. We expect that the requirements of this application scenario could have been easily met by using one of these systems to model/represent the required schemes.

* is there a structured training course that experts in your area have to take?

CQ2: Field question: Is E an expert in the field that A is in?

CQ3: Opinion question: What did E assert that implies A?

CQ4: Trustworthiness question: Is E personally reliable as a source?

* Surely you will agree with me that the evaluation of leading to A is speculative?

* What is the percentage of mistake?

* The time you spent on your evaluation is not enough

* Don't you think it is arrogant to claim that in a such a short time (say "short time", whatever the time is) you can assess A?

* My client feels victimised by you

* Give me some examples of factors which support/attack the assessment A and explain why these factors are relevant and contribute.

* Is there different way of assessing A, say between Europe and America?

* Why did you ignore ... (find something to claim was ignored).

CQ5: Consistency question: Is A consistent with what other experts assert?

* Are you doing all of this alone? It is possible that someone else, had he/she been with you would have reached a different conclusion?

* There are research results/other reliable experts, which contradict what you are saying

CQ6: Backup evidence question: Is E's assertion based on evidence?

* What are the steps taken in your evaluation leading to your expert statement A

* What method do you use?

* What exactly do you check? How do you put all your findings together and get a result?

Add CQ7 Fallacies

* Do you really think that your testimony makes any difference?

* Do you realise the damage your testimony is making?

* How many previous cases you had your testimony rejected?

We now quote from [7]:

"Critical Questions for Argument from Expert Opinion

1. Expertise Question: How credible is E as an expert source?

1.1 What is E's name, job or official capacity, location, and employer?

1.2 What degrees, professional qualifications or certification by licensing agencies does E hold?

1.3 Can testimony of peer experts in the same field be given to support E's competence?

1.4 What is E's record of experience, or other indications of practiced skill in S?

1.5 What is E's record of peer-reviewed publications or contributions to knowledge in S?

2. Field Question: Is E an expert in the field that A is in?

2.1 Is the field of expertise cited in the appeal a genuine area of knowledge, or area of technical skill that supports a claim to knowledge?

2.2 If E is an expert in a field closely related to the field cited in the appeal, how close is the relationship between the expertise in the two fields?

2.3 Is the issue one where expert knowledge in any field is directly relevant to deciding the issue?

2.4 Is the field of expertise cited an area where there are changes in techniques or rapid developments in new knowledge, and if so, is the expert up-to-date in these developments?

3. Opinion Question: What did E assert that implies A?

3.1 Was E quoted in asserting A? Was a reference to the source of the quote given, and can it be verified that E actually said A?

3.2 If E did not say A exactly, then what did E assert, and how was A inferred?

3.3 If the inference to A was based on more than one premise, could one premise have come from E and the other from a different expert? If so, is there evidence of disagreement between what the two experts (separately) asserted?

3.4 Is what E asserted clear? If not, was the process of interpretation of what E said by the respondent who used E's opinion justified? Are other interpretations plausible? Could important qualifications be left out?

4. Trustworthiness Question: Is E personally reliable as a source?

4.1 Is E biased?

4.2 Is E honest?

4.3 Is E conscientious?

5. Consistency Question: Is A consistent with what other experts assert?

5.1 Does A have general acceptance in S?

5.2 If not, can E explain why not, and give reasons why there is good evidence for A?

6. Backup Evidence Question: Is E's assertion based on evidence?

6.1 What is the internal evidence the expert used herself to arrive at this opinion as her conclusion?

6.2 If there is external evidence, e.g. physical evidence reported independently of the expert, can the expert deal with this adequately?

6.3 Can it be shown that the opinion given is not one that is scientifically unverifiable?"

The difference between the Walton Expert Witness scheme and what goes in Israeli courts can be explained as follows: In Israel senior practitioners become designated as "experts"and these can testify in court. So many of the Walton questioned do not apply to them in court. They have already been screened and vetted. The only way to attack what they say is to bring in another recognised expert. The perceptive reader might ask, if this is the case why ask questions like

- What is your qualification?

- is there a structured training course that experts in your area have to take?

Perhaps it is protocol questions or for the record? The reader should realise that these questions are what is actually asked in court. We reiterate the second author is a recognised senior expert and the data is what is actually done in court.

B   Examples of tools for assessing strength of risk factors

This Appendix quotes (we copied and pasted and edited some texts, from what is available on the internet) details about various tools which evaluate the strength of various factors discussed and assumed to be available in Section 4.

We stress that the professional literature which deals with sexual risk assessment is very diverse. It is impossible to survey and compare it all in one Chapter. The present Chapter addresses three communities, two among them are argumentation communities and so we need to give them an idea about the tools used in assessing risk of sex offenders. We therefore list the most common actuarial instruments existing in the field, which are used in different ways in different countries. In Israel, the suitability of actuarial tools was not tested and therefore, existing tools are used as tools of decision support for experts and are applied mainly to gain a view of different factors. The duty of the recognised expert is to combine the existing knowledge in the world, which is reflected also in the existing instruments, with his own experience and clinical knowledge which he has accumulated following years of practice and express his expert judgement to the court. The present Chapter and Appendix does not imply that we disregard the existing tools, but intends to show the reader, especially from the argumentation community, the complexity of the current situation, in which there is a multiplicity of tools. Those members of the argumentation community well versed with numerical fuzzy argumentation and T-norms will no doubt notice that the tools listed below in this appendix give numerical valuations. The risk assessment community use naive averaging to combine them. The argumentation community can improve the tools and help.

We can offer to use logical argumentation super-tools for combining the existing tools into one simple and effective integrative tool, which can be used for decision support for experts. We recommend and ask that the argumentation community take the challenge and develop such tools. Note however, that the risk assessment community, just like the general medical community (facing a new medicine or drug) will take their time and will test the tool and conduct statistical analyses to test the suitability and success rate of of the proposed tool to the sex offender population in different countries and different populations of sex offenders.

Sex offenders are different from each other in their character as well as in the nature of their offenses. We have given a list of the main factors that influence the decision of the expert regarding the risk assessment of the offender, and basically all of these factors are measurable, (as we have pointed out and presented). Each of the factors is addressed when it is relevant to the to the sex offender or if it helps characterize his offense and if it helps determine his degree of risk. As we wrote, we do not claim that tools presented can over-ride the discretion of the expert or his risk estimates, but the tool is used only to be decision support tool. (This is the case in Israel but there are other countries where these international tools carry much more weight and cannot be easily disregarded by the expert.) Of course, if we see in a sex offender case a factor which are not quantifiable, or such that no existing tool seems to apply to them, then it is up to the expert to decide what weight to give to such factors

On the practical level, the expert writes his opinion and presents it to the court and refers to all the relevant factors. The opinion contains several key parts.
A. Details information about the offenders past and personal life (school, social relationships, marriage, military, work, etc.).
B. Description of the offenses.
C. Description of the interview with the sex offender, his attitude to his offenses and a description of his sexual habits, which includes fantasies, whom he is attracted to, the frequency of sexual intercourse, the intensity of his sexual impulses.
D. Clinical impressions, which includes an explanation by the expert of his view of why the offender offended and what underlies his offenses.
E. Risk evaluation, including references to such factors of static and dynamically nature relating to the offender.

F. In conclusion the expert summarizes all relevant factors and gives a final risk assessment of the offender.

The judge or defense lawyer is allowed not to agree with some specific this or that factor addressed by the expert or indeed can disagree with any of the expert's conclusion and then the expert to explain any factor and a factor and the debate continues in the form explained in Section 3

Tool 1: Static-2002 (Hanson, Helmus, & Thornton, 2010)

Static sex offender risk assessment instrument

The Static-2002R is based on static (unchanging) risk factors which predict the potential for sexual re-offending. This risk assessment instrument is required by law to be used by the California Department of Corrections and Rehabilitation to assess every eligible sex offender prior to release on parole; by Probation, to assess every eligible sex offender pre-sentencing and on a probation case load; and by Department of State Hospitals, prior to release of an eligible sex offender from a DSH institution. A validation study of the Static-99 risk assessment instrument, which is used to score risk of sexual re-offence by California sex offenders, shows that Static-99 assessments are very accurate in predicting sexual re-offence by a diverse California sex offender population. The first 5 years of this study is described in the Journal of Threat Assessment and Management (2014), Vol. 1, No. 2, at pp. 102-117. The California Department of Justice partnered with the SARATSO Committee to do this study. The second five years of the study will be published in 2017. The Static-99 was found to be very accurate in predicting who would reoffend in California, predicting who would commit a new sex offence in about 82% of cases. High risk offenders had a recidivism rate of over 29%, while low risk offenders had a recidivism rate of only 1.6%. The article also describes a California SARATSO inter-rater reliability study which shows that inter-rater reliability in scoring the Static-99R is good.

Example of scoring from static- 2002 sexual assessment tool[14]

CATEGORY I: AGE (Score -3 points to 1 point)

Age at Release. The Basic Principle: The rates of almost all crimes decrease as people age (Hirschi & Gottfredson, 1983; Sampson & Laub, 2003). Sexual offending does not appear to be an exception. Most studies have found that older sexual offenders are lower risk to reoffend than younger sexual offenders (Barbaree & Blanchard, 2008; Hanson, 2002,). 2006 Research has found that the original Static-2002 did not fully account for age at release and that a new age weighting had greater predictive accuracy (Thornton, Helmus, & Hanson, 2009). With the new age weighting (used in this item), age at release no longer significantly contributed to the prediction of sexual recidivism.

Information Required to Score This Item. To complete this item the evaluator should confirm the offender's birth date from official records if possible or have other knowledge of the offender's age through collateral report or offender self-report. The Basic Rule: Score -3 to 1 point depending on the age of the offender, referencing the table below.

AGE		SCORE
18 to 34.9	=	1
35 to 39.9	=	0
40 to 59.9	=	-1
60 or older	=	-3

Under certain conditions, such as anticipated release from custody, the evaluator may be interested in an estimate of the offender's risk at some specific time in the future. Static-2002R may be scored months before the offender's release to the community and the offender may

---

[14]Static-2002R: Revised Age Weights Helmus, L., Babchishin, K. M., Thornton, D., & Hanson, R. K. (2009-10-08) Replaces Age Item in Official Static-2002 Coding Rules (Phenix, Doren, Helmus,Hanson, & Thornton, 2009.

advance an age scoring category by the time he is released. For assessing risk in the future consider what his age will be on the date of release. In this case, you calculate risk based upon age at exposure to risk. Sometimes the offender's release date may be uncertain. For example, he may be eligible for parole but does not qualify for release due to an inadequate release plan. In these cases it may be appropriate to use some form of conditional wording indicating how his risk assessment would change with a delayed release date.

STATIC-2002R CODING

ITEMS

1. Age at Release

$$
\begin{array}{lcl}
18 \text{ to } 34.9 & = & 1 \\
35 \text{ to } 39.9 & = & 0 \\
40 \text{ to } 59.9 & = & -1 \\
60 \text{ or older} & = & -3
\end{array}
$$

PERSISTENCE OF SEXUAL OFFENDING

2. Prior Sentencing Occasions for Sexual Offences.

No prior sentencing dates for sexual offences	= 0
1 Sentencing Occasions	= 1
2, 3 Sentencing Occasions	= 2
4 or more Sentencing Occasions	= 3

3. Any Juvenile Arrest for a Sexual Offence and Convicted as an Adult for a Separate Sexual Offence

   No arrest for a sexual offence prior to age 18 = 0
   Arrest prior to age 18 and conviction after age 18 = 1

4. Rate of Sexual Offending.
   Less than one sentencing occasion every 15 years = 0
   One or more sentencing occasions every 15 years = 1
   Persistence Raw Score (subtotal of Sexual Offending)

Rate of Sexual Offending		Label
0	=	0
1	=	1
2, 3	=	2
4, 5	=	3

DEVIANT SEXUAL INTERESTS

5. Any Sentencing Occasion For Non-contact Sex Offences:

$$No = 0$$
$$Yes = 1$$

6. Any Male Victim:

$$No = 0$$
$$Yes = 1$$

7. Young, Unrelated Victims
Does not have two or more victims age less than 12 years, one of them unrelated = 0
Does have two or more victims age less than 12 years, one must be unrelated = 1.

RELATIONSHIP TO VICTIMS

8. Any Unrelated Victim:

$$No = 0$$
$$Yes = 1$$

9. Any Stranger Victim:

$$No = 0$$
$$Yes = 1$$

GENERAL CRIMINALITY

10. Any Prior Involvement with the Criminal Justice System

$$No = 0$$
$$Yes = 1$$

11. Prior Sentencing Occasions For Anything:
    0-2 prior sentencing occasions for anything = 0
    3-13 prior sentencing occasions = 1
    14 or more prior sentencing occasions = 2

12. Any Community Supervision Violation:

$$No = 0$$
$$Yes = 1$$

13. Years Free Prior to Index Sex Offence:

    • More than 36 months free prior to committing the sexual
      offence that resulted in the index conviction AND more
      than 48 months free prior to index conviction = 0
    • Less than 36 months free prior to committing the sexual
      offence that resulted in the index conviction OR less than
      48 months free prior to conviction for index sex offence =
      1

14. Any Prior Non-sexual Violence Sentencing Occasion:

$$No = 0$$
$$Yes = 1$$

General Criminality raw score (subtotal General Criminality items)

Arrest/charges/Convictions		label
0	=	0
1, 2	=	1
3, 4	=	2
5, 6	=	3

General Criminality SUBSCORE
TOTAL -3 to 12
TRANSLATING   STATIC-2002R   SCORE   INTO   RISK
CATEGORIES

Score	Label for Risk Category
-3 through 2 =	Low
3, 4 =	Low-Moderate
5, 6 =	Moderate
7, 8 =	Moderate-High
9 plus =	High

Tool 2 : STABLE-2007/ACUTE-2007 (SARATSO Instrument for Dynamic Risk Assessment)

The Stable-2007/Acute-2007 was adopted by the SARATSO Committee in September 2013 as the new dynamic risk assessment instrument. The Stable-2007/Acute-2007 is scored by certified treatment providers working with sex offenders on probation or parole. (Pen. Code, sec. 290.09.) These tools measure dynamic (changing) risk factors which are empirically related to the risk of re-offence, and are evidence-based risk assessment tools. Dynamic risk assessment supplements the static risk assessment now done in California using the Static-99R, and gives a better picture of the overall risk of re-offence presented by sex offenders on supervision. The STABLE is predictive of the risk of future sexual offending.

Tool 3 The RRASOR (Hanson, 1997)

This is an actuarial instrument designed to measure risk of sexual recidivism. Scores range from 0 to 6, with a higher score indicating greater risk of sexual recidivism. It has four items: (1) prior sexual offences, (2) any unrelated victims, (3) any male victims, and (4) offender is less than 25 years of age. For the current study, the items of Static-99 were used to compute the RRASOR. The coding rules for the items of the RRASOR and Static-99 are identical with the exception of prior sexual offences. Specifically, unlike the RRA-SOR, the coding rules of Static-99 do not count pseudo-recidivism as prior sexual offences. Pseudo-recidivism is estimated to affect approximately 5% of offenders (Phenix, Doren, Helmus, Hanson, & Thornton, 2009), and hence, the difference between using the item scoring of Static-99 rather than RRASOR is expected to be minimal. In the development study, the RRASOR differentiated sexual recidivists from nonrecidivists with an Area Under the Curve (AUC) of .71 (Hanson, 1997). A recent meta-analysis conducted by Hanson and Morton-Bourgon (2009) found that the RRASOR showed similar, although slightly smaller effects, when averaged across 34 diverse follow-up studies (weighted mean d = 0.60, 95% CI = 0.54 to 0.65, N = 11,031, k = 34; which translates to an AUC of .66, 95% CI = .65 to .68).

Tool 4 :MnSOST-R

This item describes the development, reliability, and validity of the Minnesota Sex Offender Screening Tool – Revised (MnSOST-R), as well as recommended risk levels and cut. scores. Variables from multiple dimensions, both static and dynamic, were reviewed for inclusion in the MnSOST-R. Final items were selected and scored empirically based on clearly defined criteria. The resulting 16 items that comprise the MnSOST-R maximize the positive predictive power of the tool, and perform significantly better than previous versions of the MnSOST. This newest version correlate achieves impressive hit rates with rapists and extra-familial sex offenders, the population for which the instrument was developed. Very high true positive rates were achieved depending on the selected cut score.

MnSOST-R Item Scores

Number Item Description Item Score
Static/Historical Items

1. Number of sex/sex-related convictions (including current conviction):

$$\begin{array}{rcl} \text{One} & = & 0 \\ \text{Two or more} & = & +2 \end{array}$$

2. Length of sexual offending history:

$$\begin{array}{rcl} \text{Less than one year} & = & 1 \\ \text{One to six years} & = & +3 \\ \text{More than six years} & = & 0 \end{array}$$

3. Was the offender under any form of supervision when they committed any sex offence for which they were eventually charged or convicted?

$$\begin{array}{rcl} \text{No} & = & 0 \\ \text{Yes} & = & +2 \end{array}$$

4. Was any sex offence (charged or convicted) committed in a public place?

$$\begin{array}{rcl} \text{No} & = & 0 \\ \text{Yes} & = & +2 \end{array}$$

5. Was force or the threat of force ever used to achieve compliance in any sex offence (charged or convicted)?

$$\begin{array}{rcl} \text{No force in any offence} & = & \text{-3} \\ \text{Force present in at} & = & 0 \\ \text{least one offence} & & \end{array}$$

6. Has any sex offence (charged or convicted) involved multiple acts on a single victim within any single contact event?

$$No = -1$$
$$Yes = +1$$

7. Number of different age groups victimized across all sex/sex-related offences (charged or convicted):
   [] Age group of victims: (check all that apply)
   [] Age 6 or younger
   [] Age 7 to 12 years
   [] Age 13 to 15 years and the offender is more than five years older than the victim
   [] Age 16 or older
   No age group or only one age group checked 0
   Two or more age groups checked +3

8. Offended against a 13- to 15-year-old victim and the offender was more than five years older than the victim at the time of the offence (charged or convicted):

$$No = 0$$
$$Yes = +2$$

9. Was the victim a stranger in any sex/sex-related offence (charged or Convicted)?

$$No victims were strangers = -1$$
$$At least one victim was a stranger = +3$$
$$Uncertain due to = 0$$
missing information

10. Is there evidence of adolescent antisocial behaviour in the file?

$$No indication = -1$$
$$Some relatively iso- = 0$$
lated antisocial acts
$$Persistent, repetitive pattern = +2$$

11. Pattern of substantial drug or alcohol abuse (12 months prior to arrest for instant offence or revocation):

$$
\begin{aligned}
\text{No} &= \text{-1} \\
\text{Yes} &= \text{+1}
\end{aligned}
$$

12. Employment history (12 months prior to arrest for instant offence):

Stable employment for one year or longer	=	-2
Homemaker, retired, full-time student in good standing, or officially disabled	=	-2
Part-time, seasonal, unstable employment	=	0
Unemployed or significant history of unemployment	=	+1
File contains no information	=	0

Dynamic/Institutional Items

13. Discipline history while incarcerated (does not include discipline for failure to follow treatment directives):

No major discipline reports or infractions	=	0
One or more major discipline reports	=	+1

14. Chemical dependency treatment while incarcerated:

No treatment recommended / Not = 0
enough time / No opportunity

Treatment recommended and = -2
successfully completed or in
program at time of release

Treatment recommended = +1
but offender refused,
quit, or did not pursue

Treatment recommended = +4
but terminated by staff

15. Sex offender treatment history while incarcerated:

No treatment recommended / Not = 0
enough time / No opportunity

Treatment recommended and = -1
successfully completed or in
program at time of release

Treatment recommended = 0
but offender refused,
quit, or did not pursue

Treatment recom- = +3
mended but terminated

16. Age of offender at time of release:

Age 30 or younger = 1
Age 31 or older = - 1

Presumptive Risk Levels and Associated MnSOST-R Cut Scores
Presumptive Risk Level MnSOST-R Score

Low = 3 and below
Moderate = 4 to 7
high = 8 and above
Refer to county attorney = 13 and above.

T00l 5 The Sex Offender Risk Appraisal Guide (SORAG)

This is a 14-item actuarial scale designed to predict violent, including hands-on, sexual recidivism among men who have committed at least one previous hands-on sexual offence.

Table 1. Items and basic coding rules of the SORAG (Quinsey et al., 2006)

Item	Risk Factor	Coding Rule
1	Lived with both biological parents to age 16 (except for death of parent) – Score *no* if offender did not live continuously with both biological parents until age 16, except if one or both parents died. In case of parent death, score as for yes	Yes = -2 No = +3
2	Elementary school maladjustment (up to and including Grade 8)	No problems = -1 Slight or moderate problems = +2 Severe problems = +5
3	History of alcohol problems – Allot one point for each of the following: alcohol abuse in biological parent, teenage alcohol problem, adult alcohol problem, alcohol involved in a prior offence, alcohol involved in the index offence	0 = -1 1 or 2 = 0 3 = +1 4 or 5 = +2
4	Marital status (or lived common law in the same home for at least 6 months) – At time of index offence	Ever married = -2 Never married = +1

5	Criminal history score for convictions and charges for nonviolent offences prior to the index offence (from the Cormier-Lang system)	Score 0 = -2 Score 1 or 2 = 0 Score of 3 or above = +3
6	Criminal history score for convictions and charges for violent offences prior to the index offence (from the Cormier-Lang system)	Score 0 = -1 Score 2 = 0 Score of 3 or above = +6
7	Number of convictions for previous sexual offences (pertains to convictions for sexual offences that ocurred prior to the index offence) – Count any offences known to be sexual, including, for example, indecent exposure 0 = -1	1 or 2 = +1 [3] 3 = +5
8	History of sex offences against girls under age 14 only (includes index offence; if offender was less than 5 years older than victim, always score +4)	Yes = 0 No = +4
9	Failure on prior conditional releases (includes parole violation or revocation; breach of or failure to comply with recognizance or probation; bail violation; and any new charges, including the index offence, while on a conditional release)	No = 0 Yes = +3

10	Age at index offence (at most recent birthday)	$\geq 39 = -5$ $34\text{-}38 = -2$ $28\text{-}33 = -1$ $27 = 0$ $18\text{-}26 = +2$
11	Meets DSM-III criteria for any personality disorder	No $= -2$ Yes $= +3$
12	Meets DSM-III criteria for schizophrenia	No $= -3$ Yes $= +1$
13	Phallometric test results	*All* indicate nondeviant sexual preferences $= -1$ *Any* test indicates deviant sexual preferences $= +1$
14	Hare Psychopathy Checklist – Revised score (PCL-R; Hare, 1991)	[2] $4 = -5$ $5\text{-}9 = -3$ $10\text{-}14 = -1$ $15\text{-}24 = 0$ $25\text{-}34 = +4$ [3] $35 = +12$

Tool 6 : Sexual Violence Risk-20

Douglas R. Boer, PhD, Stephen D. Hart, PhD, P. Randall Kropp, PhD, and Christopher D. Webster, PhD. The SVR-20 is a 20-item checklist of risk factors for sexual violence that were identified by a review of the literature on sex offenders; factors assessed include psychosocial adjustment, history of sexual offences, and future plans.

- Specifies which risk factors should be assessed and how the risk assessment should be conducted.

- The list of risk factors is empirically related to future sexual violence, useful in making decisions about the management of sex offenders, nondiscriminatory, and comprehensive without being redundant.

- Appropriate for use in cases in which an individual has committed, or is alleged to have committed, an act of sexual violence, including pretrial release decisions, presentence assistance to judges, development of treatment programs at correctional intake, prior to discharge to assist in post-release management, custody/access assessment, determination of need for a community warning, quality assurance or critical incident reviews, and education and training.

The SVR-20 is probably the most commonly used SPJ instrument for the risk assessment of sexual offenders5. Boer and Hart (2009) stated that 'the SVR-20 has been evaluated by a variety of researchers in a variety of sites and is the best-validated SPJ for the risk assessment of sexual offenders' (p. 346). The SVR-20 is a structured clinical guideline for the assessment of risk for sexual violence in adult sex offenders designed by a group of forensic scientists who had already done research on SPJ for other offender subgroups. The SVR-20 was developed from a thorough research of the empirical literature and using the clinical expertise of a number of clinicians. In order to identify relevant risk factors, there were three general principles: The risk factor has to be (a) supported by scientific research, (b) consistent with theory and professional recommendations, and (c) legally acceptable, that is, consistent with human and civil rights. The SVR-20 consists of 20 items, divided into three domains (see Table 1). The authors developed a manual and worksheets, in order to support a reliable application of the instrument. The administration of the SVR-20 can be divided into three general steps of the risk assessment process: First, the 20 items, as well as any additional case-specific risk factors have to be coded by an experienced forensic clinician. The items are rated using a 3-point ordinal rating scale as definitely present, possibly or partially present, or absent. In the second step, the evaluator indicates for each present risk factor whether there has been any recent change in the status of that factor within a flexible time

Domain	Risk Factor
Psychological	1. Sexual deviance
Adjustment	2. Victim of child abuse
	3. Psychopathy
	4. Major mental illness
	5. Substance use problems
	6. Suicidal/homicidal ideation
	7. Relationship problems
	8. Employment problems
	9. Past nonsexual violent offences
	10. Past nonviolent offences
	11. Past supervision failure
History of	12. High density
Sexual offences	13. Multiple types
	14. Physical harm
	15. Weapons/threats
	16. Escalation in frequency or severity
	17. Extreme minimisation/denial
	18. Attitudes that support or condone
Future Plans	19. Lacks realistic plans
	20. Negative attitude toward intervention

Table 2: The Risk Factors and Items of the Sexual Violence Risk-20 (SVR-20; Boer et al., 1997)

frame. Changes are also coded on a 3-point ordinal rating scale in terms of exacerbation, no change, or amelioration. In the final step, users make a final judgement about the risk of future violence using again a 3-point ordinal rating scale. The final risk judgement should be rated as low, moderate, or high which is also indicating the degree of intervention required in this individual case. For example, a final judgement of high risk would indicate an urgent need to develop and start a comprehensive risk management plan for the individual which would feature more resources than in case of moderate or low risk.

Tool 7 : STABLE-2007

This is a structured scale for identifying factors useful in the treatment and community supervision of sexual offenders. This presentation will offer an overview of the development of STABLE- 2007, and go on to review the research concerning its reliability and validity. Evidence concerning rating reliability has been mixed. The rater reliability for the total scores is high (ICC > .90) among trained evaluators working on the same team. However, poor rater reliability has been observed when (continued on next page) raters lacked opportunity for common training and calibration. Even among well-calibrated teams, exact agreement is rare. The standard error of measurement is about 1.5 points, meaning that raters are expected to be within about 4.0 points 80% of the time. Evaluators need to consider this measurement error when interpreting individual results, and base their conclusions on the plausible range of "true" scores. Of the 13 content areas assessed by STABLE-2007, 9 can be considered empirically-supported risk indicators for sexual recidivism, as defined by $d > .15$ when aggregated across 3+ studies (Mann, Hanson & Thornton, 2010). Three of the STABLE-2007 content areas are promising (aggregated $d > .15$ based on 1 or 2 studies). One factor, Social Rejection/Loneliness, was a significant predictor in the STABLE-2007 development study but not in other research. In addition, there is one factor that was non-significant in the STABLE-2000/2007 development study, but should now be considered empirically supported, namely Child Molesters Attitudes ($d = .46$, based on 5 studies, n = 781 child molesters). To date, there have been 3 independent replications of STABLE-2000 and 1 independent replication of STABLE-2007. Overall, these studies have found levels of predictive accuracy similar to those observed in the validation study. However, a re-analysis of the original validation study indicates that STABLE- 2007 did not work well for sexual offenders of Aboriginal heritage (AUC = .58 vs. .71 for non-Aboriginal). Finally, recent research on the construct validity of the STABLE-2000 and STABLE-2007 items (e.g., Nunes & Babchishin, 2012), with a particular focus on the interpretation of Emotional Identification with

Stable 2007	Assessment ID: DOCH-STABLE-84			
	Assessed: 7/11/2008			
	Name: John Doe (SID ♯: A0980003)			
DOB: 06/04/2057	**Sentence Date** 10/09/1982		**Unit:** Intake Svc. Center	
Gender: Male	**offence Type:**	Indecent exposure **County:** Hawaii		
Assessor: Susan@cyzap		**Assessment Staus:** Pre-trial		
Purpose: Initial Assessment		**Disposition:** Pre-trial		
Case number: 733272				
Scoring Form			**Score**	
1.	**Significant Social Influences**			
	A. Number of positive influences (max 8)			
	B. Number of negative influences (max 7)		2	
	C. Number of neutral influences			
	D. Total significant social influences			
2.	**Capacity for stable relationships**			
	A. Ever lived with an intimate partner	No		
	for at least two years?	Yes	2	
	B. Currently living with an intimate partner	with concerns		
3.	**Emotional identification with children**			
	Any child victims less than 14 years?	No	N/A	
4.	**Hostility toward women**			
	Notes: Extreme hostility		2	
5.	**General social rejection**			
	Notes: anti-social		2	
6.	**Lack of concern for others**		2	
7.	**Impulsive**		2	
8.	**Poor problem solving skills**		2	
9.	**Negative emotionality**		2	
10.	**Sex drive/sex preoccupation**		2	
11.	**Sex as coping**		1	
12.	**Deviant sexual preferences**		0	
13.	**Cooperation with supervisor**		1	
		Total score	20	
		Risk category	High	

Table 3: Interpretive Ranges: $0 - 3 = $ Low, $4 - 11 = $ Moderate, $12+$ $=$ High

Children (McPhail, Hermann et al., 2011) will be presented.

CHAPTER 9
THE USE OF LOGIC AND ARGUMENTATION IN THERAPY OF SEX OFFENDERS

1 Background and orientation

1.1 Some discussion and orientation

This Chapter is intended first for the formal argumentation community. This community develops logics and systems modelling argumentation and dialogues. The community is in search of major applications areas for their models. One such application area, for example, is Law.

The message of this Chapter is that there is another major application area for formal argumentation. There is an international community of sex offender therapist which is well established, well funded, and their therapy methods use (methods which can be modelled by) formal argumentation and logic. This community presents a natural application area for formal argumentation. We thus describe in this Chapter how the sex offender therapists work, to give the formal argumentation researcher a view of this application area. What is especially important about this application area is that in order to model it and learn from it, the formal argumentation community have to evolve their formal methods and adapt to this new application. Part of this enhancement is to modify and import certain methods from other areas of Logic, for example, from Non-Monotonic logic. The members of the formal argumentation community are not familiar, on average, with other areas of logic, and so we also describe in this Chapter, what we need from neighbouring logics.

This makes this Chapter of interest also to sex offender therapist as well. They may be already familiar with their own practices but the additional logics described will be of interest to them.

There are three independent international communities dealing with reasoning and arguments, namely ((1), (2), and (3) below), (1) and (2) never interacted and have not been interacting until recently and in fact have not even been aware of their common interest, while (2) and (3) were aware of one another. These are:

1. The logic and argumentation community, studying and modelling reasoning and argumentation [5, 6].

2. The sex offender forensic therapy community, using reasoning and argumentation methods in their therapy.

3. The Logic Based Therapy (LBT), see [14, 15, 16].

The integration of logic with clinical practice is not new and already exists in the professional literature, for example in Cohen, E.D. 2022 and others [13, 19, 32, 50], which proposes the integration of logic in cognitive behavioral intervention and Therapy, (LBT). LBT recognises that the need for therapy arises from cognitive distortions in the mind of their patient which in turn causes various maladies. see the LBT list of Fallacies quoted in [10].

From the point of view of Formal logic, the fallacies (and there are maybe 400 of them) can be treated/corrected by pointing out the formal mistakes. This is pure theory . The Therapist cannot just tell the patient he/she are logically/mathematically wrong. The therapist needs to use different more human sensitive approach to help the patient. These Cognitive Distortions/Fallacious reasoning may be different in the LBT patient case as compared with the sex offender patient case and the respective therapy approach may be different. The present article recognises the achievement and wealth of experience of LBT. We can see that LBT deals with Cognitive Distortions and we refer the readers from the formal logic communities and the sex offender communities to look at for example book [13] and paper [12] but seeks to focus on the treatment of sex offenders.

The challenges facing this Chapter in its attempt to interest the formal argumentation community in the sex offender therapy area are fourfold:

1. There is no formal logical model for universal reasoning distortions.[1]

---

[1]This footnote is intended to clarify the our use of the concept Universal Distortion in

Having become aware of this phenomena we need to study it in general logic, the way we study other reasoning mechanisms like Fallacies, Fake news, bias, attack, support, etc., etc.

2. We need to identify the kind of distortions sex offenders practice and compare them with other kind of "distorted" reasoning.

3. We need to collect data from sex offender therapists to figure out how they try to correct/address sex offender distortion. This is important to General Logic. (If a good reasoner puts forward an argument which is a fallacy, our response is to point out that it is a fallacy, but doing this to a sex offender may not be the correct response.

4. We need to give a simplified exposition/survey of distortions and therapy in such a way that we get the formal argumentation and the sex offender therapy communities to work together.

Reasoning. By a universal reasoning distortion (URD) we mean here a complete distortion of a reasoning system across all of its component. We do not regard a local error of reasoning as a system distortion. For example a reasoner using one or two fallacies such as generalisation or exaggeration is not distorted in his thinking. He is just making an error. (This error may lead the patient to depression but still it is not a Universal distortion). By comparison, a sex offender who would dismiss any argument and commit any fallacy in order to make himself look good and reject any criticism, is universally distorted in his thinking. More examples of a Universal distortion is a person who is drunk. Such a person is distorted in his functionality across the board. His ability to think straight, his reaction time, his digestion, his balance, etc.

Sex offenders have typical thinking distortions for their group.

There are egosyntonic thinking distortions (acceptable to the ego) and the individual is sure that these perceptions are correct and sees no need to change them. Thus, for example, a person with pedophilic sexual deviation often believes that his thoughts are correct and true. For example, the child enjoys sexual contact with an adult. He knows what he wants, and he agrees. This is an example of a universal distortion of a sex offender who will reject any argument, facts, assumptions, etc which threatens his egosyntonic belief. By comparison in the egodystonic case the sex offender has no reasoning distortion. He knows he is wrong, and his mind is open to listening to therapists.

One of the roles of the sex offender therapy is to change thinking distortions from egosyntonic (ego-acceptable) to ego dystonic (ego alien). Following treatment, which will focus not only on trying to change thought patterns, but also with the help of the patient will try to find the source of the this universal distortion and what they serve for him. After the perceptions become the egodystonic, the individual will be able to continue the process of change and correction.

The structure of our Chapter for the case of sex offenders is as explained above:

1. a quick description of correct reasoning model;

2. a description relative to the model in (1) of the sex offender distortion;

3. description of therapy in terms of (1) and (2).

To sum up our approach, we start with a model/logic without distortion, then we identify a variety of psychological defense mechanisms and thought distortions (note that distortions occur regardless of our cognitive ability; some very intelligent people use thought distortions to justify problematic behavior). Then we investigate therapy to be used.

In the case of the sex offender, challenging his thought distortions helps change cognitive habits and reduce dangers.

Remark 1.1 (Methodology). *Before we continue, we want to give an explanatory warning to the reader. We stated that our aim is to attract and explain to the argumentation community why it is in the interest of all parties (argumentation modelling and sex therapy practitioners) to talk to each other. So this Chapter is not an argumentation research paper in the traditional sense, but more like a description of what is going on in sex offender therapy using intuitive logic in the practiced therapy of sex offenders. It is more like a scientific survey, such as may be published in maybe a geological journal, with hints of how various parts can be modelled.*

It is helpful to compare this Chapter with [26].

[26] was more ambitious. We identified the idea of "distortion" from the sex offender therapy practice and then introduced "distortion" into argumentation and wrote a full technical paper about distortion in argumentation. So in the minds of the members of the argumentation community, [26] is just another extension of argumentation, along with similar extension papers on probabilistic argumentation, bipolar argumentation, numerical argumentation, etc.

The fact that [26] contains appendices about sex offender therapy unfortunately made only slight impact.

We hope the present "survey like" Chapter will have more success.

Most crimes defined as sex offences are committed by males and so this Chapter, for convenience, will refer to the perpetrators in the masculine. Also, the corrective path to which we refer here is the one used in Israel, [20, 58].

Dealing with sex offenders is a high profile area of activity in any society. After a sex offender is given a prison sentence, remission for good behaviour depends upon the convict expresses motivation to join therapy whereupon he is offered the opportunity to join a therapy group in prison.

Of course it is not surprising that many sex offenders join a therapy groups. What is more surprising, is that the therapists use logic and argumentation to treat these offenders. The community is not explicitly aware of this connection with the logic and argumentation community. Group therapy focuses on changing behaviour pattern and cognitive schema. current article focus is only cognitive schema and cognitive distortions.

They regard the sex offender as suffering from reasoning distortions (caused possibly by physical drives) and proceed to use intuitive logic and argumentation methods to try to correct such distortions and reduce the temptation to reoffend. Once we, the authors, realised this, we were motivated to write the current Chapter and other related papers and study reasoning distortions in general. When you think about it, it is of great value to the logic community to have a very high profile medical community using logic and argumentation. If the argumentation community could observe and model case studies from such practice, this could immensely benefit both communities, as well as society in general. We envisage the argumentation community helping to improve such therapy. Currently 30% of participants show significant improvement and perhaps this success rate can be improved, [54].

So the challenge for this Chapter is that it needs to be written in such a way that a non-expert reader can understand both how sex offender therapy works and how it can be viewed, using logic. If we succeed in addressing this presentation challenge we will make both communities aware of each other. Fortunately this is possible since the non-expert is aware both of sex offenders as well as common sense every day reasoning and so we can write our Chapter using minimal technical logical structuring. The more technical reader can consult Chapter 1.

2 Introduction and overview of therapy methods

This Section is intended for the formal argumentation reader who seeks important applications areas for his formal methods.. We need it to provide sex offender data and context for the argumentation reader, following our strategy as discussed in Remark 1.1.

2.1 The theoretical basis of group treatment for persons with sexual offense histories

Most therapy groups for persons with sexual offense histories are integrative, meaning that they combine theory and interventions from various approaches. These typically include some combination of psychoeducation, Logic Based Therapy (LBT), the Good Lives Model (GLM) [46, 55], and motivational interviewing (MI) [46]. Notably, each of these approaches has at least some empirical support demonstrating their efficacy or effectiveness with this population when used on their own and delivered in a group format [4] or in combination with one another [47]. Group therapy goals are divided into outcome goals and process goals [48]. Outcome goals are behavioral changes that individuals seek to achieve by participating in group therapy, and in this case, it refers to reduction of sexual recidivism by understanding cognitive distortions and developing coping skills with stressful thoughts and situations that might lead to sexual offending. Process goals are those that relate to the process of understanding personal concerns and relating to other individuals during a group session [36]. Therefore, our work is also influenced by psychodynamic group psychotherapy. Group leaders may consider that open and closed groups for persons with sexual offense histories have a clear outcome goal of the reduction of sexual recidivism, and that process goals, such as increased openness between group members, learning to confront members assertively and not aggressively, or focusing upon interpersonal relations between group members, are more appropriate for a psychodynamic group therapy in the community. However, a study among incarcerated adult males found that they focus primarily on interpersonal interactions with staff and other inmates [48]. Therefore, the interpersonal model of group psychotherapy which emphasizes the critical nature of peer interactions and consequent dynamic interpersonal learning

[41] is highly relevant for group treatment of persons with sexual offense histories.

It should be noted that most theories and research concerning sex offender treatment focus on Anglo-American populations, and it is essential for therapists specializing in sex offender treatment to acquire cross-cultural competence. A clinician's lack of understanding can result not only in cultural insensitivity but also in the stripping away of the offender's positive elements of identity and self-esteem [11]. In this context, it is important to note that treatment in forensic settings relies on various models as one of the major ones is the RNR (Risk-Need-Responsivity) model [1]. This model outlines the basic principles of risk, need, and responsivity to generate effective interventions for persons with offending history with the ultimate goals of improving treatment for these populations and reducing recidivism [2]

The need principle suggests treatment programs should focus on criminogenic needs, meaning, to dynamic risk factors which directly relate to offending behaviour that are amenable to change [61]. A few of these factors are antisocial cognitions, antisocial associates, history of antisocial behaviour, family/marital circumstances and substance use [2]. The responsivity principle provides guidance on treatment provision and holds that it should be consistent to the individual's culture and learning style [40]. Furthermore, consideration of cultural factors can influence the individual's engagement in rehabilitation from building therapeutic relationship, to executing therapeutic strategies, and implementing appropriate intervention programs [60].

Our data is based on the accumulated knowledge of all sex offenders in Israel, which shares knowledge and experience systematically and on a regular basis and relies on internal protocols (written in Hebrew) [49]. One of the authors of the article is a therapist with rich experience who has worked and is working with many other therapists and based on the accumulated knowledge of many therapists he has built a theoretical model of thought distortions and possible reactions to these distortions based on the experience and knowledge of dozens of therapists of sex offenders.

The sex offender therapy practice uses therapy in groups [57]. It allows for two types of dedicated therapy groups for sex offenders and examines the pros and cons of each method of therapy by addressing different aspects of each and examining its logical and practical characteristics. Different groups have different strength and can address different types of distortion.

We focus on two types of groups: "closed groups" (groups with fixed membership) and "railway groups" (open groups with a non-fixed, variable membership).

In Israeli jails, the treatment groups usually include about 14 adult sex offenders which is an enormous number and there is definitely a price to pay for dealing with such a large group.

In the international community the optimal number of participants in groups is no more than 8.

2.2 Nature of therapy groups

Therapists have frequently asked themselves what are the advantages and disadvantages of each type of group and what is the preferred method. To answer this question first of all we explain what a dedicated group is for sex offenders.

Remark 2.1. *We list some background material needed to explain this subsection.*

1. *The subsection is written from the point of view of the sex offender therapist. We include it in this Chapter to provide context for the argumentation reader.*

2. *We ask the argumentation reader to recall papers by Bench-Capon about argumentation with audience, such as [29]. The audience in our case are the therapy groups.*

3. *There is a difference between audience as perceived in [29] and audience as members of a therapy group.*

   (a) *In therapy groups the members of the group all have the same kind of distortion. However, any individual in the group x can see the distortion in other member's reasoning y ≠ x reasoning even though he cannot see the same distortion in his own x reasoning. This allows the therapist to get help in pointing out*

>   *distortion in x because his colleagues y ≠ x also agree with the therapist.*
>
> *(b) The arguments between the therapist and any group member x are not pure reasoning arguments as in [29]. They also intended to extract information and character understanding of x. So some use of fallacies is present in the debate. [29] does not consider the effect of fallacies.*

We are talking about a therapy group with cognitive behavioural distortion [39]. The patients are treated by this method because other methods were not found to be effective enough [44]. Probably the reasons are:

1. Lack of motivation

2. Sex offenders perceive themselves as victims. they often talk incessantly about their victimisation.

3. Therapists also believe that in dynamic group therapy they get insight and this is important, but often not enough. Insight is only a part of the therapeutic process, but at this point we can say that the sex offender therapy has a lot of dynamic elements, which we shall discuss later on.

Why group therapy?

The simple answer is that statistically it is the most effective treatment [37]. The logic behind it is that sex offenders feel exceptional. They know they are deviants and in group therapy the feeling that they are not alone and that there are other people with similar problems often allows them to become more open and receptive ("openness"). The group helps mitigate the embarrassment and helps the therapist to confront patients. Note that sex offenders are much more receptive than other people and so if the group works well, things that patients say to each other are sometimes more important and have more influence than the words of the therapist.

The treatment, based on a model of Bengis (1986) [7] is called "relapse prevention". The goal of the treatment is to prevent future attacks on other victims.

Main goals:

A. raising the level of awareness of the sex offender about the range of behaviour options available to him.

B. developing coping skills and strategies of self-control.

C. creating a sense of personal control in the sex offender

The Therapist response is not only to confront the sex offender and correct his logic, but mainly to show him that his world view is not disqualified or judged, and to help him manage himself better and not get involved in trouble as before. The responses are not judgmental, but motivated out of concern and expression of emotion.

Basic assumptions

A. There is no cure for the disorder, (therapists are aware they are not magicians and do not know how to effect a complete cure for such things as perverted fantasies. Neither can they completely heal mental disorders) however we can teach strategies to avoid a repetition of unacceptable behaviour.

B. When a perverse thought appears, the individual can still choose to avoid violence.

C. Model SUD = seemingly unimportant decisions. Sometimes even taking seemingly unimportant decisions may lead to an offence.

For example, a sex offender with a paedophilic disorder who works as a guard is given a temporary posting at a kindergarten. The seemingly unimportant decision to accept the job may lead to abusive behaviour.

D. The offence is planned and not impulsive [59]

E. The offence cycle has many steps in it. The forensic therapy community distinguishes a stage called "lapse", when the sex offender puts himself in a situation where he might offend. If he indeed does offend then he is said to "relapse".

For example, if a pedophile starts fantasising about child and plans an offence-, it is still a "lapse" and he can decide not to offend; in comparison, a "relapse" is a situation in which there is a new offence.

We make the assumption that Not every "lapse" is a "relapse" [43]

Thus even if the patient made the wrong decision, and got himself into a dangerous situation (lapse), he can recognise it and withdraw, and thereafter can learn to avoid it, and not say to himself, "I stumbled, what can I do?"

Not all therapists agree with this form of therapy but in the opinion of this Chapter, it is believed that simple language is best and without the expectation of drastic and quick changes in emotional ability of the patient, bearing in mind where these people came from and whence they return. We consider it more important that the therapeutic content will survive in day-to-day life and situations. Also, we do not consider we have a mandate to change the culture of the person and have no right to tell the patient whether or not to get engaged, whether or not, to marry several women or one.

The therapists involved in this study believe they are allowed to intervene only if they see a direct correlation with the offender's approaches to future possible offences.

The character of groups

"Closed Groups" means that you cannot get new patients to join during the group process and the group therapy therefore has a clear beginning, middle and end. Open (railway) groups are groups where patients can join at different stages. The generally accepted practice is that the group has two therapists: a man and a woman. It is good for modelling and shows how two can communicate, and can even disagree but can still respect each other without using stereotyped behaviour. The underlying idea of this approach is to reduce extremes of disagreement. It should be pointed out that research on the juvenile probation service found no difference in the success of the group according to the gender of the practitioner [30, 35].

2.3 Advantages/disadvantages of choice of groups

After our brief review of the nature of the groups and the basic assumptions, let us examine the advantages and disadvantages of open and closed groups.

We will divide this as follows:

1. The nature of the group,

2. Planning

3. Sorting of candidates

4. Therapist

5. Integration of new participants

6. Completion of the therapy.

1. The nature of the group. A closed group is a set with linear characteristics, where the group members are supposed gradually to change, reach enlightenment, to gain knowledge (although this word is not good enough to explain the process because it becomes experiential work, requires emotional and mental involvement, etc.). The patients should identify major patterns within their personalities and learn new ways to conduct themselves. This is circular work. The group works on a particular subject and only after all the members finish working on a specific theme are they allowed to move to the next item.

An open group however, is spiral and the group moves inconsistently. The progress often moves backwards and forwards through the beginning, middle and the end. This method allows us to examine the same issue in different periods and to attack it at different times. The patient is able to do so in a different developmental stage of the group and of himself. While a closed group is relatively regular, the open group is characterised by much lability. In order to understand the movement and development of the open group we must remember the "moods" of Bion theory [8].

The open group "mood" swings could be dependent, aggressive, avoiding observation, watching, rational, emotional, controlled, uncontrolled, shrinking, and expanding. Sometimes the movement of the group resembles a sort of tango, where the progress is three steps forward and two steps to the side, but it has no distinct development phases. In the open group the therapist must hold all the required information and all the time must try to combine all the puzzle pieces together and pick up on necessary issues [33, 34].

2. Planning. Due to the nature of the closed group the therapists in the planning stage can construct a series of gradual and consistently logical contents and to stick to them, adjusting is required only when you need to make changes to the content in the working group. Before the therapist works in depth on a particular topic it is best to start with less intimidating content, to create links between the participants and gradually to go from easy to hard. For example, in prison we allow patients to work on trust, to describe experiences of their lives. The therapist should express empathy, in such a way as to reduce defensiveness and allow the offenders to develop direct reactions of empathy towards the victims. In the case of a closed group we do not have a created situation in which a new patient drops in and might be asked to relate to risk factors before working on the assault cycle and before he figures out the trigger, or manipulation and planning of his act. Such a situation certainly can happen in an open group.

3. Sorting of candidates. In a closed group, sorting and choosing the candidates must be very rigorous as if we make a mistake, nobody else can fill the space until the end of the group process and we cannot allow someone else who needs the treatment to take advantage of it. Regarding a decision about the maximum number of participants in a group we think there IS room for some flexibility. Because of the need to meet demand we have increased the number of participants in our therapeutic groups and have noted that larger groups mean slower progress and a longer time for the group therapy to reach its end The average treatment lasts for almost two years (in the past, with fewer participants, the process was shorter).

In an open group, we can be less meticulous in choosing the candidates or in cases of doubt we can consider allowing a candidate temporarily to join the group. We can and do, tell him that he goes into a two months trial period, and then later on we make the final decision regarding suitability.

Of course, this situation involves difficulties. The patients have problems with frequent changes, which challenges their need for a sense of order and regularity. We should also mention that letting a person into therapy and then removing him could further undermine his self-image, increase a sense of failure (many patients have a very low self-image) and of course it must be noted that for a patient, the retirement from dedicated treatment (whatever the reason) also increases the sexual risk level.

Despite all of that, at least it is better for a patient's self esteem to give him an opportunity. We feel that a therapist sorting candidates for an open group can make a more relaxed assessment and can take more time for important decisions.

The closed group does not have this advantage, even if we set up a preparatory short group prior to the closed group meetings in order to examine who is ready for the process and who is not.

4. Role of the therapist. When dealing with an open group the therapist must be much more alert and make sure that the complicated group dynamics, with people coming in and out and going around in circles in treatment will ensure the proper treatment of all.

In the case of an open group the therapist does not have the luxury of a gradual and systematic progress. A study conducted in 2006 by Dr. Avraham Ofek [52] which explored youth sex-offending groups says: "this railway method (open group) is good for youth and hard for therapists" the difficulties he found being those of regression and repetition. Research indicates that stress on the therapists is much heavier in open groups than in the closed groups. In open groups the therapists have to become acrobats ensuring that all the patients, although they joined the group at different times, will be treated on all the basic issues of a dedicated sex offenders group. Already at this point we can say that one way to ensure that most of the required content has been addressed is to perform a number of assault cycles for every patient (it is difficult to believe how circles seem different at different times).

In our estimation, closed groups are more appropriate for training new therapists at the beginning of their careers, because with the closed group, the new therapist has the opportunity to observe and learn consistently about intervention, and identifying the logic behind the gradual construction of the contents and development of a group. Probably learning from an open group can be confusing and overwhelming. For that reason, a trainee therapist might consider joining the open group during later stages after viewing closed groups.

5. Integration of new participants. About joining new patients in open groups, there is the fear that the veteran (i.e. senior members) would get bored, having to repeat procedures they have already learned. The therapist must be very creative and try to reach the same goal using various techniques and adjusting the techniques to form the group. Yalom claims that, adding a new member successfully, depends partly on timing: there are better and less good times to add members [45]. During a crisis and struggles it is harder to integrate a new member or if he is admitted, the fear is that the energies are directed to the new member which may disturb the flow of conversation about a burning issue. The most convenient time for admitting a new member, according to Yalom, is when the group is not really moving forward.

A social worker named Tamir Ashman adds that the number of joiners is important. According to Ashman [35], it is important to induct two members simultaneously, in order to facilitate the process. He says that when the group is experiencing a period of crisis, it is hard to absorb a new single member but if two or three join, they are able to create leverage and start a fresh viewpoint. Even if the burning issue is forgotten, the process will return to it in another way, with the perspective of the new people. This can create dynamics that allow a more productive viewpoint. Ashman claims that 2 patients joining at the same time is the most optimal number.

In the experience of our forensic expert (the second author – Dr Gadi Rozenberg), he has not been able to find a definite formula definitively to answer the question about what is the optimal number. It depends greatly on the members' personalities and the state of the group.

The expert states that if the group is in crisis, with no trust between members and no significant progress, it is a mistake to insert a new person. He believes that it is necessary to focus on the obstacles and only then, to induct newcomers.

The fear is that the entrance of a new patient will delay the group work and hurt the trust and intimacy which has been created, but certainly it can be said that the integration of a new person at an appropriate time allows "freezing" of the image of the group and gives an opportunity to review the progress. For example, with the joining of new participants all are asked to describe their offences and we can see a difference in description, with the adding of relevant information, etc.

One of the significant advantages of an open group is that it allows for a change of status and position in the group. In a closed group almost every-one establishes a position in the hierarchy of the group and it is very difficult for an individual to change status thereafter. In open groups every patient has more opportunity for a change of status. A 'back-bench' member of the group might very well become a veteran who can give the newer members the benefit of his experience. The very important thing is that the new patients constitute for the veteran a kind of mirror of where they were to where they are now (we cannot count the number of times we have heard a patient say "In the past, we thought the same as you...") and this is certainly a very new experience for both. One sees where he should strive to get and the veterans have a chance to see the progress they have made, which helps improve their self-image. This lets the veterans look "sideways" on themselves.

In the past the therapist had very clear positions on specific situations, for example that new patients are not allowed to join during the process of the departure of an old patient. Today he is not so sure. Life is unpredictable and all kinds of different situations develop. Instead of the introduction of a newcomer during the separation of a veteran being unfair to both, giving neither the requisite attention, he found that the handover can make a sub-stantial contribution to the treatment of all the patients in the group. Letting a person enjoy telling about the process he just finished and giving him the responsibility of introducing the new one to the group. This actually helps to reduce anxieties and fears, while the new member sees that it is possible to finish the treatment and to change himself. Mostly, new patients enter natu-rally into an existing commitment and there are fewer power struggles with the therapists. This is the type of barometer that shows the patient where he wants to go and the others to demonstrate what they have done.

Although it is possible to concentrate on how the nature of the group affects patients, we must also note that for therapists trying to assess whether or not it is time to add a new patient is very stressful. What did we miss? How much time must we devote to each subject? Etc. Also, due to the nature of the group the therapist must decide on the recommendations to shorten or extend the duration of the stay.

6. Completion of the therapy.   Completion will be dealt with after the group has been described.

3   The therapy process

Sections 3 below describes the therapy process . It is intended to familiarize the argumentation reader with the universal logical distortions of the sex offender.

There are two aspects to the process. One is to discover the distortions of the patient and second to try and fix them.

The Therapist relies on many years of accumulated experience, with constant examination and deliberation of what works for years and what does not work. The Therapist community uses brainstorming and learning of professionals that leads to conclusions of which responses help and which do not

For closed groups, Treatment focuses on the emotional, cognitive and behavioural aspects and the steps upon which the Group focusses are as follows

1. **Familarisation**
   Every one gives his name and important facts about himself

2. **"I and the other"**
   = Patients are asked to use drawings of themselves and of significant figures in their lives and the work on drawings allows focussing on the life history of the individual.

   Other methods at this stage are guided visualisation, therapy cards (known as "Anibi")

3. **Empathy or identification of emotions**
   work on the different feelings and emotions of the patient and the others. This can be done using pictures with facial expressions, writing a letter to a victim and a letter from a victim, reading the testimonies of the victims and so on.

4. **List of sex offender's arguments**
   See section 4 where it is discussed in detail

   This list was obtained as a result of accumulated experience of many years of professional therapist who have learnt in practice how to react to sex offenders distortions

5. **Addressing emergency pressures**
   Aggressiveness, assertiveness, passivity. The patients are asked to describe such challenges and how they dealt with them.

6. **Sexual functioning**
   It is imperative that it is conveyed to the patient not only what is prohibited, but also what are the alternatives.

7. **Offence cycle**
   This can be identified as a summary of the therapy progression [42].

   (a) trigger: this is what sets the pattern of behaviour. The trigger need not have any sexual connection.

   (b) feelings and thoughts (cognitive distortions)

   (c) belief-decreasing inhibitors: alcohol, drugs or pornography

   (d) planning

   (e) focus on the offence

   (f) reconstruction of the process.

   Finally the therapist works on how to deal with risk situations, using films and other means of impressing all senses.

8. **The closed groups make several "stations" in the process**
   These stations involve stopping and getting feedback which allow the

patients an opportunity to test themselves while the therapists can use this to assess the progress and identify points to re-inforce and improve.

We now come to the last part:

9. Finishing closed groups

Closed groups have a set time to finish the process.

Treatment can be described as linear, and at the end of it we can look at the path we have taken and then disengage. As a group we wind up the process and go our separate ways.

This is a very exciting process of termination and hope. Professional literature [38, 56]talks about how the patients can enhance the effectiveness of the treatment by constantly reviewing the content they worked on in the group. Therefore each client receives a file containing all of his homework, confessions and other things he worked on. This therapist organised a reunion for one group of graduates from a prison programme and found the renewed contact both exciting and hopeful as well as being very significant for both significant for us and for them. We would definitely recommend and he recommends this as standard practice.

In open group therapy the situation is different. According to Tamir Ashman, one of the disadvantages of an open group is that it does not deal with the group "here and now" and there is no processing of a separation experience because in contrast to closed groups the patients do not say goodbye to the group, but the group says goodbye to the patient.

In a closed group the therapist can say goodbye to all the patients in a winding-up ritual and thereby has the chance not only to review the progress that the participants have made but to take satisfaction from seeing that the patients have more insight and from the anticipation that they can be better, more well-integrated members of society than before they started the treatment.

In railway groups the problem is that there is no such opportunity for a formalised parting nor will any parting take effect with all participants simultaneously. The therapist merely says goodbye to one and then immediately receives a new patient. Sometimes this makes it very difficult for the therapist to gain a sense of achievement and job fulfilment.

10. **Advantages and disadvantages.** Finally, this therapist adds that there advantages and disadvantages to both methods and various solutions have been proposed to reduce the disadvantages of each method, but the therapist in an open group is in a particularly difficult and stressful situation. In a closed group, the therapist handler finishes with patients, then has a break before taking on a new group. However, in an open group the therapist works through the process, seeing patients grow but instead of finishing the treatment, the therapist then has to start work with other patients in an unending flow with no break, no grand finale, but only a continuous conveyor belt of patients.

4   Our model for sex offender arguments

This section is intended for both the argumentation reader and the sex offender therapist reader. It gives some background and examples for common sense reasoning. Common sense reasoning is needed to enhance formal argumentation to enable it to model sex offender reasoning. (Sex offender reasoning suffers from universal distortion, so to model that we need to distort also common sense reasoning.) Since many members of the argumentation community do not know the ins and outs of common sense reasoning /non-monotonic logics, we include this section here. We hope some sex offender therapists will find this Chapter/section interesting.

To understand the schematics of the model, let us describe the principles involved and then consider from Appendix A Example A.1.

Let us begin by putting forward two principles [P1], and [P2]:

[P1] Non-monotonic Common sense reasoning is based on the notion of expectation (not in the probability sense but in the sense of social behavioural and social rules conventions). The basic relation involved is that of:

(*) 'on the basis of A one expects B, following knowledge context Δ of how things are in our world'. This basic relation is denoted by $A \to B$.

To give but a few examples, see [21].

For the argumentation readers, whom we can assume that they know logic, please recall the theory of formal systems of non-monotonic consequence . Please specifically recall the approach where we have a formal database Δ which determines the context of the discussion between two people and any claim C introduced will be added to Δ. So if we want to claim that $A \to B$, then we must show that

(**) $\Delta \cup A$ non-monotonically proves B.

A distortion affecting this claim can be executed by changing Δ to a slightly different Δ^*, causing the result that $\Delta^* \cup A$ no longer non monotonically proves B.

This is not the same as denying B, or attacking the importance or relevance of B. It is an attempt to distort the background information(context) Δ.

The above approach is we hope well understood by the formal logician.

For the non-logician sex offender therapist, we need to explain the above approach in an intuitive way. We can explain the non-monotonic consequence through the notion of Expectation, as it manifests itself in natural language by the use of the word "but".

The connection is as follows: Let Δ describe a day to day intuitive common sense context. Then "$\Delta \cup A$ non-monotonically proves B" corresponds more or less to saying statement (***) below:

(***) In the context described by Δ, the linguistic statement "A but not B" sounds odd.

The rest of this section explains (***) to the non-logician.

First let us give a few examples of Expectation, in ordinary day to-day context, see [21]:

1. When receiving a letter one is expected to reply to the letter.

   John received a letter $\to$ John replied.

2. When a traffic accident occurs an ambulance is expected to arrive.

3. John is offered more money for doing a job →ďŤ John accepts it.

4. John leaves home for the weekend → ďŤ John locks his door.

[P2] The English word "but" can be used to test expectation. Consider the following (XX indicates that the sentence is not acceptable) examples:

5. (a) He liked the food but didn't ask for more.

   (b) He liked the food but asked for more.

6. (a) He married her but did not sleep with her.

   (b) He married her but slept with her.

   (c) He did not sleep with her but married her.

7. (a) John was offered the money but did not take it.

   (b) John was offered the money but took it.

   (c) John did not take the money but was offered it.

8. (a) John is handsome but has a terrible temper.

   (b) John has a terrible temper but is handsome.

9. (a) Jones is very rich but is chronically ill

   (b) Jones is chronically ill but is very rich.

   There are other words which involve expectation:

10. (a) He is fast despite being overweight

    (b) He is tall despite being religious

11. Though she did not know it, she trusted him.

12. (a) He bought a lottery ticket but he didn't win.

    (b) He bought a lottery ticket but he won.

13. Not a day went by but brought us news of yet another calamity.

14. Even Scrooge was not so dreadfully cut up by the sad event, but that he was an excellent man of business on the very day of the funeral.

15. She wanted to make a speech but did not know how to begin.

16. XX She was already engaged but might have accepted him as a lover.

17. She was already engaged or she might have accepted him as a lover.

18. No goals were scored, though it was an exciting game.

19. Although Britain considers itself an advanced country, it has a very old fashioned system of measurement.

20. He borrowed my mower, even though I told him not to.

21. Even if you dislike ancient monuments, Warwick Castle is worth a visit.

22. If he is poor, at least he is honest.

23. XX If he is poor, at least he is ugly.

24. Though well over eighty, he can still do it.

25. Naked as I was, I braved the storm.

26. Whether or not he finds a job, he is getting married.

27. She looks pretty, whatever she wears.

28. No matter how hard I try, I can never catch up with him.

29. (a) Although we never interviewed him we were willing to offer him the position.
   (b) Although we never interviewed him we were not willing to offer him the position.

30. Even if you pay me, I will not like you.

[P3] It is obvious from the above examples that the word "but" and some other key words can be used to test expectation.

So given two statements of ordinary day today life and we want to test whether $A \rightarrow B$ is a common sense acceptable rule, we form the sentence "A but not B". If this sentence is not odd then we conclude that on the basis of A we can expect B.

Let Ω be the set of all acceptable rules in some context (say a context of discussion between a normal person and a sex offender) then Ω is the set of rules which are used in the discussion. Ω is not listed formally because it is infinite and because whenever a candidate rule of the form $A \rightarrow B$ is used it can be tested.

The distortion of a sex offender is that he uses rules that test negative and the challenge of the therapist is to make him see that.

Example 4.1. *This example illustrates further the idea of expectation and its connection with legal reasoning and with the test with the word "but".*

A logician and a linguist are sitting together in a pub drinking beer.

At the next table, two workers are sitting, W1 and W2, drinking beer and talking. W1 keeps on goading W2 about how bad his workmanship is, and what a failure he is not only in his work but in his performance with his wife (who by the way is the sister of W1).

W2 gets more and more agitated.

The logician and the linguist observe the exchange and on the basis of normal human behaviour, they judge that W1 is goading W2 beyond normal endurance and they expect W2 to lash out violently at W1.

The common sense rule here is

- *W1 goading W2 as heard $\rightarrow$ W2 strikes W1*

Consider three scenarios:

(S1) *The linguist and logicians decide to intervene and de-fuse the situation and they invite W1 and W2 to free beer at their table (this scenario shows how people capable of taking action, actually take the action following their expectation).*

(S2) *The linguist and logician do nothing but W2 is able to exercise control and there is no violence.*

The linguist says

W1 goaded W2 beyond endurance but W2 did not resort to violence (this shows how language describe what happened and the word "but" shows there was expectation)

(S3) W2 did strike W1. W2 sued W1 for assault. The judge in court says

There are mitigating circumstances. W1 goaded W2 beyond normal endurance; W2 could not have been reasonably expected to control himself. (This shows how future law sees what happened and the expectation is transformed into mitigating circumstances. See Example 4.2 below for a real life example.)

Example 4.2. *We want to point out an example in real life. Recently there was a trial of a sports doctor (this is W1)who had, over a period of years, been sexually abusing the young female athletes he was treating. At the end of the trial the father (this is W2) of three of the victims leapt across the courtroom and attacked the abuser (this is case (S3)). He was pulled off and subsequently returned to the court and apologised. The judge said that his apology was accepted and that, considering the provocation, she could see no reason to punish him (this is the mitigating circumstances of (S3)) provided he did not act badly in the future. (Actually, you might disagree with her, but that's is another matter).*
See `https://www.youtube.com/watch?v=Bhplg8YCu-M` *and* `https://www.youtube.com/watch?v=tFLEym5aBzA`.

Example 4.3. *This example is for the benefit of those among the sex offender therapists who happen to know a little bit more about logic. We refer to Appendix A where there is a long example, Example A.1.*

Let us revisit the exchange in Example A.1 and make the Ω used in the exchange explicit. Please note the meaning of the logical symbols used

$\wedge$ *stands for and*
$\rightarrow$ *stands for implies (if . . . then)*
$\neg$ *stand for not*

The theory Ω is:

$$\Omega = \{(B' \wedge X'' \rightarrow X'), (B' \rightarrow \neg X'), (B \wedge X' \rightarrow X), (B \rightarrow \neg X)\}$$

And the inference rules are modus ponens (namely from the pair $X, X \rightarrow Y$ we infer Y)and the defeasible principle that a more specific rule wins over a less specific rule (namely of the two rules do not use the less specific one).

By more specific rule we mean that the rule 1,

1. $X \wedge Y \rightarrow Z$
   Is more specific than the rule 2

2. $Y \rightarrow \neg Z$
   And if both X and Y are agreed to hold, we have a clash between Z and $\neg Z$, because Z can be deduced from rule 1 and $\neg Z$ can be deduced from rule 2. But since the rule 1. Is more specific than rule 2, it has the upper hand, and we accept/ deduce only Z.

   We might have another rule 3

3. $W \wedge X \wedge Y \rightarrow \neg Z$
   Which is even more specific than rule 1, and so now rule 3 is the most specific and so we do not use rules 1 and 2 and use rule 3 and therefore we reject Z and accept $\neg Z$.

   Note that if we have two rules say 4. and 5.,

4. $U1 \rightarrow V$

5. $U2 \rightarrow \neg V$ which clash and neither is more specific than the other then we are stuck and cannot deduce anything (neither V nor $\neg V$).

The argument of Example A.1 makes use of more and more specific rules from Ω.

Let us now write the argument formally

(A) $\{\top \rightarrow X\}$,
   In truth (i.e., $\top$) I am not a sex offender

(B) $\{(B \rightarrow \neg X), B\}$,
   I saw you groping your secretary and we have the rule that

(C) { groping $\rightarrow$ sex offender}.
   This rule is more specific than the rule $\{\top \rightarrow X\}$.
   Therefore you are a sex offender (i.e. we deduce $\neg X$).

(D) $\{(B \wedge X' \to X), (B \to \neg X), B, X'\}$.

The secretary is my wife, and we have the rule groping secretary $\wedge$ secretary is wife $\to$ not sex offender, and this rule is more specific and therefore we deduce now X, i.e. I am not a sex offender

(E) $\{(B' \to \neg x'), B \wedge X' \to X), (B \to \neg X), B, X', B'\}$.

The secretary not registered as wife, therefore the secretary is not your wife and you do not have available the more specific rule you used in (D).

(F) $\{(B' \wedge X'' \to X'), (B' \to \neg X'), (B \wedge X' \to X), (B \to \neg X), B, X', B', X''\}$,

The secretary not registered as my wife is a tax dodge. Therefore we use the rule Secretary not registered as my wife and it is intended as a tax dodge $\to$ secretary is my wife So we can still deduce X, namely that I am not a sex offender.

5 Discussion of the actual sex offender universal distortions

5.1 Preliminary discussion

Here we discuss some simple logical principles/model, which can explain/ annotate the items in the list which follows.

The sex offender's reasoning is distorted. He will input replies in an argument that are not relevant, but are what he thinks are proper relevant arguments. To illustrate, imagine I say to a person,

"you did not accept the offer of a cup of coffee from our host and he was offended".

The person replied

"coffee makes my heart go faster, it is bad for me"

to which we can further say,

"well, you could have explained your reasons"

and the person could reply

"I was embarrassed".

The above is a normal possible conversation. However imagine the following distorted conversation:

"you did not accept the offer of a cup of coffee from our host and he was offended".

To which the person replies:

"Donald Trump is the president of the USA".

There are several challenges here:

1. What is the connection between Trump and this person drinking coffee? To clarify, had the person said "I am trying to stop smoking", we could probably think that having coffee makes him want to smoke. But there is no "public" connection between Trump and drinking coffee.

2. Having realised that this person has a reasoning "distortion" or some private ideas about the world, what do we say to him? Do we say "what is the connection between Donald Trump and coffee?". Perhaps the person has a private explanation, maybe Trump tweeted something about coffee and fake news, or that coffee is a fake drink. But what if the person says and thinks:

"Of course nobody drinks coffee when Trump is a president".

What do we say in this case?

The above is an illustration of what therapists face when they deal with sex offenders. If we were to ask a sex offender, "why did you force this woman to have sex with you" one of the possible answers is, "She sleeps with everybody any way so she must sleep with me". As far as he is concerned, "If she sleeps with everybody, she should sleep with him". We have the problem of how to respond. It is no use telling him he is wrong, because he absolutely thinks that he is not wrong and he is fed up of hearing that he is wrong from many people, like the police, the judge and many others.

A new approach is required.

The new approach relies on four obvious principles.

1. We have to understand and get into the mind of this person and understand the structure of his distortion and speak to him in his own language. A mother might be able to do this with her child, but , all we can do is ask for more information . We can ask what else he drinks, what else he eats, what he thinks about Donald Trump, or his opinions about what Donald Trump does or says. If we are lucky, we might discover that according to this person, Trump does not eat ice cream but this person likes ice cream and eats it. There is an obvious inconsistency in this person's behaviour and we can point it out. Why do you not drink coffee, but do eat ice cream. Of course the person might ignore the inconsistency but fortunately sex offenders tend to behave differently.

2. They are sensitive to inconsistency

3. All sex offenders more or less share similar distortions.

4. Two sex offenders, will frequently put forward the same distortion, and although neither will see his own misapprehension they will both identify the fault in the other's reasoning. This is key in group therapy because each member of the group can be criticised by all the other members who can see all of his distortions. By the way, this is true of normal people as well. One often cannot see one's own problem, but can see with clarity the same problem in others.

> "And why beholdest thou the mote (i.e: the small speck) that is in thy brother's eye, but considerest not the beam (i.e: the tree trunk) that is in thine own eye?" (Jesus: The sermon on the mount. King James version)

5.2 List of offender's arguments with commentary

To the trained logical reader it may seem that some of the reasoning arguments of the therapist are not logically sound. This is intended as such, to either extract a reaction from the patient or as a preliminary to make the patient aware of his wrong reasoning, by playing at his own game. So the dialogue you see between the therapist and the patient is not a rational dialogue being unfolded in logical argumentation [53, 18].

The dialogue below is not an actual therapy dialogue but a compilation of the kind of questions and answers used in years of dealing with sex offenders during Therapy.

The logician reader should bear in mind that in the group therapy the dialogue is carried out with one of the sex offenders (say S1) when all the others are listening. S1 cannot see the distortion in in his own thinking, but all the others, say S2, S3,..., can see it. However if we have the same conversation with S2, then S2 will exhibit the same distortion as S1 and will not see it, but S1 will see it.

5.2.1 Exaggeration

A simple insult can become a major attack which requires a serious countermeasure.

> "I beat the victim up because he talked to me in a disrespectful way"
> *Logical test sentence*
> *X was disrespectful to y but y did not beat him up*

Possible reply: You are quite right. However why didn't you kill the bastard? This way he certainly will not disrespect you anymore.

Analysis: Criticism does not work here. The offender will switch off. We need to sympathise but at the same time show him that he has exaggerated in his reaction. Why did he not kill the victim? Why was killing too much but not beating up?

5.2.2 Generalisation.

"Why did you attack/rape this girl, a stranger¿'

> "One girl rejected me and so I have no chance with girls and my only option is to take one by force."
> *Logical test sentence*
> *One girl rejected him but he did not take another girl by force*

Possible reply: So you are going to rape girls the rest of your life? This seems to be expensive. Maybe there is another way to solve your problem?

5.2.3 Misinterpretation of facts

> My wife smiles at someone and I am sure she is having an affair.
> *Logical test sentence*
> *My wife smiled at someone but I did not think she was having an affair with him.*

Possible reply: You say that you are "sure", would you bet a million dollars on this? What if you are wrong? Will you take the risk? Perhaps you want to say "maybe" and discuss the possibilities. Also, have you ever seen your mother or sister smile at someone? Does it mean they are sleeping with him?

5.2.4 Unfounded deduction

> A woman accepts my invitation for coffee, which means she agrees to have sex with me
> *Logical test sentence*
> *The woman accepted my invitation for coffee but she did not agree to have sex with me.*

Possible reply: I believe that this is what you thought, but maybe she was not yet clear about what she wanted. Maybe she liked you but eventually decided you are not for her.

Even if she knew your intentions can she not change her mind?

When you say at a job interview you want the job, are you not allowed to change your mind? Do you maintain that you are then committed to take the job and do whatever they tell you to do?

5.2.5   Extreme opinions

a. My wife says she wants a divorce, but if I force sex on her she will stay mine.

b. Children love sex with grownups.

c. I must have sex with this woman, otherwise my life is not worth living

Logical test sentences

a. *My wife says she wants a divorce, but I am not going to force sex on her so she will stay mine.*

b. *They are children but they do not love sex with grownups.*

c. *I did not have sex with this woman but my life is still worth living*

Possible respective answers

a. OK so if someone in jail beats you up then you would love him and love jail?

b. you can love or not love someone you know. If the child does not yet understand SEX, then he cannot say how he/she feels about it.

Moreover, as a child you want to tell me that you would have liked a grown up do with you what he wants simply because he thinks you would like it?

c. You do not really have to have sex with the woman. You think you need to, but it is not the case. Imagine that you have to make a choice between the life of your mother or having sex with this woman, what will you choose?

5.2.6 Reasoning distortions

The sex offender distorts the system in order to feel more comfortable with what he is doing. In the annihilator model he would change *V* in a way which puts anything having to do with himself in the highest *V* value. The following list gives samples of sex offenders' rationalisations:

Kindheartedness

    a. I was not attacking, I was only trying to help.

    b. I did not do anything.

    c. I exhibited myself in order to teach the children about sex, or the child was sad and I only amused him.

    Logical test sentence

    a. He attacked the woman but it did not help her

    b. If the meaning is denial of the action then it may be true or false but in principle it is logical. If the meaning is

        "I did it but what I did is not wrong"

    Then the test sentence is

        "I raped her but she thought it was wrong"

    c. I exhibited myself to the child but he did seem to learn about sex

Possible answers:

    a. Would you have liked your own daughter to receive such "help"

    b. Is there anything wrong in your personality or anything you would have preferred changed? If not then I want to to know how this is done. You are the first person I have ever met who feels their personality is perfect. I am also sure that you could still learn new tricks to further perfect your personality, to be even wiser and more careful.

   c. Did the child ask you to learn in this way?
      If this is not wrong why did you run and hide?
      Is this the way you yourself learned about sex?
      What is so interesting/amusing/instructive in a naked man?

Helplessness

   I cannot stop myself. My drive controls me.

Logical test sentence
I cannot stop myself; My drive controls me but I am blamed for it.

Possible response: If a policeman were standing next to you, would you still have done it?...So you see you still have control and choice

Projection-blaming

   a. She made me do it.

   b. I was drunk.

   c. My friends started it, I was just swept along by them.

Logical test sentence

   a. See our comment to 4.6.1 b.
      Maybe she held a gun to his head?

   b. I was drunk when I did it but I was held responsible

   c. My friends started it, I was just swept along by them but I was held
      responsible

Possible answers:

a. Are you so weak in character that someone can force you to do what you do not want?

b. Most people fall asleep after consuming alcohol. Some however let out what they already have inside them!

c. And if your friends would jump off a high roof would you also jump with them?

I have the right to...

a. I spent money on her, she owes me.

b. She is my wife, I have the right.

c. She is my daughter, I created her.

d. My wife denies me sex, so her daughter takes her place.

e. It is ridiculous. A man cannot be accused of raping his wife any more than he can be accused of stealing his own radio.

Logical test sentence

a. *I spent money on her, she owes me. But nevertheless I did not rape her*

b. *She is my wife, I have the right but nevertheless I did not rape her*

c. *She is my daughter, I created her but nevertheless I did not rape her*

d. *My wife denies me sex, but I did not let her daughter take her place.*

e. *this is a wrong analogy and one needs to give a similar analogy which looks very wrong "So the finance minister in charge of the treasury can take money for himself?"*

Possible answers:

a. How much would you charge the man sitting next to you to let him have sex with you?

b. So if you were born a woman, your husband could force sex on you even if you happen to be unwell and in pain?

c. Is the duty of a father to raise kids and educate them or is it to destroy them?

d. Are there any other possible ways to react? I am just curious if your wife would catch you cheating with other woman, she should rape your son?

e. So, your wife can do to you the same things you did to her?

Minimalisation

a. It did not bother her.

b. Other people do worse.

Logical test sentence

a. *She was raped but it did bother her*

b. *Other people do much worse than rape but I did not myself rape.*

Possible answers

a. You sit and I will ask friends to keep standing still next to you without touching you, I bet you will feel under pressure and uncomfortable. Now imagine how terrible terribly painful is what you did to your victim

b. A person who was raped is in distress and pain he/she will not be comforted by thinking it could have been worse.

Justification

a. She annoyed me. She deserves it.

b. Youngsters nowadays know more about sex than grown ups. They want sex. So what if she is only 12 years old?

c. I had a hard day and was a long time without sex.

d. She sleeps with everybody, why pick on me?

Logical test sentence

a. *She annoyed me but I did not rape her as she deserved.*

b. *Youngsters nowadays know more about sex than grown ups, but I did not sexually offend them.*

c. *I had a hard day and was a long time without sex, but I did not rape anyone.*

d. *She sleeps with everybody, but I did not rape her.*

Possible answers:

a. So I can rape you if you annoy me right now?

b. This is a strange argument because emotionally young people have not changed from earlier times compared with today

c. Why did you not masturbate or go to a professional woman? Why rape?

d. If you like pizza, does it mean you eat 100 pizzas a day, every day?

Self importance

a. I am beyond the law.

b. All women adore me. I thought she was just playing hard to get.

c. I know what women think. I know she wanted me.

d. She contacted the police only because I stopped having sex with her and she just can't give up on me.

Logical test sentence
These cases are more difficult to analyse because they state false facts.

Possible answers:

a. I met many people who thought they were above the law but the law thought otherwise.
   As a person above the law you realise now that after you raped you are going to be at the bottom of the social ladder in prison.

b. A man whom women admire does not force himself on any one of them.
   If you thought she was playing hard to get why did you not play along ? Why rush and rape?

c. Wanting you is a reason to rape her?

d. So you think it is not you who needs treatment but that she needs to be here now?

6 Further points of discussion

This section gives further clarification by the authors to the readers concerning the List of offender's arguments and the therapist answers, using a question and answer system.

This question and answering system below is not a sex therapy dialogue but it is a simulation of what the reader of this Chapter might ask and what e we the authors answer it.

It is not part of the therapy, but a teaching exposition mechanisms between the authors and the readers.

Question 1. How can one ensure that the therapeutic group, instead of helping and criticising each other, unite and strengthen one another's distortions and commit more offences? Does the choice of heterogeneous candidates for the same group help to disperse the distortions of thinking and reduce the likelihood that these criminals will strengthen each other? For example, if we take a group of 10 people who all rape their wives for the same reason, (for example, "the woman smiled, she is definitely a traitor and should be raped"), would they strengthen each other instead of criticising and altering a distortion? Is there any data on it?

Answer 1. The groups are heterogeneous groups such as exhibitionists, pedophiles, rapists of adult women and rapists of adult men . It can be seen that perceptual delusions of paedophiles are often different from the distortions involved in offences against adults. Thus, dispersion is created and different patients hold different distortions and even contradict each other's distortions. For example, a pedophile who hurt a child might say that the worst thing is a homosexual abuser while the other tells him that the worst thing is harming a helpless little girl who did nothing wrong to anyone.

Question 2. Are the problems in question 1 related to the type of group namely closed or train?

Answer 2. In our assessment, in the train group, the chances of further narrowing the thinking of other patients are smaller because patients are in different stages of group participation, there are status differences, and they have reached different stages of insight and improvement.

Question 3. There are social groups on the Internet, interested in a particular subject. some of them deal with motorbikes and old radios and some deal with distorted topics, such as conspiracy theories, flat earth theories, fetishes of various sorts, and groups that encourage sex with children. do these groups strengthen the distortion held by other participants in the group and why does one pedophile in such a group not tell the other, "you are a pedophile, you are doing something wrong "why are you doing it?" in other words, why do they not they criticise each other?

Answer 3. Yes, the different groups tend to strengthen the opinions and proclivities of the group members rather than the opposite. Usually, people choose a group that will strengthen their perceptions (for example, a person who is attracted to children is more likely to join a group that believes that children in the current generation are already mature and want sex.) Because it creates cognitive dissonance that causes tension and discomfort such people will avoid joining groups that talk about how damaging premature sexual experience is to a child.

Question 4. What conclusions can be drawn from answer 3 about the dangers of such groups reinforcing and perpetuating distortions?

Answer 4. People want to find justifications for their behaviour and the rich internet space where you can find anything may encourage reinforcement of misconceptions to the extent that an individual give a person reinforce his misconceptions and sometimes even the individual will believe that his / her misperception has some scientific validity. Flat earth theorists for example frequently cite each other as authoritative source material. So, for example, it is often possible to see patients who have committed sexual offences using the sympathetic arguments put forward by their lawyers in court, as justification for their behaviour. The patient might have a suppressed awareness that it is the lawyer's job to win the case on behalf of the client by downplaying the seriousness of the crime but the patient will seize on these arguments to support his own self-image as an innocent victim of circumstances.

7 Conclusion and comparison with cognitive behavioural therapy and logic based therapy

This concluding section summarised what we are trying to achieve in this Chapter.

The authors observed that similar logics/reasoning are used by argumentation/non-monotonic logicians and sex therapists clinical experts.

The aim of this Chapter is to get them to talk to each other for the benefit of both communities.

There are difficulties

1. The two communities use different professional languages

2. Each community is immersed in their own problems, have different attitudes towards what is important, and might have a natural resistance to new ideas.

3. In this Chapter we address two of these communities, each from their own point of view showing how the other community approaches the Logic of Sex offenders as it manifests itself in Therapy.

   The logician might see an opportunity to develop new systems for example distortion.

   The therapist might use the formal awareness of what he is doing, to develop tricks to influence the sex offender.

4. Formal awareness of the logic involved in therapy will make it easier to teach new therapy student how to practice.

5. There is another benefit of getting the the logicians and therapists to talk to each other. This is the influence it might have on the perception toward the therapy of sex offenders.

Acknowledgements

The authors would like to thank the referees for their detailed comments. We also thank the referees for bringing to our attention research done by the LBT community. This is very relevant and we have modified the Chapter accordingly.

References

[1] Andrews, D. A. (2006). Enhancing adherence to risk-need-responsivity: Making quality a matter of policy. Criminology and Public Policy, 5(3), 595-602. https://doi.org/10.1111/j.1745-9133.2006.00394.x

[2] Andrews, D. A., & Bonta, J. (2017). The psychology of criminal conduct (6th ed.). Routledge.

[3] Ashman, T. (2010). Stages in the development of a timed group [electronic version]. Nadla on 10/24/2010, article site

[4] Alphin, R. (2020). Treatment providers perceptions of effective sexual Offender Treatment Modalities. Walden Dissertations and Doctoral Studies. 8510. https://scholarworks.waldenu.edu/dissertations/8510.

[5] Baroni, P, Gabbay, D., and Giacomin, M., eds. *Handbook of Formal Argumentation*, College Publications, 2018.

[6] Baroni, P, Gabbay, D., and Giacomin, M., eds. *Handbook of Formal Argumentation, Volume 2*, College Publications, 2018.

[7] Bengis, S. M. (1986). Training series: The diagnosis and treatment of the juvenile sex offender. Holyoke, MA: *New England Adolescent Research Institute*.
Comment on Reference:
This paper describes the most common therapeutic model for treating sexually abusive people. The model helps the patient understand that it is his choice to cause harm and therefore even if he enters a tempting risky situations he can identify the risk and stop himself from dangerous behavior and avoid causing injury.

[8] Bléandonu, G. (2020). *Wilfred Bion: His life and works*. Other Press, LLC.

[9] https://psychology.fandom.com/wiki/Cognitive_Behaviour_Therapy

[10] https://psychology.fandom.com/wiki/Cognitive_distortion

[11] Carrasco, N., & Garza-Louis, D. (1997). Hispanic sex offenders: Cultural characteristics and implications for treatment. In B. K. Schwartz (Ed.), Handbook of sex offender treatment (pp. 45-1 45-10). Civic Research Institute, Inc.

[12] Elliot D. Cohen The Psychoanalysis of Perfectionism: Integrating Freuds Psychodynamic Theory into Logic-Based Therapy. *International Journal of Philosophical Practice* Volume 6, No. 1 (Spring 2020) pp. 15–27.

[13] Elliot D. Cohen. *Cognitive behavior interventions for self defeating thoughts. Helping clients to overcome the tyranny of "I can't"*. Routledge, 2022.

[14] Elliot D Cohen. *Logic-Based Therapy and Everyday Emotions: A Case-Based Approach* (Lanham, MD: Lexington Books, 2016).

[15] Elliot D Cohen. *Making Peace with Imperfection: Discover Your Perfectionism Type, End the Cycle of Criticism, and Embrace Self-Acceptance* (Oakland, CA: Impact, 2019).

[16] Elliot D Cohen. *Critical Thinking Unleashed (Elements of Philosophy)*, Row-

man & Littlefield Publishers, 2009.

[17] Elliot D. Cohen. *Caution: Faulty Thinking Can Be Harmful to Your Happiness*. Kindle Edition 2013

[18] DUrso, G., Petruccelli, I., Costantino, V., Zappulla, C., & Pace, U. (2019). The role of moral disengagement and cognitive distortions toward children among sex offenders. *Psychiatry, psychology and law*, 26(3), 414-422.

[19] Albert Ellis. *Reason and emotion in Psychotherapy*. Lyle Stuart, 1962.

[20] Embry, R., & Lyons Jr, P. M. (2012). Sex-based sentencing: Sentencing discrepancies between male and female sex offenders. *Feminist Criminology*, 7(2), 146-162.

[21] Dov Gabbay. *Logic for Artificial Intelligence and Information Technology*, Monograph, College Publication 2007, 580 pp, Chapter 11 Section 4, Case study: concessive clauses .
Comment on Reference:
Basic text in Applied Logic. Taught at King's College London.

[22] Dov Gabbay and Artur Garcez Logical Modes of Attack in Argumentation Networks, in *Studia Logica*, 93(2-3): 199–230, 2009.
Comment on Reference:
Advanced paper, but rather intuitive.

[23] Dov Gabbay. *Meta-Logical Investigations in Argumentation networks*. Research Monograph College publications 2013, 770 pp
Comment on Reference:
This is an advanced monograph

[24] Dov Gabbay and Michael Gabbay. Argumentation as information input. 2015. Submitted to the special issue of the Isralog 2017 conference D. Gabbay and M. Gabbay. Argumentation as information input. 2015.
Short version published in Proceedings COMMA 2016, Computational Models of Argument Pages 311–318
DOI10.3233/978-1-61499-686-6-311 Volume in Series Frontiers in Artificial Intelligence and Applications IOS press Volume 287:
Full version PUBLISHED in College Publication Simari Tribute Argumentation-based Proofs of Endearment, 2018, pp 145-197.
Comment on Reference:
Advanced paper, but worth the effort to get a feel of the subject.

[25] D Gabbay and L Rivlin. HEAL2100: Human Effective Argumentation and Logic for the 21st Century. The Next Step in the Evolution of Logic *IFCoLog Journal of Logics and their Applications* Volume 4, number 6, July 2017, pp.

1633–1687.

[26] D. Gabbay, G. Rozenberg and L. Rivlin. Reasoning under the influence of universal distortion. *Ifcolog Journal of Logics and Their Applications*, 2017, pp 1769-1900.

Comment on Reference:

This is an advanced paper on distortion with discussion of distortion in sex offender thinking.

[27] Dov Gabbay, Gadi Rozenberg and Lydia Rivlin; Argument, Sex and Logic, Monograph, to Appear with College Publications, 2023

[28] Dov Gabbay, Gadi Rozenberg and possible Co-authors. A study of Logic Based Therapy from the point of view of Modern Formal Argumentation, in preparation.

[29] Gannon, T. A., Olver, M. E., Mallion, J. S., & James, M. (2019). Does specialized psychological treatment for offending reduce recidivism? A meta-analysis examining staff and program variables as predictors of treatment effectiveness. *Clinical Psychology Review*, 73, 101–752.

Comment on Reference:

This is an examination of the effectiveness of dedicated treatment of sexual offenders.

[30] Gerber, J. (1994). The use of art therapy in juvenile sex offender specific treatment. *The Arts in Psychotherapy*, 21 (5), 367-374. Gerber, J. (1994). The use of art therapy in juvenile sex offender specific treatment. The Arts in Psychotherapy, 21 (5), 367-374.

[31] Charles Leonard Hamblin. Fallacies (University Paperbacks) Paperback 8 Jun. 1972

[32] Stefan G. Hofmann. *An Introduction to Modern CBT Psychological Solutions to Mental Health Problems*. John Wiley & Sons, 2012.

[33] Howard, M., de Almeida Neto, A., & Galouzis, J. (2019). Relationships between treatment delivery, program attrition, and reoffending outcomes in an intensive custodial sex offender program. *Sexual Abuse*, 31(4), 477-499.

Comment on Reference":

This paper discusses value of open-ended vs closed groups and group therapist warmth for reducing attrition/drop-out. In sum, our predictions for the current study are as follows. We hypothesized that implementation of changes to treatment delivery in a residential SOTP, including application of the rolling group format and systematic emphasis on positive therapist characteristics, would have a significant impact on offenders' likelihood of program completionNo previous research has compared open and closed groups in sex offender treatment, a small number of studies have indicated rolling groups may improve rates of dropout, while achieving similar treatment outcomes, compared with closed groups in other psychotherapeutic settings (Graham, 1999; Hoffman, Gedanken, & Zim, 1993; Tourigny & Hebert, 2007).

[34] Howard, M. & Zhigang Wei, Z. (2021) Effects of closed versus open groups on attrition and recidivism outcomes for sex offenders in custody-based treatment programmes, *Journal of Sexual Aggression*, DOI: 10.1080/13552600.2021.1905894.

Comment on Reference:
There is growing recognition for the importance of group processes in treatment for sexual and other offenders. However, there is little evidence about the influence of programming factors that affect the composition of groups, including use of closed or open groups. The current study examined how group format was associated with programme attrition and reoffending outcomes for sex offenders attending custody-based treatment programmes. Cohorts of offenders who attended a programme as it transitioned from a closed group format to an open group format were compared to cohorts of offenders in a second programme that remained in closed group format over the study period (nǍL'=490). We found that post-intervention cohorts in both programmes showed reductions in rates of programme non-completion which were not significantly associated with group format. Group format was not a significant predictor of reoffending outcomes. The results suggest that both open and closed groups may be viable alternatives for achieving treatment outcomes in intensive sex offender programmes.

[35] http://news.sky.com/story/newcastle-gangs-abused-adults-and-children-with-arrogant-persistence-review-finds-11263201
Accessed February 25, 2018

Comment on Reference:
This is example of how the authorities in England deal with sexual delinquency.

[36] Jacobs, E. E., Maasson, R. L. L., & Hrvill, R. L. (2006). Group counseling: Strategies and skills (6th ed). Thomson Brooke Cole. Journal for Specialists in .Group Work, 30(2), 159172.

[37] Jennings, J. L., & Deming, A. (2013). Effectively utilizing the behavioral in cognitive-behavioral group therapy of sex offenders. *International Journal of Behavioral Consultation and Therapy*, 8(2), 7–13.

[38] Jennings, J. L., and Sawyer, S. (2003). Principles and techniques for maximizing the effectiveness of group therapy with sex offenders. *Sexual Abuse: A Journal of Research and Treatment*, 15, 251-267.

[39] Kim, B., Benekos, P. J., & Merlo, A. V. (2016). Sex offender recidivism re-

visited: Review of recent meta-analyses on the effects of sex offender treatment. *Trauma, Violence, & Abuse*, 17(1), 105–117.

[40] Kahn, R. E., Ambroziak, G., Hanson, R. K., & Thornton, D. (2017). Release from the sex offender label. Archives of Sexual Behavior, 46(4), 861864. https://doi.org/10.1007/s10508-017-0972-y

[41] Leszcz, M. (1992). The interpersonal approach to group psychotherapy. International Journal of Group Psychotherapy, 42(1), 37-61. https://doi.org/10.1080/00207284.1992.11732579

[42] Levinson, B. (2019). The Treatment of Sex Offenders Using a Task Centered Approach. In *Clinical Management of Sex Addiction* (pp. 457-480). Routledge.

[43] Lussier, P., McCuish, E. C., & Cale, J. (2021). The Lapse and Relapse of Correctional-Based Sex Offender Treatment and Intervention. In *Understanding Sexual Offending* (pp. 273-311). Springer, Cham.

[44] Ly, T., Fedoroff, J. P., & Briken, P. (2020). A narrative review of research on clinical responses to the problem of sexual offenses in the last decade. *Behavioral sciences & the law*, 38(2), 117-134.

[45] MacNair-Semands, R. (2021). *The Theory and Practice of Group Psychotherapy*: by Irvin D. Yalom and Molyn Leszcz. New York, NY: Basic Books, 2020. 818 pp.

[46] Marshall, L. E., & Marshall, W. L. (2017). Motivating sex offenders to enter and effectively engage in treatment. In D. T. Wilcox, M. L. Donathy, R. Gray, & C. Baim (Eds.), Working with sex offenders: A guide for practitioners (p. 98112). Routledge.

[47] McKillop, N., Hine, L., Rayment-McHugh, S., Prenzler, T., Christensen, L. S., & Belto, G. (2022). Effectiveness of sexual offender treatment and reintegration programs: Does program composition and sequencing matter? Journal of Criminology, 55(2), 180-201. https://doi.org/10.1177/26338076221079046

[48] Morgan, R. D., Garland, J. T., Rozycki, A. T., Reich, D. A., & Wilson, S. (2005). Group therapy goals: A comparison of group therapy providers and male inmates.

[49] Ministry of Health, Israel. Ethical and professional rules for assessing the dangerousness of sex offenders, July 2021. https://www.health.gov.il/Subjects/mental_health/dangerous/Pages/offenders.aspx

[50] Jerome Neu. *Emotion, thought and Therapy. A Study of Hume and Spinoza and the Relationship of Philosophical Theories of Emption to Psychological Theories of Therapy*. Routledge and Kegan Paul, 1977.

[51] P H Nowell-Smith, Ethics, Penguin Books, 1954

[52] Ofek. A. (2006). Group treatment program for adolescent sex offenses. pp.

218-228. Ramat HaSharon: Ofek Institute for Management and Research

[53] Patterson, C. (2018). Does the adapted sex offender treatment programme reduce cognitive distortions? A meta-analysis. *Journal of Intellectual Disabilities and Offending Behaviour.*

[54] Rice, M. E., and Harris, G. T. The size and sign of treatment effects in sex offender therapy. *Annals of the New York Academy of Sciences*, 989(1), 428-440, 2003.

[55] Scheory Bitton & Abulafia, 2021. Edited book of papers all in Hebrew : Sex Offenders: Current Trends in Legislation, Enforcement, Evaluation and Treatment, Karmel-Keshet, 2021.

[56] Schmucker, M., and Lösel, F. (2015). The effects of sexual offender treatment on recidivism: An international meta-analysis of sound quality evaluations. *Journal of Experimental Criminology*, 11, 597-630.

[57] Schmucker, M., & Lösel, F. (2017). Sexual offender treatment for reducing recidivism among convicted sex offenders: a systematic review and meta-analysis. *Campbell Systematic Reviews*, 13(1), 1-77.
Comment on Reference:
This is a summary of studies on dedicated care of Sex Offenders.

[58] Shanee, N. S. (2018). Sex Offences in Israel-Public opinion, Risk Management and Treatment. *Sexual Offender Treatment*, 13(1/2).

[59] Simmons, M. L. (2019). Evaluating the legal assumptions of Victorias Sex Offender Registration Act 2004 from a psychological perspective. Psychiatry, *Psychology and Law*, 26(5), 783-796.

[60] Tamatea, A. J., Webb, M., & Boer, D. P. (2011). The role of culture in sexual offender rehabilitation: A New Zealand perspective. In D. P. Boer, R. Eher, L. A. Craig, M. H. Miner, & F. Pfäfflin (Eds.), International perspectives on the assessment and treatment of sexual offenders: Theory, practice, and research (pp. 313329). Wiley-Blackwell. https://doi.org/10.1002/9781119990420.ch16

[61] Viglione, J. (2018). The risk-need-responsivity model: How do probation officers implement the principles of effective intervention? Criminal Justice and Behavior, 46(5), 655-673. https://doi.org/10.1177/0093854818807505

[62] Zhumalai, S., Muralidhar, D., Dhanasekarapandian, R., and Nikketha, B. S. (2018). Group interventions. *Indian journal of psychiatry*, 60(Suppl 4), S514.

A Appendix

Some semi-formal logic

Let us begin with a familiar story, which will help us introduce some structure: A celebrity, X, is accused by two independent alleged victims A and B, of sexually inappropriate behaviour. X denies the accusations and counter attacks, claiming that A and B are lying and are just seeking attention.

The logical structure of this situation can be presented as follows:

1. We have a set $S = \{A, B, X\}$ of the individuals involved.

2. We have an attack/accusation relation R. The attack/accusation relation is denoted by an arrow, $\rightarrow$, ($\rightarrow$ is R).

We have

$$A \leftrightarrow X \text{ and } B \leftrightarrow X.$$

According to current formal argumentation theory (see for example [23]), the basic simplest argumentation model gives the following logical options for our situation:

(O1) We believe X and do not believe A and B (the technical term for it is X is "in" and A and B are "out").

(O2) We do not believe X (X is "out") and therefore A and B are believed and are "in".

(O3) We cannot decide who is telling the truth (the technical term for this is all "undecided").

It is clear that this basic logic model does not allow for the possibility, say, that we believe B but do not believe A and so we have (O4) below:

(O4) A is "out", B is "in" and X is "out".

Similarly, we do not have (O5):

(O5) A is "in", B is "out" and X is "out".

There are fortunately more advanced logical models (see [23]) which can allow for (O4) and (O5). One such model allows for a valuation function V. We give, using the function V, numerical valuations to arguments (say in percents or using values between 0 and 1) (think of it as strength or credibility) where 0 is no strength at all and 1 is 100% strength.

So if we set

$$V(A) = 0, V(B) = 1 \text{ and } V(X) = 1,$$

then we practically eliminate A from the situation. This gives us (O4).

Similarly we can get (O5), by giving

$$V(A) = 1, V(B) = 0 \text{ and } V(X) = 1.$$

These values can represent what those trying the case in court think of the credibility of the people involved. Of course once we use a mathematical representation of such a model we have to say what we do in the most general case of V, namely how we propagate these V values.

For example in the case we have for example the values of V below

$$V(A) = V(B) = 0.5 \text{ and } V(X) = 0.6.$$

We need to say what are the new values of V (say V^* is the new value) for A, B, X respectively, (namely what are $V^*(A), V^*(B)$ and $V^*(X)$) and further say for what values of $V^*(X)$ do we consider that X is "out"?

There are various logical models for calculations of the new V^*, see [23].

The above is a formal model of the form (S, R, V), of arguments with a notion of attack and a notion of strength which we may consider in applying to some aspects of the sex-offender situation when he is attacked by alleged victims.

However, two aspects of this proposed model are immediately clear:

(a1) It is in the interest of X to lower the values V of all his attackers and increase the value V for himself. X would offer arguments intended to create new values V^*, favourable to him.

(a2) The most obvious and crucial inadequacy of the proposed model is that it considers the units A, B, X as atomic. In real life A and B are stories telling the (alleged) circumstances in which A and B were victimised and X tells his own version of events. The attack relation is the input of criticism and additional details and arguments about each other's versions. The numerical values of credibility is imposed on top of all of the above interaction.

There is no clearly defined such model in the argumentation literature, though some model components and ideas are available [26, 22, 24].

We have developed such a model and see how such arguments (as used by sex offenders) fit into the model. All this is done in Section 4.

To give the reader an idea of how this works, consider a case where B is attacking X saying he is a sex offender, because she saw him groping his secretary A.

A does not say anything, but the incident was caught on B's camera.

The next Example A.1 documents the exchange, (there is no need to use the numerical valuation V in this example):

Example A.1. *We start with the statement:*

> *X: I (Mr. X) am not a sex offender*
> *B: X was seen groping his secretary*

Clearly in the above we have that B attacks X
> *X defends himself, saying X′:*

> *X′: My secretary is my wife*

Clearly X + X′ attack B.
> *However, B does not let go and checks the employment records and continues with B′:*

> *B′: the secretary is not registered or reported as X's wife*

Clearly B′ attacks X′.
> *To this X answers with X″:*

*X″: The secretary is indeed my wife. I kept it a secret because
it is a tax dodge, there is less tax deduction for one's wife's
employment. So I am not a sex offender, just a tax dodger.*

Clearly X″ attacks B′.

*The above exchange of attacks is a case of information input. Each of the
statements is further information supplied in order to affect the conclusion
of whether X is true or not, see [24].*

*Note that X + X′ + X″ + B + B′ are all consistent together, and they
create the details of the story/reality of the case.*

The above Example A.1 is a rational exchange by normal people in
Western Society. The attacks and the information inputs are relevant to the
narrative.

A sex offender's reasoning is distorted and he might supply unreasonable inputs (the distortion manifests itself in that the input is unreasonable in
normal society). If the offender comes from a different cultural background
he might not even understand that there is a problem/offence. He might
misinterpret signals and behaviours.

To clarify, for example, it might be acceptable by the majority of common sense ordinary people that groping one's wife in the office, although
distasteful, is not a sex offence (though it might be grounds for a divorce
claim). So X' is a reasonable answer to the attack B.

A real sex offender, however, might answer something like Y below:

*Y: I am wonderful, women are crazy about me. I was doing my
secretary a favour by groping her, releasing her from her misery
in admiring me.*

The reasoning distortion of the sex offender is in thinking that Y is a proper
answer to B. See such distortions/answers in Section 4.

The logical part of the therapy is to make the sex offender understand
that his thinking is distorted.

The reader will benefit from consulting the following very recent article on the internet, accessed on February 25, 2018 1400 hours UK time. It
clearly and painfully demonstrates the problem. Newcastle gangs abused
adults and children with arrogant persistence, review finds

```
http://news.sky.com/story/newcastle-gangs-abused-adults-
and-children-with-arrogant-persistence-review-finds-
11263201
```
The key passage in the article is the following:

> At the final trial, 17 men and one woman were convicted of charges including rape, supplying drugs and inciting prostitution. The report author invited all of those convicted to be interviewed in his review. Only one agreed. "There was a complete lack of remorse," said Mr Spicer, who cannot identify which abuser he interviewed. "In fact he didn't accept that he'd done anything wrong... He felt the victims were responsible for their own abuse."

B Some information on the logic based therapy (LBT) for the formal argumentation reader

B.1

We became aware of the LBT community from the referee, also looking at cognitive distortions and it is our intention to study the field in the same way we approached the sex offender community. No doubt the LBT methods are being used implicitly by the sex offender therapist. We do not know however how the use of logic is reflected. We believe and urge the formal argumentation community to take interest in LBT and see what new ideas and enhancement to formal argumentation we can get from observing LBT practice.

We need to work directly with a LBT therapist expert co-author to improve our logic service to the communities and patients. We identified Elliot D Cohen as a Logic Based researcher most compatible with our logic approach, and we intend to write a paper entitled: A Study of LBT from the point of view of Formal Argumentation. See [28].

The present article advised care-givers and therapists of sex offenders to try and avoid the accepted narrative in general society which is based on criticism, justified of course, towards sexual offenders emphasising the legal aspect and the seriousness of their offences during therapy. there are various bodies involved in this kind of narrative/therapy. Sexual abusers who come to treatment usually know society's attitude towards them and in treatment the goal is not to recycle and represent the known narrative, but to try and deal with their vulnerability and distortions of thinking using the sex offenders own point of view and own narrative. This point of view is more appropriate and better adapted to the language of the abusers. This practice is expected to achieve good results and lead to a perceptual change in the sex offender and consequently to behavioral change of the offenders.

One of the examples that can be offered to emphasize the intervention is the following example.

Example B.1. *A man caught his wife cheating on him. He got angry, cried, and most of all, he felt that his whole world had been destroyed. He was not able to confront his wife, talk to her, examine whether and how to proceed from this point. He felt a complete failure and chose to leave the house. he walked around like a sleepwalker through the streets of the city feeling miserable and conflicted. On the one hand he felt anger at his wife and on the other hand felt it was his own failure to keep his wife with him.*

At this point the patient is exhibiting some reasoning distortion. He can benefit from a LBT therapist who will show the patient the flaws in his reasoning and perhaps help him accept what has happened as not his failure/fault and deal with the situation rationally.

To compare the LBT therapy with the sex offender therapy with the sex offender case, let us continue the story.

The patient decided to go into the pub and got himself drunk. When a woman sat down not far away from him, the patient in his drunken state approached the woman in an inappropriate and disrespectful way. The woman responded to him dismissively and rejected all attempts at his courtship. The man was even more angry, did not stop with his comments and continued drinking. At one point the woman went to the bathroom and he followed her, entered her bathroom cubicle before she could lock the door and began to touch the different parts of her body, telling her how beautiful and sexy she is and how much she must be interested in sex. The woman tried to drive him away and resist, but he forced himself on her.

We now have a case of a sex offender and we can apply sex offender therapy.

In therapy, great emphasis was placed on his male self-image on finding what other factors exist that pushed him to attack the woman. To continue the example, the therapist might find out that, from an early age, the patient grew up alongside a neutered mother who constantly felt intrusive and over-bearing and often destroyed his attempts at relationships with other women. The father of the patient was weak and even afraid of his wife/mother and all decisions in the house were made by her. More than once the father would choose to sit with his son and complain about his bitter fate and complain that his wife did not respect him.

The treatment revealed that the patient's mother was the one who chose his wife for him and today, during the treatment, he identifies the similarity between his mother and his wife and recognizes that quite a few of the difficult feelings he had in front of his mother are the same feelings he experiences in front of his wife.

The subject felt impotent in his feelings, but was unable to share his feelings and desires with his wife and when he caught his wife cheating on him he felt a lot of unpleasant emotions, but did not know how to express them and during the treatment he felt that the fragile male image he had completely shattered and he felt "not a man" and the appeal to alcohol and sexual abuse of a woman is an expression of a distorted attempt to "prove" his masculinity and he projected frustration with his feelings from the mother and later, from his wife which led him to hurt another woman.

So to compare LBT with Sex offender therapy, both use logic, but LBT address faulty reasoning resulting in patient condition while Sex offender therapy also has to cope with more serious distortions arising from much deeper causes as well as follow the sex offender into prison and court cases and assess how dangerous the offender is to society if released from prison, etc., etc.

Let us consider another example, to compare LBT Therapy with Sex Offender Therapy

Example B.2. *Consider a case of an immigrant pupil (13 years old) who keeps failing in school studies and who does not believe in himself.*

A LBT therapist will apply his method to correct the patients reasoning and bolster his confidence. However, the LBT Therapist might discover that the patient was sexually abused and that this is the cause of his failures. The reasonable course of action is for the LBT Therapist to redirect the patient to a Sex Offender Therapist, as the causes are much more serious and the case is embedded in a larger social context.

The LBT community lists the following main distortions leading to illness [10].

Cognitive therapy and its variants traditionally identify ten cognitive distortions that maintain negative thinking and help to maintain negative emotions. Eliminating these distortions and negative thought is said to improve mood and discourage maladies such as depression and chronic anxiety. The process of learning to refute these distortions is called "cognitive restructuring".

- *All-or-nothing thinking*. thinking of things in absolute terms, like "always", "every" or "never". Few aspects of human behavior are so absolute. (See false dilemma).

- *Overgeneralization*. taking isolated cases and using them to make wide, usually self-deprecating generalizations. (See hasty generalization).

- *Mental filter*. Focusing exclusively on certain, usually negative or upsetting, aspects of something while ignoring the rest, like a tiny imperfection in a piece of clothing. (See misleading vividness).

- *Disqualifying the positive*. continually "shooting down" positive experiences for arbitrary, ad hoc reasons. (See special pleading).

- *Jumping to conclusions* assuming something negative where there is actually no evidence to support it. Two specific subtypes are also identified:

  - *Mind reading*. assuming the intentions of others
  - *Fortune telling* guessing that things will turn out badly. (See slippery slope).

- *Magnification and Minimization*. exaggerating negatives and understating positives. Often the positive characteristics of other people are exaggerated and negatives understated. There is one subtype of magnification:

- *Catastrophizing*. focusing on the worst possible outcome, however unlikely, or thinking that a situation is unbearable or impossible when it is really just uncomfortable.

- *Emotional reasoning*. making decisions and arguments based on how you feel rather than objective reality. (See appeal to consequences).

- *Making should statements.* concentrating on what you think "should" or ought to be rather than the actual situation you are faced with, or having rigid rules which you think should always apply no matter what the circumstances are. (See wishful thinking).

- *Labelling.* related to overgeneralization, explaining by naming. Rather than describing the specific behavior, you assign a label to someone or yourself that puts them in absolute and unalterable terms.

- *Personalization (or attribution).* Assuming you directly caused things when that may not have been the case. (See illusion of control).

- *Fairness Fallacy.* Expecting life to be fair is a cognitive distortion since, unfortunately, life is often not fair.

- *Karma Fallacy.* Similar to the fairness fallacy, the karma fallacy is the belief that people will get what they deserve. A belief that karma will level the playing field in short order is a cognitive distortion. This is sometimes called the heaven's reward fallacy.

- *Fallacy of change.* There are two parts to this: one is that by manipulating others, we can change them. The others is that by changing others, we can become happy.

- *Always being right.* In this cognitive distortion being right is more important than anything else including other peoples feelings.

- *Blaming.* This distortion is the flip side of personalization. Instead of feeling that one is responsible for everything, one blames others for everything that happens.

B.2

The formal argumentation reader should look up Fallacies, for example [51, 31].

Elliot D. Cohen in his book [14] offers the following table:

	Cardinal Fallacy	Fallacy Type	Guiding Virtue
1.	Dutiful Worrying	Emotional	Prudence (in confronting and resolving moral problems)
2.	Demanding Perfection	Emotional	Metaphysical Security (security about reality)
3.	Damnation (of self, others, life, or the universe)	Emotional	Respect (for self, others, life, or the universe)
4.	Awfulizing	Emotional	Courage (in the face of evil)
5.	The-World-Revolves-Around-Me Thinking	Emotional	Empathy (connecting with others)
6.	Oversimplifying Reality	Reporting	Objectivity (in making objective, unbiased judgments)
7.	Distorting Probabilities	Reporting	Foresightedness (in assessing probabilities)
8.	Blind Conjecture	Reporting	Scientificity (in providing explanations)
9.	Cantstipation	Behavioral	Temperance (self-control)
10.	Bandwagon Thinking	Behavioral	Authenticity (being your own person)
11.	Manipulation	Behavioral	Empowerment (of others)

Table 1: Fallacies Table from Eliot T. Cohen [10]